Quality in the Era of Industry 4.0

Quality in the Era of Industry 4.0

Integrating Tradition and Innovation
in the Age of Data and AI

Kai Yang
Wayne State University
Michigan, USA

Published by John Wiley & Sons, Inc., Hoboken, New Jersey.
Published simultaneously in Canada.

For general information on our other products and services or for technical support, please contact our Customer Care Department within the United States at (800) 762-2974, outside the United States at (317) 572-3993 or fax (317) 572-4002.

Wiley also publishes its books in a variety of electronic formats. Some content that appears in print may not be available in electronic formats. For more information about Wiley products, visit our web site at www.wiley.com.

Library of Congress Cataloging-in-Publication Data Applied for

Hardback ISBN: 9781119932444

Cover Design: Wiley
Cover Image: © chombosan/Alamy Stock Photo metamorworks/Getty Images

Set in 9.5/12.5pt STIXTwoText by Straive, Pondicherry, India

SKY10061352_113023

Contents

Preface

Brief Description of the Book

Overview of the Primary Focus and Themes of the Book

Quality in the Era of Industry 4.0: Integrating Tradition and Innovation in the Age of Data and AI seeks to explore the complex interplay between emerging technologies that are behind Industry 4.0 such as artificial intelligence (AI), big data analytics, smart manufacturing, and the multifaceted domain of quality management. The primary focus of this book is to examine how these disruptive technologies, particularly the rapid advancements in AI, are altering the paradigms of quality, from design to delivery and beyond.

Key themes covered in this book include:

1) **Historical Evolution of Quality:** Tracing the evolution of quality management over time, culminating in the contemporary transformation driven in part by Industry 4.0 and AI.
2) **Changing Paradigms:** Exploring how conventional quality management and quality engineering must be complemented, and in some instances replaced, by new realities and technology-driven approaches.
3) **Design and Innovation:** Charting the evolution of quality by redirecting its focus toward the early stages of the product or service lifecycle. This involves harnessing the capabilities of cutting-edge technologies like connectivity and AI to streamline product design and service delivery, resulting in superior customer value through dynamic, real-time iterations among stakeholders, including customers.
4) **Predictive Quality and Data:** Investigating how AI-driven analytics and machine learning models are becoming foundational to predictive quality management, allowing businesses to anticipate and address issues before they escalate into critical failures. Data quality emerges as a critical component in today's digital age; practical guidance on ensuring data integrity is provided.
5) **Risk Management:** Examining the integration of AI into risk assessment, utilizing data analytics to predict potential future risks, and devising proactive strategies for mitigation. In the rapidly evolving landscape of the 21st century, effective risk management has become an indispensable cornerstone of quality assurance, helping organizations navigate the complexities of modern operations.

6) **Future of Organizational Structures:** Analyzing the impact of technological advancements such as Industry 4.0 and AI on organizational structures, subsequently influencing mechanisms for quality management.

Explanation of the Significance of Discussing Quality in the Context of Industry 4.0 and AI

In an era marked by rapid technological changes, with AI standing at the forefront, the concept of quality is expanding beyond traditional realms like defect control and customer satisfaction. Industry 4.0 and AI together introduce disruptions and opportunities that necessitate a reassessment of quality management practices. Smart manufacturing systems, real-time data analytics, interconnected business ecosystems, and AI technologies are not only enhancing but also revolutionizing how we define, implement, and evaluate "quality."

The recent impressive developments in AI, such as advanced neural networks, natural language processing, and robotics, are adding layers of complexity and capability to quality management systems. These advancements make it not just possible but imperative to incorporate AI into discussions of quality in Industry 4.0. Issues like data quality, cybersecurity, and workforce reskilling are becoming increasingly intricate due to AI's growing role in organizational processes.

By focusing on these themes and exploring their intersection, this book aims to provide quality professionals, academicians, and industry leaders with the tools and insights needed to navigate the evolving landscape of quality management.

Target Audience

The content and insights provided in *Quality in the Era of Industry 4.0: Integrating Tradition and Innovation in the Age of Data and AI* are specifically designed to cater to a diverse audience with varying levels of expertise and interest in quality management and AI. Following are the key target audience groups:

Industry Professionals: This book aims to serve as a comprehensive guide for quality managers, engineers, data analysts, and other professionals working in fields disrupted by Industry 4.0 and AI. It offers actionable insights and methodologies for adapting to new paradigms of quality management.

Academicians and Researchers: Professors, scholars, and students in disciplines such as industrial engineering, data science, and business management will find a wealth of research material, case studies, and future research avenues. The book also serves as an up-to-date academic resource that reflects the latest trends and technologies, including advancements in AI.

Corporate Leaders and Decision-Makers: CEOs, CTOs, and other executives can benefit from this book by gaining an understanding of how Industry 4.0 and AI are transforming quality management. The book provides strategic insights into integrating technology-driven quality measures and risk assessments into organizational processes.

Consultants and Advisors: Those involved in advising firms on quality management and technological adoption will find this book to be a valuable resource. It covers the breadth and depth of quality management changes brought on by AI and Industry 4.0, providing a robust framework for consultation services

Policy Makers: With Industry 4.0 and AI ushering in a new regulatory environment, policy makers can gain insights into the complexities and challenges posed by these technologies, helping them craft more informed and effective policies for quality control and data management.

Technology Enthusiasts and General Readers: Anyone interested in the intersection of technology and quality will find this book to be an accessible and enlightening read. It aims to demystify the complexities surrounding Industry 4.0 and AI, making the content relatable to those without a specialized background.

Data Scientists and AI Practitioners: Given the significant focus on predictive quality and data analytics, professionals in these fields will find relevant methodologies and algorithms that can be directly applied to AI-driven quality management systems.

Non-Profit Organizations and Activists: Those interested in ethical manufacturing, consumer protection, and the societal implications of AI will find the book's sections on risk management and ethical considerations particularly relevant.

By addressing the needs and curiosities of such a varied audience, this book aims to be a comprehensive, go-to resource for anyone looking to understand the transformative impact of Industry 4.0 and AI on the world of quality management.

Motivations to Write This Book

The decision to write *Quality in the Era of Industry 4.0: Integrating Tradition and Innovation in the Age of Data and AI* was spurred by multiple motivating factors, each contributing to the overall purpose and drive behind this endeavor. Here are the key motivations:

1) **Stagnation in Quality Theories and Methods**: One of the primary motivators was the observation that the field of quality management has experienced a significant stagnation in terms of theory and methods. The core principles and methodologies that have served as the backbone of quality management have seen little innovative evolution in the past 60 years. This static nature presented an urgent call to rejuvenate the field and explore new avenues and frameworks.

2) **The Advent of Industry 4.0, AI, and Big Data**: The emergence of technologies under the Industry 4.0 umbrella, particularly AI and big data, has radically altered the operational landscape across industries. These advancements have not only opened up opportunities for redefining quality but also made it imperative to reevaluate traditional quality management frameworks. The book aims to bridge this gap by providing cutting-edge insights into how Industry 4.0 and AI are transforming the field.

3) **Rich Experience in the Field of Quality**: With several published books in the quality domain—including works on Design for Six Sigma (DFSS), Voices of Customers, and Applied Statistical Methods, and my research and consulting experiences, I bring a deep-rooted understanding and expertise in quality management. The extensive experience allows for a nuanced understanding of the field's intricacies and forms the backbone of the actionable insights offered throughout the book.

4) **Community Consensus on the Need for Change**: There is a growing consensus among professionals and governing bodies about the urgent need for change in quality management paradigms. Prominent organizations like the American Society for Quality (ASQ) have introduced concepts such as "Quality 4.0," acknowledging the transformation brought about by Industry 4.0. This book aims to contribute to that discourse, offering a comprehensive look at evolving paradigms and practical implementations.

The synthesis of these motivating factors lends this book its focus and urgency. It is an endeavor to not only highlight the challenges and opportunities introduced by Industry 4.0 and AI but also act as a catalyst in propelling the field of quality management into a future teeming with possibilities and innovations.

Book Structure and Chapter Summaries

To ensure the reader gains a holistic and nuanced understanding of quality management in the age of Industry 4.0 and AI, the book is organized into eight comprehensive chapters, each serving a specific purpose and building upon previous chapters. Here is an overview:

1) **Chapter 1: Evolution of Quality Through Industrial Revolutions**
   This chapter serves as the foundational block, tracing the journey of quality management through various industrial revolutions. It not only sets the historical context but also introduces the reader to the concept of Industry 4.0 and its potential impact on quality management.
2) **Chapter 2: Evolving Paradigm for Quality in the Era of Industry 4.0**
   After establishing the historical backdrop, the book delves into the current shifts in quality paradigms, made more complex and challenging by technologies like AI and super connectivity. The chapter scrutinizes the emergence of "Quality 4.0," as outlined by the ASQ.
3) **Chapter 3: Quality by Design and Innovation**
   This chapter explores the role of design in quality, especially in a landscape transformed by AI and Industry 4.0 technologies. Key industry case studies like Apple and Samsung provide actionable insights.
4) **Chapter 4: Quality Management in the Era of Industry 4.0**
   Focusing on practical applications, this chapter goes deep into how quality management is changing in various sectors from manufacturing to service, thanks to smart technologies and AI-driven data analytics.
5) **Chapter 5: Predictive Quality**
   This chapter dives into one of the most cutting-edge applications of AI in quality—predictive analytics. With a focus on AI-powered predictive models, the chapter serves as a how-to guide for implementing predictive quality measures across various sectors.
6) **Chapter 6: Data Quality**
   Data serves as the backbone of AI and Industry 4.0, making this chapter crucial. It covers everything from the basics of data quality to advanced data governance strategies, all the while emphasizing its importance in modern quality management.
7) **Chapter 7: Risk Management in the 21st Century**
   Taking a holistic view, this chapter explores how risk management is being redefined by AI and data analytics. It provides a comprehensive discourse on both traditional and emerging frameworks in risk management, making it highly relevant in today's volatile environment.

8) **Chapter 8: Emerging Organizational Changes in the 21st Century**

The final chapter delves into the human and organizational aspects of quality management in the age of AI and Industry 4.0. It explores how roles, skills, and organizational structures are evolving and what that means for quality professionals.

By systematically addressing these various facets, the book offers a multi-dimensional view of quality in the age of Industry 4.0 and AI. It is designed to be not just a resource but a guide that challenges, educates, and inspires its readers to think differently about quality.

Acknowledgments

I am profoundly grateful for the invaluable advice and encouragement I have received from a host of individuals throughout the development of this book. My heartfelt thanks go to Gregory Watson, Larry Smith, Nicole Radziwill and Ron Atkinson of the American Society for Quality (ASQ), as well as Dr. Jack Feng from the Institute of Industrial and Systems Engineers. I am also deeply appreciative of my colleagues on ASQ's Quality 4.0 Technical Program Committee: Christiana Hayes, Wendy Diezler, Kristine Bradley, Nicole Johnson, Lisa Custer and Manny Veloso.

Special recognition is extended to Mr. Lin Ben, Mr. Xueping Lin, Mr. Lingyun Dong, and Mr. Tony (Wei) Tong for their invaluable contributions, offering in-depth insights into real-world Industry 4.0 practices.

I extend my sincere gratitude to Dr. Ben Mejabi for his significant input in Chapter 2, which explores the paradigms on quality set forth by Robert Pirsig and Christopher Alexander.

My thanks also go to Dr. Zuoping Yao and Dr. Jie Hu, who enriched Chapter 3 with their expertise on Quality of Experience practices at SAIC-GM-Wuling Automobile Company.

Contributions from Yuping Liu and Hongjuan Yao have enriched Chapters 2 and 3, offering vital perspectives on super-connectivity and co-creation practices of Haier Group.

I am very grateful to Dr. Carlos Alberto Escobar Diaz, Dr. Ruben Morales-Menendez, and Dr. José Antonio Cantoral Ceballos of Tecnológico de Monterrey for their scholarly contributions to Chapter 5.

Feedback from readers is invaluable; your comments and suggestions are most welcome. Rest assured, I will give thoughtful consideration to your input for future editions of this book.

1

Evolution of Quality Through Industrial Revolutions

Industry 4.0, also known as the Fourth Industrial Revolution, is fueled by a remarkable suite of emerging technologies. These encompass everything from artificial intelligence and interconnected smart devices to groundbreaking advancements in nanotechnology. Such innovations are spawning unprecedented capabilities, fostering dynamic shifts in interpersonal communications, societal relationships, business operations, and service delivery. They are also stimulating product usage diversity and heightening production flexibility.

The momentum and breadth of this revolutionary transformation are intrinsically tied to the maturation rates of these underlying technologies, which are evolving at a swift pace. Historical precedents show us that each industrial revolution has deeply influenced management sciences and practices, including those in quality management. There is no reason to assume that the Fourth Industrial Revolution will be any different—it is poised to reshape our world in ways we are only beginning to comprehend.

Established quality management systems, such as total quality management (TQM), originated in the early period of the previous century. These systems were designed in response to the needs of the Second Industrial Revolution, which was characterized by mass production. Their primary goal was to mitigate defects and nonconformities within manufacturing processes. It is undeniable that these quality systems and methodologies have proven robust over time, consistently enhancing quality. They have undergone regular upgrades to stay relevant, and they continue to be fundamental components of our contemporary operational landscape.

However, it has had no innovative changes for nearly 60 years, and many methods and tools are outdated, so they are less and less used in the field. Figure 1.1 illustrates Google search trends for quality management and Six Sigma, and they are clearly trending downward since 2004. Obviously, pursuing good quality is human nature, and it will never go out of style. It is the current quality management theories and practices that become stagnant [1–3]. It is time for a breakthrough in quality management.

Quality is an ancient, complex, and eternal concept. Throughout history, as science and technology develop, at some point, industrial revolution will happen, as a result, new production systems may emerge, and the quality paradigm usually evolves with it to keep pace with these changes. Since the central theme of this book is "Quality in the Era of Industry 4.0," it is beneficial to see how quality has been evolving through all previous industrial revolutions. It will help us to predict what is next for quality in Industry 4.0. This is the key topic of this chapter.

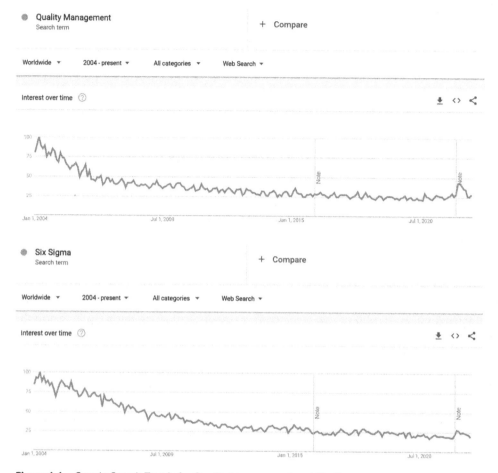

Figure 1.1   Google Search Trends for Quality Management and Six Sigma

1.1   Quality Before Industrial Revolutions

Throughout the course of history, quality has consistently manifested as an elegant, profound, yet elusive concept. The term "quality" has its roots in ancient Rome, derived from "qualis," signifying a "degree of excellence" in things or human values [4]. This concept echoes the philosophical musings of Greek thinkers such as Plato and Socrates on the subject of epistemology. Further exploration by philosophers, including John Locke, introduced the duality of quality: the objective aspect, which is inherent in the items themselves, and the subjective aspect, reflecting individual perceptions of these items [4].

With the burgeoning consumer markets that began in the 1600s, products started reaching a mass audience. This meant a diverse array of consumer expectations and needs, compounded by a wide spectrum of economic capabilities. Consequently, the interpretation of quality began to transition from an "absolute excellence" toward the concepts of "value for money" and "cost-effectiveness."

In the business realm, "value" has emerged as a critical measure of quality. In this context, value is defined as the ratio of a product's benefits to its cost. Prior to the era of industrialization, goods were predominantly crafted through craftsman production systems. In such setups, a craftsman or a team would oversee the entire design and production process, from inception to completion.

Craftsmen typically possessed high levels of skill, dedication, and self-discipline. They approached their work with an artisan's mindset, finely attuned to the client's subjective evaluation of quality or their esthetic sensibilities. These highly esteemed craftsmen were, in essence, the arbiters of quality for their products. As highlighted by Juran [5], they meticulously monitored every detail of design and production.

During this era, quality assurance relied heavily on the integrity of the craftsmen who valued their reputation highly. Consumers, too, played a role by carefully inspecting the products. In many parts of Europe, guilds implemented self-inspection and certification protocols, significantly enhancing the branding and quality assurance of artisans and their workshops. An official standard verification model was also prevalent, with government-led verification and oversight mechanisms being particularly advanced in China since the Qin and Han dynasties [5].

Throughout this period, the craftsman held primary responsibility for the product. This role was sometimes played by an individual entrepreneur, at other times by a team, albeit typically a small one. The craftsman managed all aspects of the product's design, production, and delivery. When required, craftsmen liaised closely with clients to ensure a high standard of work. Given their comprehensive oversight of the entire process, they could ensure full production control, preclude coordination issues, and maintain a tight rein on product defects.

This approach remained in vogue until the onset of the Second Industrial Revolution.

1.2 Quality in the First Industrial Revolution

The First Industrial Revolution unfolded in Great Britain between the 18th and 19th centuries. The driving forces behind this revolutionary period included the use of fossil fuels for power generation and the advent of steam engines for trains and ships. These developments were bolstered by significant breakthroughs in the iron, steel, and chemical industries and the rise of machine-driven textile industry.

The First Industrial Revolution incited transformative changes across economies, societies, and lifestyles. While it did not fundamentally alter the craftsman model, it did raise the complexity of many products well beyond that of traditional handicraft items. Consequently, the teams of craftsmen expanded, giving rise to a workshop-based supplier system—commonly known as the "cottage industry." This new system provided parts, materials, and subsystems for complex products. To coordinate this growing supplier system, the inception of early versions of industrial standards was necessary [6].

The craftsman model is characterized by its practitioners being "jacks-of-all-trades"—highly skilled individuals who manage all aspects of production. However, a notable drawback of this model is the extensive training period required for craftsmen.

In the craftsman production model, the manufacturing of complex products lacks streamlining. Lengthy process transitions and preparation times lead to lower productivity, extended production duration, and consequently, higher costs. It is also important to note that the First Industrial Revolution did not significantly impact quality control methods.

1.3 The Second Industrial Revolution and the Birth of Modern Quality Management

The Second Industrial Revolution predominantly originated in the United States, spanning from the late 19th century to the early 20th century. This revolution was marked by the extensive adoption of electrical energy and the implementation of a mass production system. This system was grounded in moving assembly lines and the use of standard, interchangeable parts.

The influence of the Second Industrial Revolution on production methods and working practices was incredibly profound. It has shaped modern quality management significantly and has even influenced the definition of quality itself, a testament to its enduring impact that persists to this day.

1.3.1 Mass Production System Is a Game Changer

Since the First Industrial Revolution, scientific and technological advancements have accelerated rapidly, significantly boosting the production capacity of industrial materials. This led to the emergence of various complex new products, such as automobiles. Crafting such products via traditional craftsmen and workshops was prohibitively time-consuming and expensive, making them accessible only to a select few.

The breakthroughs in electrification infrastructure presented opportunities to expedite the production process and reduce costs. Among those seizing this opportunity was Frederick Taylor, who proposed the "Scientific Management Method," also known as *Taylorism*. This method proposed several core principles [7]:

- Break down large, complex tasks, such as automobile assembly, into numerous smaller steps—potentially in the thousands.
- Divide labor among workers, assigning each to a specific process step, with each worker performing the same task repetitively.
- Utilize stopwatches and motion analysis to determine the most efficient work practices and use these as a standard for training workers.
- Abandon the craftsman production system completely by segregating management, design, and production sectors, and by instituting an extensive division of labor and professions, marking a clear distinction between "blue collars" and "white collars."

Another pivotal figure in the Second Industrial Revolution was Henry Ford, founder of the Ford Motor Company and the originator of the moving assembly line operation and the mass production system [8]. His core contributions include:

- Full utilization of standardized, interchangeable parts across the entire industry—preceding Ford, the American firearms industry had already implemented the practice of standardized parts in the late 19th century.
- The first electrically powered moving assembly line: the assembly line comprised numerous workstations, each executing a single operation performed by one worker.
- Simplified, standardized, and interchangeable worker skills.

At its inception, the moving assembly line exhibited the following characteristics:

- The product (automobile) had a single design.
- The assembly line was a rigid yet precise process, capable of producing only one specified product through a sequence of pre-determined process steps.
- The work of product design was also highly specialized and separated from production.

Compared to the craftsman production model, this mass production model drastically reduced, if not entirely eliminated, setup and change-over times between consecutive process steps. Consequently, it markedly improved productivity and production capacity, substantially reduced the cost and sales price of products, and significantly broadened the consumer base. However, the product variety in the mass production system became quite limited, with monotonous styles and typically unimpressive esthetics.

(a)

(b)

Figure 1.2 Comparison Between Cars Produced from Craftsman and Mass Production Systems. (a) Mass Production: Ford Model T, 1922 Price: U$319, 1.3 million produced per year. *Source:* Shawshots / Alamy Stock Photo. (b) Craftsmen Production: Rolls-Royce Twenty £1600, 2940 cars produced from 1922 to 1929. *Source:* Barker / Wikimedia Commons / CC BY-SA 2.5

Figure 1.2 contrasts two types of cars—one produced by a mass production system and the other by a craftsman production system. The vast differences in cost, productivity, and style between these two cars are clearly discernible.

The mass production model, or the Taylor model, had a disruptive and profound impact on the economic activities, business models, lifestyles, and corporate organizations of that era, effects that largely persist today, particularly in Western countries.

Some key effects of the mass production system include the following features:

- **The decline of the craftsman/artisan class**: Craftsmen, highly skilled and versatile individuals, once constituted the backbone of key economic activities in ancient and pre-modern times. With the proliferation of the mass production model, this class has become nearly extinct, now appearing predominantly in the luxury goods and handicraft industries.

- **Realignment of Society**: A highly specific division of labor and professions has become a common practice, extending beyond manufacturing to government, service, medical, and scientific research sectors.
- The widespread adoption of well-designed processes and standardization in all disciplines.

The mass production system also exerted a paradigm-shifting, disruptive influence on quality control. In ancient times, quality represented "the degree of excellence," and with the maturity of the mass consumer market, the concept of quality evolved to encompass "benefits versus cost" and "customer value." Compared to handcrafted luxury products, when the mass production system substantially reduced product prices, consumers, driven by cost considerations, were willing to forgo all luxurious features and decorations in favor of basic functions. For instance, the Ford Model T, available only in black color, was inexpensive, functional, and reasonably reliable, making it the most popular car model of its time.

However, the early Taylorism mass production system was highly fragmented. Workers operated separately at different stations along the assembly line without communicating with each other, contrasting starkly with the craftsman model, where one craftsman would complete the entire order from start to finish.

1.3.2 The Start of Modern Quality System

Amid the surge in complexity attributed to the large number of components and steps involved in the mass production system, the susceptibility of products to errors and defects intensified. Consequently, the critical aspect of quality assurance shifted toward managing defects by guaranteeing conformance from every supplier and each stage of the production process. This need served as a catalyst for the evolution of modern quality management systems pioneered by distinguished figures such as Shewhart, Deming, and Juran [9].

Several fundamental quality systems and methodologies, such as statistical quality control (statistical process control [SPC]) [10] and total quality management (TQM) [11], took root during the pinnacle of the Second Industrial Revolution. Notably, following the widespread adoption of Taylorism's mass production model, the working definition of quality underwent a transformation. It evolved from notions of "excellence" and "customer value," to the more tangible metrics of "defect free," "low scrap rate," and "low failure rate." This interpretation of quality has persisted to this day, with scrap and failure rates remaining the sole quantifiable quality indicators being assessed.

In conclusion, the emergence of modern quality management and quality engineering can be attributed to the pressing issue of handling, controlling, and eradicating defects and failures within mass production systems.

The transition in defining quality, from a "degree of excellence" to "free from defects and adherence to standards," is a drastic shift. To make sense of this "definition of quality," it is essential to explore the numerous studies that have debated this issue. This brief summary provides an overview. In their article, Reeves and Bednar [12] argue that perceptions of quality vary based on individual perspectives. For instance, from a consumer's perspective, quality may signify "excellence," "customer value," or "surpassing expectations." In contrast, producers perceive quality as "conformance to specifications" and low rates of scrap, failure, and defects. In the initial phases of the mass production model, a seller's market prevailed due to product scarcity, which likely influenced this differentiation.

Garvin's 1987 article [13] outlines eight dimensions of quality: performance, features, reliability, conformity, durability, service, esthetics, and perceived quality. This comprehensive overview contemplates multiple perspectives, encompassing both "objective quality" and "subjective quality." However, from then until the present, the quality community has prioritized aspects such as reliability, conformity, and durability.

Since the advent of the mass production system, the initial method of quality control involved inspecting products prior to shipping, repairing or discarding defective products, and retaining the quality ones. Large repair shops were commonplace in early automobile factories. Sampling inspection [14], involving random or full inspection to prevent defective supplier parts from reaching factories, was another common practice. If a batch failed this inspection, it would be returned to the supplier. Inspection was also used to weed out inferior products within factories before shipping them to customers.

However, this inspection approach had clear disadvantages. First, the inspection was labor intensive and time consuming, with the manual inspection proving less than foolproof. Should an issue be detected during inspection, the root cause could potentially lie anywhere within the thousands of process steps, due to worker errors, defective incoming parts, or machine failure. Identifying and eliminating these problems via end-of-production inspection proved challenging, hence their recurring nature. The production process, as shaped by the Taylor system, was fragmented and compartmentalized, complicating the swift identification and resolution of quality issues. For instance, in early 20th-century automobile plants, the repair workshop, managed by the quality department, aimed to eliminate defects. This objective clearly conflicted with the assembly plant management's goal of swiftly dispatching manufactured cars.

Since quality issues typically arise during the production process, and inspection is the last step, some argued that "quality control should move upstream." Consequently, it was deemed necessary to control quality at critical upstream process steps. In 1931, American statistician Walter Shewhart's book "Economic Control of Manufactured Product" formally defined quality control as a statistical issue and introduced the concept of the statistical process control (SPC) [15]. This method of managing key processes marked a significant shift, with numerous subsequent quality methods and indicators (such as process capability indices) deriving from it and remaining in use today.

Before World War II, the usage of control charts was not widespread among manufacturers; they were primarily utilized in Bell Labs factories. However, Edward Deming and Joseph Juran [9], who were involved with these early initiatives, later emerged as key pioneers of modern quality management.

It is noteworthy to observe the gradual evolution of the mass production system between 1920 and 1950. In the 1910s, the Ford production system epitomized the first version of the mass production system. The assembly line was virtually inflexible, producing solely the Ford Model T. Initially, due to its low price and reasonable reliability, Model T dominated the market. However, post-1920s, General Motors (GM) employed several strategic measures to outmaneuver Ford:

1) They deviated from a single product line setup, introducing several brands catering to different price points to meet varying consumer needs. Each brand operated akin to a small company, all under the GM umbrella.

2) They initiated the practice of introducing new models annually. While there may not have been significant changes each year, the presence of new features was ensured. This marked the beginning of the mass production system's gradual transition toward a flexible production system [16].

1.4 The Third Industrial Revolution and the Maturity of Modern Quality Management System

In the 1950s, Toyota's lean production system emerged, enabling greater flexibility in production and an increased variety of products without sacrificing efficiency [17]. The advent of the Third Industrial Revolution in the 1960s, powered by computer and information technology, introduced further flexibility and control over the production process [18]. It also significantly improved product design capabilities. These advancements underscored the emphasis on another critical aspect of quality: a robust focus on customer satisfaction. This series of events precipitated the comprehensive evolution of quality management systems and quality engineering.

1.4.1 Contributions of Japan to Quality Management

The initial wave of modern quality management unfolded in post-World War II Japan. Although Japan had industrialized prior to the war, its reputation for consumer goods' quality was unimpressive. The label "Made in Japan" was equated with cheap and low-quality products during this period. The Korean War transformed Japan into a critical strategic hub for the US military, necessitating the rectification of their problematic, defect-ridden, and subpar telephone communications within a stipulated timeframe. As a measure to "urge Japan to improve quality," American quality experts Edward Deming and Joseph Juran were invited to deliver open classes on quality management to Japanese corporate executives. The US military mandated the attendance of top Japanese company executives at these sessions [19].

Intriguingly, Deming and Juran had previously conducted similar lectures in the United States, but their insights garnered far less attention from American executives. However, their presentations in Japan led to significant impacts and spurred a series of follow-up activities:

- As per historical documents, the Japanese executives attending these sessions expressed significant interest in the Plan-Do-Study-Act (PDCA) or Plan-Do-Check-Act (PDSA) work process. This process entailed "identifying and defining a critical problem and persisting with it until it was resolved" [20].
- At that time, Japan was grappling with severe economic hardships in the post-war period. The nation needed to export consumer goods to alleviate these economic difficulties and sought to enhance the quality of its products to alter the reputation of Japanese goods. Consequently, many companies demonstrated considerable interest in this approach.

Following the absorption of American methodologies, several Japanese experts began propagating these principles via broadcast lectures tailored to the needs of Japanese business individuals. This marked the beginning of many Japanese experts and business professionals enhancing these quality methodologies originating from America and developing their own unique approaches.

1.4.1.1 Total Quality Control

Local expert Ishikawa Kaoru, a professor at the University of Tokyo, played a significant role. He translated and expanded on Deming and Juran's teachings. In 1950, in collaboration with front-line industry workers, he developed the cause-and-effect diagram (also known as the fishbone diagram). Post-1960, he posited that not only middle and upper-level enterprise managers should learn quality management, but first-line supervisors and on-site workers should as well. This belief led him to pioneer the "total quality control (TQC)" activity model and training method, which subsequently became the first practical version of the TQM system, underpinned by Ishikawa's 11 points [21].

1.4.1.2 Taguchi Method

Another local expert, Genichi Taguchi, joined Nippon Telegraph and Telephone Company (NTT) in 1950 with the aim of enhancing the quality and reliability of Japan's telecommunications industry. He later developed the Taguchi method to improve durability, stability, and reliability during product design—a method which was adopted seriously by Toyota and promoted worldwide post-1980 [22].

1.4.1.3 Quality Function Deployment

Quality function deployment (QFD) is a unique design planning methodology originating from Japan, which is focused on incorporating the customer's product expectations into the design process. The inventor of QFD, Yoji Akao, saw his concept first applied in Japan in 1966. QFD gained initial popularity in Japan before being adopted globally [23].

1.4.1.4 Kano Model

Professor Noriaki Kano from the Tokyo University of Science and Technology introduced the renowned Kano Customer Satisfaction Model (Kano Model) in the 1970s and 1980s. This model categorizes customer needs into three types (or five, according to some studies): must-have, one-dimensional, and attractive quality. Different categories of customer needs are addressed in various ways. Kano also developed a method to identify these three types of customer needs through customer surveys [24].

1.4.1.5 Affinity Diagram

Japanese anthropologist Jiro Kawakita proposed the affinity diagram methodology in the 1960s [25]. This method helps analyze data from large volumes of scattered, rambling words collected from customer surveys or brainstorming sessions. The technique can discern patterns and underlying structures within these words. It is now widely used to analyze and synthesize customer needs, serving as a fundamental technique for brainstorming and one of the seven tools of TQM.

1.4.1.6 Kansei Engineering

Originating in Japan, Kansei Engineering is a product design methodology that stems from the observation that consumers possess conscious or subconscious preferences or dislikes for specific product features based on their personal feelings. These feelings can be evoked by a comprehensive psychological state induced by the customers' five senses—sight, hearing, taste, smell, and touch. As such, designers must comprehend the correlation between these consumer perceptions and the features of the design, aiming to enhance the design's positive impression. Kansei Engineering has evolved several robust methods to capture, quantify, and analyze sensations. For instance, it uses techniques such as analyzing the gaze, attention, facial expressions, and smiles captured in video recordings of visitors during a product exhibition. The outcomes of this human perceptual analysis are translated into product design elements, and products are manufactured to align with people's preferences. Pioneering the introduction of perceptual analysis into the realm of engineering research were researchers from the Faculty of Engineering at Hiroshima University, Japan. Starting in 1970, with a focus on residential design that takes into account occupants' emotions and desires, they investigated how to embody occupants' sensibilities into engineering techniques used in residential design. Renowned expert Nagamachi Mitsuo has contributed significantly to this field, authoring works such as "Kansei Engineering of Automobiles" and "Kansei Engineering and New Product Development" [26].

1.4.1.7 Poka-Yoke

Shiego Shingo, a technical pioneer of the Toyota production system, devised key processes and techniques for quick die change and rapid process changeover to achieve zero inventory and flexible production [27]. As the Toyota production system underscores "one piece flow," conditions such as "zero defects" and "zero failures" are indispensable. This led Shingo to extensively investigate quality control, arriving at several critical observations:

1) Shingo recognized the human tendency to make mistakes, acknowledging that perfection is unattainable. However, he believed that an effective system to automatically correct errors could prevent human mistakes from turning into significant defects or flawed products.
2) He categorized the prevalent post-production inspection to separate good from bad products, often done manually, as "judgmental inspection." Shingo pointed out several drawbacks to this method, such as the inevitability of human error and its general inability to identify the root cause of a quality issue.
3) Shingo found that SPC, which originated in America and was popular in Japan, had limitations. Statistical process control relies on sampling data from production for quality-related judgment, requiring production to halt and root cause identification when the SPC charts indicate an "out of control" situation. Shingo argued that due to its reliance on sampling inspection and time lag in problem identification, SPC might miss defects and prove ineffective for retrospective investigation of root causes.
4) He advocated for three inspection methods to effectively guarantee zero defects: (i) step-by-step inspection, wherein the downstream process checks the upstream's semi-finished products, investigating any problem's source immediately; (ii) self-inspection, which involves checking the semi-finished product before it leaves the process, and tracing the source of any issue promptly; (iii) source inspection, which finds and blocks the source of a problem to prevent its occurrence.

After an exhaustive study of existing quality inspection and control methods from the United States, Shigeo Shingo proposed his Poka-Yoke (fool-proof) quality assurance method. He outlined the following guiding principles:

- To achieve zero defects, 100% inspection is necessary.
- Judgmental testing, if needed, should be objective, and not manual.
- Inspection must be low cost and automated.
- On detecting faults and defects, the root cause should be identified immediately.
- All hidden root causes should be identified and eliminated one by one.
- Upon finding the source of the problem, it needs to be blocked at the source to prevent its occurrence using an automatic detection device, termed a Poka-Yoke device.

Shingo emphasized the characteristics of the Poka-Yoke device, which should be inexpensive, capable of 100% inspection, produce real-time results, and be implemented either through technical means or process design.

Interestingly, Shingo also shared his thoughts on statistical methods and SPC. Initially, he viewed SPC and statistics as the ultimate solution to quality issues. However, he later saw SPC as a means of estimating and maintaining the current process, not necessarily improving it. He regretted that his early reverence for statistics and SPC might have delayed the perfection of his Poka-Yoke system. Despite Shingo's somewhat controversial views, both non-statistical and statistical methods have their places in quality management, varying case by case and over time.

In practical terms, Shingo Shigeo has undeniably triumphed. The modern digital Poka-Yoke system, governed by automatic sensors, facilitates automatic real-time 100% inspection and online real-time control, becoming the mainstay of online quality control [28].

1.4.2 Total Quality Management (TQM)

Starting from the 1970s, the remarkable success of Japan in multiple domains posed significant competitive pressure and challenges to industries in Europe and America. This led to introspection across various sectors in the United States. In 1980, a conversation was sparked by an NBC TV show titled "If Japan can, why can't we?" which led to heated discussions about "Japan's success and the United States' response."

Edward Deming and Joseph Juran, who were acknowledged for introducing modern quality management to Japan, were then solicited by many large Western corporations, such as Ford and General Motors, to guide their operations. In the process, Deming and Juran learned about many Japanese practices like the total quality team, which they had greatly influenced during their involvement in Japan's quality initiatives. From the 1970s onward, they published numerous works on quality management. Some examples are Deming's "Out of Crisis," "The New Economics for Industry, Government, Education" [29], and Juran's "Quality Planning and Analysis," "Upper Management and Quality" [30].

They asserted that in a mass production system, the assurance of quality can only be achieved through comprehensive control and continuous improvement in all key process elements throughout the entire product lifecycle. Since all actions are performed by people, everyone needs to receive proper quality training and be empowered to take responsibility. They saw quality as a meticulously planned and executed systematic endeavor by the whole organization, a concept they referred to as TQM.

This concept of TQM received varying degrees of response and support from the governments and industries of Europe and America:

- Beginning in 1984, the United States, initially with the military and with the Department of Defense, and then with the US federal government, embraced the concept of TQM.
- From 1987 onward, the United States has annually evaluated and bestowed the Malcolm Baldrige National Quality Award [31] upon select outstanding companies and entities.
- Since the 1990s, many European countries have standardized TQM, leading to the creation of ISO 9000 certification, which is now issued to companies worldwide.

1.4.3 The Third Industrial Revolution and Its Impact on Quality Management

The Third Industrial Revolution [18] marked a transition from mechanical and analog electronic technology to digital electronics. This shift, which began in the latter half of the 20th century, was characterized by the adoption and proliferation of digital computers and related information technology. These advancements had profound impacts on the manufacturing industry as sensors, computers, and information technology were continuously integrated into the production process. Consequently, important parameters in industrial processes, such as workpiece dimensional measurements, temperatures in chemical reactors, and real-time pressure measurements in containers, could be collected in real time.

The technological developments of the Third Industrial Revolution had several significant impacts on quality management and quality engineering, particularly within the manufacturing industry:

1) **Widespread Application of Automatic Sensors, Detectors, and Digital Poka-Yoke Devices/Systems in the Production Process**: Starting from the 1980s, numerous manufacturing companies invested billions of dollars in installing these devices and systems in their factories. Their implementation significantly reduced production failures and assembly quality issues, leading to a marked improvement in production quality.

2) **Computer-Aided Design (CAD) and Computer-Aided Analysis (CAE)**: A multitude of quality issues are attributed to poor design, affecting not only performance, esthetics, and features but also resulting in malfunctions, hidden safety concerns, and low reliability. Addressing these problems typically requires a significant investment of manpower, materials, and time. However, from the 1980s onward, the extensive use of CAD and CAE for simulation testing, product modeling, and virtual reality exercises on computer platforms allowed engineering teams to detect and rectify design issues quickly and inexpensively. Consequently, CAD and CAE significantly improved both the quality and speed of design.

3) Advances in computer capabilities facilitated the widespread application of numerous rigorous scientific methods. These included scientific modeling, statistical analysis, ultrafast computing, and the handling of vast digital data storage. These advances simplified the use of powerful quality tools such as SPC, design of experiments (DOE), and statistical modeling and analysis. Some advanced industrial companies were even capable of integrating sophisticated statistical methods with engineering disciplines such as mechanics and materials science, along with other specific sciences, to implement in-depth and accurate process modeling. This process facilitated the monitoring, control, and optimization of manufacturing processes. These advancements in computer capabilities also served as a key enabling factor for the Lean Six Sigma initiatives in the quality community.

1.4.4 Lean Six Sigma

1.4.4.1 Overview of Lean Six Sigma

Six Sigma [32], which was first pioneered by Motorola, is a business operating system initially aimed at eliminating manufacturing defects. The term "Six Sigma" originates from the statistical field of process control, signifying the capacity of manufacturing processes to generate a substantial proportion of output within specification. In short-term operations, processes running with "six sigma quality" are projected to yield long-term defect rates below 3.4 defects per million opportunities (DPMO).

The concept and movement of Six Sigma were initiated as early as 1986 at Motorola and later spread to various manufacturing companies in the 1990s. In 1995, General Electric (GE) officially launched the Six Sigma movement. Due to the immediate benefits it brought about such as quality improvement, cost reduction, and profit enhancement, Six Sigma rapidly expanded across the entire Western manufacturing industry. The primary components of Six Sigma include:

1) The concept of continuous improvement through reducing scrap rate, costs, and waste.

2) **Establishing an Organization and Team**: The company's top management and administrative team must shoulder the responsibility of leadership and implementation, while also selecting and recruiting technical teams. An improvement project team, the administrative team, and the technical team collaboratively select projects for the project team.

3) Each project must establish specific goals and timelines, such as "reducing the paint shop scrap rate by 50% within three months." Each project must also undergo the Six Sigma (DMAIC)

process (a digitized PDCA). Upon completion, the project's cost and return are reviewed by the finance department.

4) Both the administrative and technical teams should receive bespoke training. The executive team learns the concepts of TQM and Lean, while the technical team should grasp more comprehensive quality methods and lean techniques.

It is noteworthy to mention that Six Sigma's initial goal in the early 1990s was to minimize the "cost of poor quality" with an extremely low scrap rate, signifying 3.4 defects per million (3.4 ppm). From the 1990s to the 2000s, several influential books on the Toyota production system and lean manufacturing were published in the West, including "The Machine that Changed the World," "Lean Thinking," and "Toyota Way" [33–35]. The Toyota system gained popularity due to its clear concept, limited mathematical components, wide applicability, and immediate impact on efficiency improvement and waste reduction. Given that Lean and Six Sigma address different problems—efficiency and quality, respectively—they complement each other. Incorporating Lean into an already established Six Sigma team enhances its effectiveness. Thus, Lean Six Sigma has become a standard business improvement operating system. After manufacturing, Lean Six Sigma quickly spread to other industries, including services, medical, financial, and government sectors.

In summary, Six Sigma has evolved into an unprecedentedly large-scale, high-intensity, and far-reaching mass quality movement, which continues to progress.

1.4.4.2 Limitations of Lean Six Sigma

In 2004, the number of Google searches for "Six Sigma" peaked, and has been gradually decreasing year on year since then. Similarly, the number of individuals mentioning Six Sigma in their LinkedIn job search biographies has declined each year thereafter. The stock value of GE, a cornerstone of Six Sigma, hit US$600 billion in the mid-2000s, making it the world's largest at that time. It now stands at less than US$60 billion. Since 2010, GE no longer designates Six Sigma as its "company-wide guiding philosophy." While Lean Six Sigma remains popular and is still utilized by many companies, its peak popularity has waned. Are there any limitations or flaws in Six Sigma? [36] We identify the following points:

- Six Sigma processes and methodologies originate from traditional manufacturing industries (mechanical, electrical, etc.) and are effective within these contexts. However, they may not be fully applicable to certain industries, such as the emerging software industry and the internet sector, where these working methods may be seen as too rigid.
- Six Sigma heavily depends on statistical analysis and its existing quality methods. While these methods can address some issues, they are not infinitely applicable. Take GE as an example. After implementing Six Sigma for five years, the model's effectiveness reached its zenith within GE. Its utility started to decline thereafter.
- Many companies rigidly adhere to Six Sigma, enforcing uniform processes and providing identical training across the board (a potential issue that might also plague ISO 9000). Many businesses report that this rigid adherence to Six Sigma stifles creativity. For instance, Mikel Harry, one of the founders of Six Sigma, proposed the "three iterations of Six Sigma": "The first is to reduce waste (which we have done), the second is to reduce costs (which GE has accomplished), the third iteration should be a system of *Value Creation* that benefits everyone." However, Mikel Harry admitted that the existing Six Sigma model could not achieve this goal. Mikel Harry passed away in 2017.
- All in all, as an increasing number of companies succeed in reducing their scrap rates to minimal levels, low scrap rates no longer serve as a competitive edge. Now, other quality aspects such as cost-effectiveness and innovation have gained greater importance.

1.5 Current Challenges and Difficulties for Quality Management

1.5.1 Industry 4.0 Is Coming

The Fourth Industrial Revolution, also known as Industry 4.0, [37, 38] is currently transforming the world. This technology-driven industrial revolution relies heavily on the maturity and rapid advancement of various digital technologies, as depicted in Figure 1.3. The breadth of Industry 4.0, in terms of both its core technologies and its areas of application, is vast. These technologies span from the Internet of Things (IoT) to highly interconnected smart machines and devices, powered by tools such as 5G wireless technology, artificial intelligence, scalable analytics (Big Data), and augmented reality.

The integration of these technologies into sophisticated cyber-physical systems signifies the present state of Industry 4.0 applications. We are still progressing toward a mature, steady state of implementation that offers a return on investment for businesses and society at large. The applications of Industry 4.0 are not confined to manufacturing. They reach into every sphere of human activity, encompassing energy, agriculture, healthcare, government, and services, and extending to all facets of social interaction.

Among all the techniques associated with Industry 4.0, such as cloud computing, big data, and additive manufacturing, the IoT plays a crucial role. The two key components of IoT technology are IoT devices and IoT platforms. The IoT devices are everyday objects and physical devices equipped with internet connectivity through the embedding of nonstandard computing devices. Examples of IoT devices include smart TVs, wearable technology, and smart appliances. Unlike

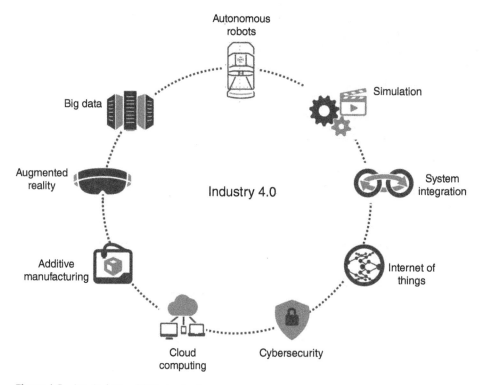

Figure 1.3 Key Industry 4.0 Technologies

traditional devices, the IoT devices can transmit data, communicate, and interact over the internet, and can be remotely monitored and controlled by IoT platforms.

With the advancement of Industry 4.0 technologies, an increasing number of products, as well as their production equipment, infrastructure, and tools, are becoming IoT embedded, evolving into smart products, smart factories, and smart suppliers. Moreover, all stakeholders of products, such as customers, producers, and suppliers, are equipped with internet-enabled computers, smartphones, tablets, and the like. In summary, our business world is progressively transforming into numerous internet-embedded dynamic ecosystems that thrive, adapt, and interact within various market environments, which is vastly different than in the past.

1.5.2 Customers in Industry 4.0 Age and Their Expectations

In today's world, customers have access to an abundance of information about any product they are interested in, including its functionality, price, and reputation, thanks to online resources. They engage in various social networks and interact with each other. What's more, customers' desires and needs continually evolve along with social and technological trends, as well as personal preferences. Figure 1.4 illustrates the evolution of customers and manufacturers over the past 150 years [39].

Before the advent of the Second Industrial Revolution, the craftsman production model was predominant. This model was based on individual customer orders and production, characterized by high costs and low efficiency. The Second Industrial Revolution, which began in the early 20th century, heralded the golden age of mass production from 1913 to 1955. During this period, product variety was low, and volume was high. Henry Ford famously said, "Any customer can have a car painted any color that he wants so long as it is black."

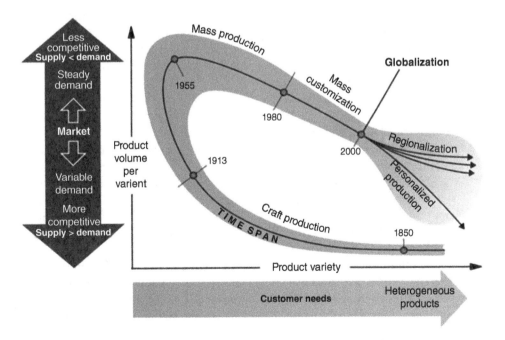

Figure 1.4 Evolutions of Customers and Manufacturers

With the gradual advent of the Third Industrial Revolution, characterized by digital control, sensors, and information technology, the production of small batches with greater variety became economically feasible. Currently, as we progressively embrace the Fourth Industrial Revolution, the trend is moving toward an even greater variety of products, more customization and individualization, but with higher efficiency and lower cost. This corresponds to the last segment of Figure 1.4.

1.5.3 Challenges for Modern Quality Management Brought by Industry 4.0

Historically, every industrial revolution has brought about significant changes in quality management practices, and Industry 4.0 is expected to be no exception. As we discussed in previous sections, the current quality management and quality engineering systems were developed to meet the demands of the Second Industrial Revolution, which focused on managing, controlling, and eliminating defects in mass production systems. However, two industrial revolutions (Industry 3.0 and 4.0) have since unfolded, bringing with them considerable changes and posing significant challenges to the quality community. Some of these challenges are discussed in the following subsections.

1.5.3.1 The Limitations of Traditional Quality Management Practices

The "traditional" quality management focuses on "defects reduction." As previously discussed, traditional quality management was initiated to address the need for controlling defects, also referred to as conformance quality. However, high conformance quality has now become commonplace and widely accessible, due to technological advancements as we discussed earlier. Here the automotive industry can provide a useful example.

Figure 1.5 shows J.D. Power's initial quality rating and market capitalization data for the automotive industry in 2022.

J.D. Power, an American consumer research, data, and analytics firm established in 1968, has become a gold standard for the automotive industry. Conducting nearly 200 benchmarking studies annually, J.D. Power derives insights from consumer behavior and market data across various industries. Since the 1980s, their quality ratings have been featured in over 350,000 television commercials and two billion print advertisements.

Recently, an apparent paradox has arisen. Data from February through March 2020 showed Tesla ranking last in J.D. Power's initial quality survey among 32 major brands. The J.D. Power initial quality rating is based on the number of problems experienced per 100 vehicles (PP100) during the first 90 days of ownership, with a lower score indicating higher quality. Essentially, this is a measure of conformance quality. At the same time, however, the company topped the list in J.D. Power's APEAL study, which measures the emotional attachment and excitement of customers toward a new car [39, 40]. This paradox was addressed in an October 2020 editorial in *Automotive News* by Professors Siegel and Yang [41]. They posited that the J.D. Power initial quality survey measures traditional manufacturing conformance quality and is thus blind to the actual subjective customer experience (emotional appeal).

This leads us to a critical conclusion: the traditional relationship between manufacturing conformance and consumer-based measures of quality is starting to break down. These divergent rankings, coupled with Tesla's market capitalization surpassing all other automakers, are indicative of the limitations of traditional quality management—the minimization of defects—in enhancing perceived business competitiveness.

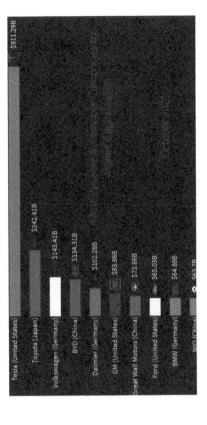

J.D. Power
2022 U.S. Initial quality study^{SM}
Brand ranking
Problems per 100 vehicles (PP100)

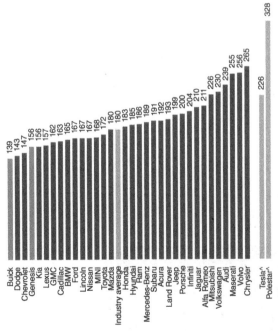

Buick — 139
Dodge — 143
Chevrolet — 147
Genesis — 156
Kia — 156
Lexus — 157
GMC — 162
Cadillac — 163
BMW — 165
Ford — 167
Lincoln — 167
Nissan — 167
MINI — 168
Toyota — 172
Mazda — 180
Industry average — 180
Honda — 183
Hyundai — 185
Ram — 186
Mercedes-Benz — 189
Subaru — 191
Acura — 192
Land Rover — 193
Jeep — 199
Porsche — 200
Infiniti — 204
Jaguar — 210
Alfa Romeo — 211
Mitsubishi — 226
Volkswagen — 230
Audi — 239
Maserati — 255
Volvo — 256
Chrysler — 265

Tesla^ — 226
Polestar^ — 328

Buick ranks highest Overall and among Mass Market brands, and is noted by a gold bar.
Genesis ranks highest among Premium brands, and is noted by a gold bar.
Note: 'Brand is not rank eligible because it does not meet study award criteria.

Source: *J.D. Power 2022 U.S. Initial Quality Study^{SM}*

*Charts and graphs extracted from this press release for use by the media must be accompanied by a statement identifying
J.D. Power as the publisher and the study from which it originated as the source. Rankings are based on numerical scores,
and not necessarily on statistical significance. No advertising or other promotional use can be made of the information in this
release or J.D. Power survey results without the express prior written consent of J.D. Power.*

Figure 1.5 Initial Quality Rating and Market Capitalization of J.D. Power for Automotive Industry

1.5.3.2 Changing Realities in a Connected World

In our increasingly connected world, fundamental aspects such as customers, products, and business environments have significantly transformed. Customers today are well-informed and interconnected, possessing personalized, rapidly evolving needs influenced by iterative experiences. This is particularly true for Generation Z, the cohort born between 1995 and 2010, who are digital natives deeply influenced by the internet, instant messaging, and social media. A McKinsey report in 2019 revealed that 77% of Gen Z consumers prefer products and services tailored to their personal needs. A Target Group Index (TGI) (Figure 1.6) study further disclosed that superior user experience is the most crucial factor in Gen Z's purchasing decisions.

The advent of the Third Industrial Revolution introduced products with embedded electronic and IT components, replacing the one-time, buy-and-sell transactions of the past. The ongoing Fourth Industrial Revolution has further integrated IoT technologies into products, which includes two critical elements—IoT devices and IoT platforms. The IoT devices—like smart TVs, wearables, and smart appliances—are physical objects with internet connectivity through embedded nonstandard computing devices. These devices can transmit data, communicate and interact over the internet, and can be remotely monitored and controlled by IoT platforms.

With the proliferation of such technologies, an increasing number of products are evolving into smart products, with functionalities largely defined or delivered by software (as illustrated in Figures 1.7 and 1.8).

123   128   124   111   134

Quality   Price   Brand   Fashion   Experience

Figure 1.6 Target Group Index for Gen Z Consumers

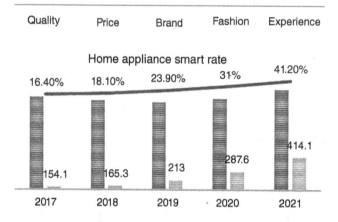

Home appliance smart rate

16.40%   18.10%   23.90%   31%   41.20%

154.1   165.3   213   287.6   414.1

2017   2018   2019   2020   2021

Figure 1.7 Increasing Shares of Smart Products in Appliance Market

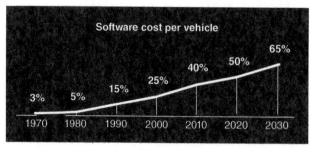

Software cost per vehicle

3%   5%   15%   25%   40%   50%   65%

1970   1980   1990   2000   2010   2020   2030

Figure 1.8 Increasing Software Contents for Automotives

These smart products have drastically reshaped the relationships between producers and consumers:

- The relationship is no longer a single transaction; producers can now access vast amounts of product usage data after the sale. These data can inform after-sales services, consumer behavior analysis, and product evaluations for future improvements. Producers can also provide over-the-air software updates to continually enhance existing products without the need for replacement or recall. Customers can interact wirelessly with producers to provide feedback and request personalized after-sales services. Consequently, the relationship between producers and customers extends throughout the product's lifecycle.
- Beyond the connection between producers and consumers, customers are more connected to other entities like social media, related services, and businesses, while producers are more integrated with other partners. These constant interactions have created an evolving business ecosystem, breaking away from previously siloed professions and business sectors. For example, clothing producers and consumers might interact with lifestyle social groups and sports clubs.

1.5.3.3 Smart Producers, Old Quality Management

As a result of the rapid development of Industry 4.0 technologies, world-class producers are now able to work with increased speed and flexibility. Pervasive sensing and connectivity allow for real-time data generation, capture, aggregation, and sharing from any location. Furthermore, IoT platforms facilitate collaboration among stakeholders—including innovators, engineers, suppliers, and customers—in iterative and concurrent product development. This enables innovative ideas from anywhere in the world to be quickly incorporated into the product. Smart factories are now capable of producing a wider variety of products with nearly zero changeover time. All of these advancements have significantly shortened the product development cycle compared to a decade ago.

However, many current quality management practices, such as certification, accreditation, and regulatory approval, have remained too slow and rigid, thereby becoming bottlenecks in the quest for efficiency.

1.5.3.4 Quality and Innovation

Quality and innovation must increasingly function in tandem in today's rapidly evolving business landscape. Evidence of this is ubiquitous, as observed in the tremendous success of highly innovative companies such as Google, Amazon, and Tesla Motors. In our earlier discussion, we mentioned the perception among many organizations that Six Sigma practices have had a stifling effect on creativity. Furthermore, the current quality management and quality engineering systems do not possess a clear framework for seamless integration with innovation activities. This represents a significant gap in modern quality practices that needs to be addressed.

1.5.3.5 Quality and Risk Management

As technology evolves, so do the inherent risks. A key responsibility of the quality management system is to guarantee the reliability and dependability of products or services. Various effective methodologies, such as Failure Mode and Effects Analysis (FMEA) and fault tree analysis, have been developed and implemented to meet this goal. However, with the advancement of Industry 4.0—the emergence of embedded products and superconnectivity—systems have become exponentially more complex and vulnerable. Consequently, the need for improved risk management strategies has never been more critical.

Table 1.1 Evolution of the Quality Concept in Step with the Different Industrial Revolutions

Timeline	Production System	Quality System	Types of Products
Pre-industrialization Ancient time to late 18th century	**Craft production system** Craftsman or team does all the work	Craftsmen interact with users to know their needs, have a deep commitment to excellence, work like artists, apply workmanship	Custom made, high quality, very expensive
Industry 1.0 From 1784 to late 19th century, the steam engine, and mechanization	**Craft production system** Cottage industry (suppliers), some industrial standards	Similar to craft production, but products become more complex and supplier chains must be managed	Custom made or batch produced
Industry 2.0 From 1870s to 1930s, electric power, and petroleum assembly line	**Mass production system** Standardization, rigid processes, division of labor, and silos	Top priority of quality is defect reduction, emergence of modern quality systems such as statistical process control and total quality management (TQM)	High volume, low cost, little or no variety
Industry 3.0 From 1970s to 2010s, automation, computers, and sensors	**Flexible, lean production system**	Application of TQM, customer centric design, elaborate data analysis, lean Six Sigma	Low cost, more variety some customization
Industry 4.0 From 2010 to present artificial intelligence times, IoT, and big data	**Business ecosystem** Smart, connected, innovative, and knowledge-based production	Quality 4.0. lifecycle data tracking	Smart products

1.6 Summary

In this chapter, we have traced the evolution of quality management from ancient times through the four industrial revolutions, summarizing our findings in Table 1.1. The prevailing theories and practices of Quality Management and Quality Engineering (QM and QE) were initiated after the Second Industrial Revolution, fulfilling the need to control defects and maintain conformance in mass production processes. The Third and Fourth Industrial Revolutions, however, brought about significant changes in lifestyle, business operations, and society as a whole, posing substantial challenges to our existing QM and QE practices. However, these technological revolutions also present tremendous opportunities for the evolution of QM and QE. In subsequent chapters, we will discuss the new generation of QM and QE in the era of Industry 4.0.

References

1 George, B., Hanus, B., and Scott, R. Investigating the dynamics of quality management theory evolution through text analytics. *2019, INFORMS Annual Meeting*, Seattle, WA (20–23 October 2019).
2 Zonnenshain, A. and Kenett, R.S. (2020). Quality 4.0—the challenging future of quality engineering. *Quality Engineering* 32 (4): 614–626.

3 Watson, G.H. (2019). The ascent of quality 4.0. *Quality Progress* 52 (3): 24–30.

4 Dahler-Larsen, P. (2019). *Quality: From Plato to Performance*. Springer.

5 Juran, J.M. (ed.) (1995). *A History of Managing for Quality: The Evolution, Trends, and Future Directions of Managing for Quality*. ASQ Press.

6 Hounshell, D.A. (1984). *From the American System to Mass Production, 1800–1932: The Development of Manufacturing Technology in the United States*. Baltimore, Maryland: Johns Hopkins University Press.

7 Taylor, F.W. (1911). *The Principles of Scientific Management*. Harper & Brothers.

8 Ford, H. and Crowther, S. (1922). *My Life and Work*. New York, USA: Garden City.

9 Juran, J.M. (1951). *Juran's Quality Control Handbook*. New York: McGraw Hill.

10 Barlow, R.E. and Irony, T.Z. (1992). Foundations of statistical quality control. In: *Current Issues in Statistical Inference: Essays in Honor of D. Basu* (ed. M. Ghosh and P.K. Pathak), 99–112. Hayward, CA: Institute of Mathematical Statistics.

11 Martínez-Lorente, A.R., Dewhurst, F., and Dale, B.G. (1998). Total quality management: origins and evolution of the term. *The TQM Magazine* 10 (5): 378–386.

12 Reeves, C.A. and Bednar, D.A. (1994). Defining quality: alternatives and implications. *Academy of Management Review* 19 (3): 419–445.

13 garvin, d.a. (1984). what does "product quality" really mean. *Sloan Management Review* 25: 25–43.

14 Dodge, H.F. (1928). A method of rating manufactured product. *Bell System Technical Journal* 7 (2): 350–368.

15 Shewhart, W.A. (1931). *Economic Control of Manufactured Product*. van Nostrand.

16 Kuhn, A.J. (1988). *GM Passes Ford, 1918–1938: Designing the General Motors Performance-Control System*. University Park, PA: Pennsylvania State University Press.

17 Ohno, T. (1982). How the Toyota production system was created. *Japanese Economic Studies* 10 (4): 83–101.

18 Haidegger, G. (2017). Evolution of technology and users requirements of factory communication systems from the 3rd to the 4th Industrial Revolution. *2017 2nd International Conference on Telecommunication and Networks (TEL-NET)*, Noida, India (10–11 August 2017), pp. 1-1. IEEE.

19 Landesberg, P. (1999). In the beginning, there were Deming and Juran. *The Journal for Quality and Participation* 22 (6): 59.

20 Koiesar, P.J. (1994). What Deming told the Japanese in 1950. *Quality Management Journal* 2 (1): 9–24.

21 Ishikawa, K. and Loftus, J.H. (1990). *Introduction to Quality Control*, vol. 98. Tokyo: 3A Corporation.

22 Taguchi, G. (1985). Quality engineering in Japan. *Communications in Statistics-Theory and Methods* 14 (11): 2785–2801.

23 Chan, L.K. and Wu, M.L. (2002). Quality function deployment: a comprehensive review of its concepts and methods. *Quality Engineering* 15 (1): 23–35.

24 Mikulić, J. and Prebežac, D. (2011). A critical review of techniques for classifying quality attributes in the Kano model. *Managing Service Quality: An International Journal* 21 (1): 46–66.

25 Iba, T., Yoshikawa, A., and Munakata, K. (2017). Philosophy and methodology of clustering in pattern mining: Japanese anthropologist Jiro Kawakita's KJ method. In: *Proceedings of the 24th Conference on Pattern Languages of Programs*, October, 1–11. Vancouver, Canada: The Hillside Group.

26 Nagamachi, M. (1995). Kansei engineering: a new ergonomic consumer-oriented technology for product development. *International Journal of Industrial Ergonomics* 15 (1): 3–11.

27 Dillon, A.P. (1986). *Zero Quality Control: Source Inspection and the Poka-Yoke System*. Productivity Press.

28 Widjajanto, S., Purba, H.H., and Jaqin, S.C. (2020). Novel POKA-YOKE approaching toward industry-4.0: a literature review. *Operational Research in Engineering Sciences: Theory and Applications* 3 (3): 65–83.

29 Deming, W.E. (2018). *The New Economics for Industry, Government, Education*. MIT Press.

30 Suarez, J.G. (1992). *Three Experts on Quality Management: Philip B. Crosby, W. Edwards Deming, Joseph M. Juran*. Arlington VA: Total Quality Leadership Office.

31 Winn, B.A. and Cameron, K.S. (1998). Organizational quality: an examination of the Malcolm Baldrige national quality framework. *Research in Higher Education* 39 (5): 491–512.

32 Tjahjono, B., Ball, P., Vitanov, V.I. et al. (2010). Six Sigma: a literature review. *International Journal of Lean Six Sigma* 1 (3): 216–233.

33 Womack, J.P., Jones, D.T., and Roos, D. (2007). *The Machine that Changed the World: The Story of Lean Production—Toyota's Secret Weapon in the Global Car Wars that Is Now Revolutionizing World Industry*. Simon and Schuster.

34 Womack, J.P. and Jones, D.T. (2010). *Lean Thinking: Banish Waste and Create Wealth in Your Corporation*. Simon and Schuster.

35 Liker, J.K. (2004). *Toyota Way: 14 Management Principles from the World's Greatest Manufacturer*. McGraw-Hill Education.

36 Staley, O. (2019). Whatever happened to six sigma? *Quartz* (3 September). https://qz.com/work/1635960/whatever-happened-to-six-sigma (accessed 29 June 2023).

37 Lasi, H., Fettke, P., Kemper, H.G. et al. (2014). Industry 4.0. *Business & Information Systems Engineering* 6 (4): 239–242.

38 Bai, C., Dallasega, P., Orzes, G., and Sarkis, J. (2020). Industry 4.0 technologies assessment: a sustainability perspective. *International Journal of Production Economics* 229: 107776.

39 Thomas, R. (2020). Tesla ranks last on influential JD Power quality survey. *The Verge* (June 25) 6:39 am. https://www.theverge.com/2020/6/25/21302804/tesla-ranks-last-on-influential-jd-power-quality-survey.

40 Szymkowski, S. (2022). Tesla tops J.D. Power APEAL study, Dodge ranks first for mass market brands. *Road and Show by Cnet* (22 July) 9:00 am. PT. https://www.cnet.com/roadshow/news/tesla-jd-power-appeal-study-dodge.

41 Siegel, J. and Yang, K. (2020). Tesla last in quality, but it's the industry that needs to improve. Commentary by *Automotive News* (26 October) 12:00 am. https://www.autonews.com/commentary/tesla-last-quality-its-industry-needs-improve.

2

Evolving Paradigm for Quality in the Era of Industry 4.0

In an era of unparalleled innovation, the dawn of Industry 4.0 has redefined our approach to business, manufacturing, and society as a whole. However, our understanding of quality – a concept fundamental to all areas of industry – remains largely predicated on theories and practices over six decades old. These constructs, rooted in the crucible of the Second Industrial Revolution, have served us admirably, but are they adequate for the realities of Industry 4.0 and artificial intelligence (AI)? Are they robust and flexible enough to address the seismic shifts brought about by advanced technology and the rapidly changing landscape of contemporary industry?

In the realm of quality management and quality engineering, we are standing at a critical juncture. On one hand, we have the well-established, tried, and tested paradigms that have shaped the industrial world for decades; on the other hand, we face the complexity and volatility of the Fourth Industrial Revolution, an era underpinned by data, interconnectivity, and AI. Quality's intricate and multidimensional nature necessitates careful, informed adaptation of its definitions and standards.

It is time for a reevaluation, a revolution in our understanding of quality that matches the upheavals of Industry 4.0. However, to navigate this new world, we first need to gain a deep and comprehensive understanding of what quality means in the context of this technological and industrial revolution.

This chapter aims to shine a light on the evolving paradigm of quality in the era of Industry 4.0. We will start by revisiting the current definitions and paradigms of quality in Section 2.1, delving into the profound changes brought about by Industry 4.0 in Section 2.2, and building an understanding of what constitutes "Quality 4.0" in Section 2.3. Furthermore, we will explore "hidden gems," lesser-known yet potent ideas on quality in Section 2.4, all leading up to a discourse on the evolving paradigm of quality in the context of this new industrial era in Section 2.5.

2.1 Current Quality Definitions and Paradigms

Quality is a majestic and beautiful word, yet it is laden with confusion and subjectivity. It evokes connotations of excellence and superior value, yet it remains a nebulous concept, subtly changing its contours based on context and perception. The journey to understand "quality" is akin to traversing a labyrinth, each turn revealing new complexities and insights.

In this section, we will embark on this journey, exploring the definitions and paradigms of quality offered by various communities and thought leaders. The discourse will draw from diverse

Quality in the Era of Industry 4.0: Integrating Tradition and Innovation in the Age of Data and AI,
First Edition. Kai Yang.
© 2024 John Wiley & Sons, Inc. Published 2024 by John Wiley & Sons, Inc.

sources – professional bodies such as the American Society for Quality (ASQ) and International Organization for Standardization (ISO), revered quality gurus, namely, W. Edwards Deming and Joseph M. Juran, as well as esteemed members of the academic community.

Each of these sources provides a unique perspective on quality, offering insights that have shaped our understanding and practice of quality over the years. However, these traditional viewpoints will only form part of our exploration. We will also delve into the unconventional, yet profoundly insightful, definitions, and paradigms of quality proposed by two illustrious figures—Robert M. Pirsig and Christopher Alexander.

Robert M. Pirsig, through his philosophical opus, "Zen and the Art of Motorcycle Maintenance," presented a nuanced view of quality, while Christopher Alexander, an architect and design theorist, offered a unique perspective on quality rooted in his "Pattern Language" theory. Their seminal work transcends the confines of traditional definitions, providing an expanded, deeply contextual understanding of quality.

In the following sections, we will thread together these disparate ideas, weaving a rich tapestry of understanding about what quality has been, is, and could be. This exploration will serve as the foundation for our subsequent discussions on the evolution of quality in the context of Industry 4.0.

2.1.1 Definitions from Quality Community

"Quality"—a term resplendent in its excellency and beauty—remains a deeply perplexing concept that has eluded consensus within the quality community. Despite its ubiquity, the simple question "What is the definition of quality?" begets a multitude of responses, each echoing the subjective lens of its respective interpreter.

Let us now delve into some of the prevalent definitions of quality from our traditional quality community:

The American Society for Quality gives the following definition: "Quality" is a subjective term for which each person or sector has its own definition. In technical usage, quality can have the following two meanings:

1) the characteristics of a product or service that bear on its ability to satisfy stated or implied needs.
2) a product or service free of deficiencies.

According to Joseph Juran, quality means "fitness for use"; according to Philip Crosby, quality means "conformance to requirements."

According to ISO 9000, another globally recognized authority, quality represents "the degree to which a set of inherent characteristics of an object fulfills requirements."

Renowned quality guru Dr. Deming offered yet another perspective, defining quality as "a predictable degree of uniformity and dependability with a quality standard suited to the customer."

These definitions, while seemingly divergent, share common threads—satisfying needs, meeting requirements, and defect avoidance. However, their vague and generalized nature often poses challenges in providing clear operational guidance. A notable exception is the straightforward and operationally viable definition – "defect-free product" – which continues to serve as a practical reference point in quality management.

Noticeably, ASQ also underscored the individualistic and subjective facets of quality, a notion that validates the idea that "quality" bears different meanings for different people. As we venture deeper into our exploration of quality, this subjective aspect will surface time and again, reminding us of the multidimensional essence of "quality."

2.1.2 Quality Definitions and Paradigms from Academic Community

The renowned Harvard Professor David Garvin once declared, "If you want to manage quality, you must first understand what quality is." Garvin masterfully outlined eight dimensions of quality—performance, features, reliability, conformance, durability, serviceability, esthetics, and perceived quality. Moreover, he emphasized that the definition and connotations of quality vary significantly when viewed from different vantage points. He distilled these perspectives into five distinct categories [1]:

1) **Transcendent Perspective**: In this view, quality symbolizes the "degree of excellence." However, akin to abstract concepts like beauty, quality, under this perspective, remains an experiential phenomenon that eludes strict definition and quantification.

2) **Product Perspective**: Adopting the lens of product developers and designers, quality is discerned through quantifiable technical metrics such as product functions, features, components, and durability. For instance, in the world of digital photography, image resolution serves as a critical quality measure.

3) **User's Perspective**: From the standpoint of users or customers, quality is a reflection of "how well the product meets their needs," encapsulating whether customers derive the intended utility from the product. The user's needs are individualistic, and associated supporting services also play a crucial role in customer satisfaction. While the user's judgment of quality tends to be subjective, it is an essential factor influencing product success.

4) **Manufacturer's Perspective**: For manufacturers, who can be considered as the downstream users of product developers and designers, quality is synonymous with "standard adherence." Once the design is finalized, their quality goal revolves around producing products that align with these design standards. Any product falling short of these standards is deemed defective or scrap. This view of quality partially fulfills customer needs because a product that does not meet the standards is prone to failure, and no customer favors a defective product. However, even if a product design lacks flair and the product performance fails to impress, from a manufacturer's perspective, as long as it meets the specifications, it is considered of adequate quality. This practical and straightforward interpretation of quality, easiest to implement, has served as the working definition of quality since the Second Industrial Revolution.

5) **Value Perspective**: This viewpoint factors in the product's price while assessing quality. Here, quality is perceived as the ratio of all benefits garnered from the product such as functionality, esthetics, and psychological satisfaction to the price paid.

Carol Reeves and David Bednar [2] contribute another intriguing angle to the discussion on the definition of quality:

6) **Customer Expectation Perspective**: According to this viewpoint, a product of high quality is one that either meets or surpasses customer expectations. Originating from the users' and the value's standpoint, this definition acknowledges the psychological nature of expectations and the comparative analysis customers often engage in when deciding to purchase. Outperforming competitors and exceeding expectations often emerge as decisive factors in many purchasing decisions.

Each of these perspectives not only offers unique insights but also presents limitations. Here are my observations:

- **Transcendent Perspective**: Although "degree of excellence" serves as a striking and accurate definition of quality, its operationalizability remains challenging. Questions such as "how to design products that customers perceive as excellent?" and "who defines the standard of excellence?" are complex and not easily answerable.

- **Product Perspective**: Undoubtedly, quality is chiefly determined by design—an essential aspect. However, this perspective, typical of designers, researchers, and engineers, may be biased toward "technology is everything" and "experts know best," which might not resonate with consumers. The "technology-everything" approach may surpass consumer needs and fail to gain adequate recognition. Furthermore, addressing the diversity of customer needs and their subjective nature proves to be a substantial challenge.

- **Customer Perspective**: Meeting customer needs is a crucial goal pursued by quality professionals, employing numerous methods such as voice of customers (VOC) surveys, Quality Function Deployment (QFD), Kansei engineering, and customer-centric innovation. Despite some successes, this approach is fraught with issues such as customer needs are personalized, but surveys often capture generalized, average needs. Most consumers are not subject matter experts and are unaware of the potential capabilities of a product. Dependence on customer research can sometimes yield dull, uninspiring products, as evidenced by General Motors' experience. On the contrary, Apple, Inc. under the leadership of Steve Jobs, dared to incorporate groundbreaking technologies, resulting in revolutionary products and a successful strategy. Traditional customer surveys might not keep pace with the evolving needs of customers.

- **Manufacturer's Perspective**: This perspective focuses on metrics such as compliance rate, defect rate, and process capability, providing accurate and operational data. These metrics, which are integral since the Second Industrial Revolution, have led many to equate the manufacturer's quality perspective with the overall quality. However, this perspective only represents a part of quality. A low defect rate is desirable, but a mediocre product, even with minimal defects, may not impress customers. Moreover, consumers generally do not understand or concern themselves with internal indicators such as compliance with standards. In today's era of rapidly evolving technologies and increasing correlations between products and services, service quality has gained prominence. However, service quality, being experiential and emotional, is not effectively captured by traditional metrics such as compliance and rejection rates.

- **Value Perspective**: The concept of customer value or price/performance ratio closely aligns with most customers' purchasing decisions. However, determining "how to enhance customer value through good design" is a daunting challenge. There is also a school of thought that believes quality and value cannot be wholly equated, particularly for products with low prices but average functionalities. These may sell well and offer good value for money but may not necessarily be equated with high quality.

- **Customer Expectation Perspective**: This view shares similarities with the value perspective but incorporates the subjective expectations of the customer and the emotional service experience, making it a challenging quality indicator to capture and manage.

These perspectives underscore the multifaceted nature of quality, each with its merits and limitations, together offering a comprehensive understanding of this complex concept.

This exploration elucidates why quality, as a key success factor, remains an intricate and multifaceted concept. Its understanding and definition vary among individuals. The definitions and perspectives on quality mentioned above, each based on solid rationale, play pivotal roles in comprehending this complex notion. However, it also sheds light on why, from the Second Industrial Revolution until now, "low failure rate" has persisted as the most common definition and metric of quality. Despite only representing a fraction of the overall quality concept, it offers the most operational perspective under a mass production system. Although other perspectives such as customer value and product view might carry more importance and value, the quality community has yet to devise a fully operational framework for their practical implementation. Thus, the quest to understand quality in its entirety and devise effective measures continues, necessitating an evolving approach in the wake of rapid technological advancements.

2.1.3 Robert M. Pirsig's View on Quality

2.1.3.1 Summary of Robert M. Pirsig's View on Quality

Robert M. Pirsig, the renowned author of two insightful books [3, 4], emerges as one of the seminal thinkers in the pursuit of defining quality. Although he was not a formal quality professional, his ideas have deeply influenced this field. In particular, his seminal work, *Zen and the Art of Motorcycle Maintenance: An Inquiry into Values*, first published in 1974, extensively explores his concept of Quality. Despite enduring 121 rejections before finding a willing publisher, the book quickly became a bestseller, selling at least five million copies worldwide, and remains influential to this day.

Robert M. Pirsig introduces a comprehensive quality framework, which he proposes as a trichotomy incorporating physical, subjective, and spiritual dimensions:

1) **Objective**: Also referred to as classic, physical, or intellect, this dimension focuses on the underlying science and technology that produces and maintains the physical product. It aligns approximately with design and manufacturing quality in traditional quality parlance, or broadly, the degree of "empirical goodness" of a product or service.
2) **Subjective**: Alternatively named romantic or emotional, this dimension concerns the internal value judgment or emotional reaction a customer has toward a product or service. This reaction influences how much they like or dislike the product. Various factors, including people's romantic views, emotions, prior experiences, life circumstances, and cultural values, influence this quality judgment.
3) **Spiritual**: This aspect acknowledges the innate human desire to enjoy better things and strive for superior life and goals. It can be referred to as the "spirit of excellence."

In his acclaimed book [4], Robert M. Pirsig posits that while the classic or scientific knowledge and work determine what the product is, its quality is judged by the subjective or emotional views of the users. "*Emotion* versus *intellect, technology* versus *romanticism, subject* versus *object by establishing the idea of 'quality' as the foundation for both sides*." He further elaborated that "There are two ways of looking at the world, one is classical understanding, and the other is romantic understanding. A classical understanding sees the world primarily as underlying form itself, by reason and by laws (field of science, law, and medicine). A romantic understanding sees it primarily in terms of immediate appearance. It is primarily inspirational, imaginative, creative, intuitive. Classical understanding is derived from the tradition of Western philosophy, focusing on logic and analytical thinking, while romantic understanding tends to grasp the whole of things and pay attention to beauty." He emphasizes that objective and subjective views of quality are two sides of the same coin and should harmonize rather than oppose each other. He said that "The opposition and split between these two ways of knowing makes our modern life crippled. Without romantic understanding, classical understanding will become a terrifying scientific monster, or a boring 'squareness'. Without classical understanding, romantic understanding will become mysticism. We should strive to combine classical with romantic modes. Let reason take you far enough, but let emotion bring you home."

Importantly, Robert M. Pirsig underlines a spiritual aspect of quality, manifested through the inherent human aspiration for improvement and excellence. He used the fact that even an infant will respond differently toward a beautiful person, and people have an eternal aspiration for betterness and excellence. Quality is also reflected in spiritual interactions among people. He said the following:

> Care and Quality are internal and external aspects of the same thing. A person who sees Quality and feels it as he works is a person who cares. A person who cares about what he sees and does is a person who's bound to have some characteristic of quality.
>
> In our highly complex organic state, we advanced organisms respond to our environment with an invention of many marvelous analogues. We invent earth and heavens, trees, stones and oceans, gods, music, arts, language, philosophy, engineering, civilization, and science. We call these analogues reality. And they are reality. We mesmerize our children in the name of truth into knowing that they are reality. We throw anyone who does not accept these analogues into an insane asylum. But that which causes us to invent the analogues is Quality. Quality is the continuing stimulus which our environment puts upon us to create the world in which we live. All of it. Every bit of it.
>
> [4, 1974, p. 317]

Quality, he proposes, evolves not just as a product metric, but through the entirety of the customer's experience, emotional and spiritual evolution, and the ongoing relationships among all stakeholders.

By way of Robert M. Pirsig's holistic view of quality, we discern the fundamentally individualized nature of quality. It goes beyond physical form to encompass the unique subjective and spiritual aspects of each person.

2.1.3.2 Possible Contributions for New Quality Paradigm

Robert M. Pirsig's views on quality, expressed most notably in his philosophical novel, "Zen and the Art of Motorcycle Maintenance," provide rich insights that could contribute to new paradigms of quality.

In Robert M. Pirsig's novel, he explores the concept of Quality as an undefinable force that exists independently of subjects and objects. Quality, in his view, isn't just about the superiority of a product, but is an intrinsic characteristic of reality itself that humans recognize and respond to but cannot define objectively.

Let's consider how his perspectives on quality could inform a new quality paradigm:

1) **Subjective Experience of Quality**: Robert M. Pirsig argues that "quality" is something we intuitively perceive, and this perception may vary between individuals. In the context of a new quality paradigm, this supports a user-centered approach where the consumer's subjective experience of a product or service plays a key role in determining its quality.
2) **Quality as Excellence**: For Robert M. Pirsig, "quality" is not only about meeting specifications but also about excellence. This could contribute to a new quality paradigm that emphasizes continual improvement and the pursuit of excellence rather than mere compliance with set standards.
3) **Quality as Value**: Robert M. Pirsig suggests that "quality" is equivalent to value and that we inherently recognize and seek quality in everything we do. This could shift the focus of the

quality paradigm from a purely objective measurement to a more complex interplay of subjective experience and objective criteria.

4) **Integrative Approach**: Robert M. Pirsig contrasts the "classical" view of understanding the world (analyzing and categorizing its parts) with the "romantic" view (appreciating the world in its entirety). He suggests that a synthesis of these perspectives leads to a deeper understanding of quality. This could contribute to a new quality paradigm that values both analytical and holistic approaches, recognizing the need for both rigorous metrics and subjective evaluations.

5) **Intrinsic Motivation**: Robert M. Pirsig's view of Quality underscores the importance of intrinsic motivation and personal commitment in pursuing excellence. This perspective could influence a new quality paradigm by highlighting the role of individual commitment and passion in achieving quality outcomes.

6) **Quality as Preintellectual**: Robert M. Pirsig argues that quality is a "preintellectual" experience, which means that it is recognized before it is analytically understood. This could shift the new quality paradigm to consider intuitive, experiential, and emotional aspects of quality, in addition to traditional metrics.

Robert M. Pirsig's views provide a philosophical grounding to the concept of quality, broadening its interpretation beyond objective measures, and emphasizing the interplay of objective and subjective; analytical and intuitive; and parts and whole in understanding and achieving quality. This holistic perspective can be highly valuable in shaping new paradigms of quality.

2.1.4 Christopher Alexander's View on Quality

2.1.4.1 Summary of Christopher Alexander's Work

Christopher Alexander, a renowned architect and design theorist, introduced a unique perspective on architectural design and quality. His paradigm for quality is based on the concept of "living" or "whole" structures that harmoniously fit into their context and meet the needs and preferences of the people who use them. His ideas have been influential not only in architecture but also in other fields such as software design.

His major works, *A Pattern Language* [5] and *The Timeless Way of Building* [6], introduce several key concepts as follows:

1) **Concept of Patterns**: The book introduces the idea of "patterns," or solutions to design problems that recur in different contexts. Each pattern describes a problem and offers a solution, with the understanding that this solution can be adapted to suit individual situations.

2) **Hierarchy of Patterns**: The patterns are arranged in a hierarchy, from the largest patterns (such as regional patterns) down to the smallest (such as those seen in individual rooms). This allows for an understanding of how patterns interact and influence each other, and how solutions at one scale might affect patterns at another.

3) **Human-Centric Design**: The book is centered around the belief that users of space know what is best for their own needs, a contrast to the belief that expert designers know best. The book encourages involving users in the design process, making the patterns more likely to serve their intended purpose.

4) **Holistic Approach**: "A Pattern Language" uses a holistic approach, emphasizing the importance of considering all aspects of a built environment (including social, cultural, and psychological factors) when designing spaces. It recognizes that good design is about creating places that enrich people's lives, not just about esthetics or function.

5) **Importance of Community**: Many patterns in the book deal with creating spaces that foster community, either by facilitating interaction, creating a sense of identity, or providing services and amenities that meet communal needs.

6) **Timeless Quality**: The book advocates for the "timeless way of building," which suggests that there are certain universal and enduring qualities to good design that transcend trends and cultures. The patterns presented in the book are meant to capture these qualities.

7) **Patterns as a Language**: The authors propose that the patterns can form a language that anyone can use to design and build at all scales, from regional planning down to the construction of individual elements in a space. This language helps people to design and build in a way that naturally meets human needs.

"A Pattern Language" has had a broad impact not only influencing architecture and urban design but also influencing fields such as software engineering, where the concept of patterns has been highly influential.

In the field of *quality*, Christopher Alexander contributed at least the following three powerful paradigms:

1) **The *Quality Without a Name* (QWAN)**: Christopher Alexander refers to a certain quality that exists in some buildings and not in others. This quality is hard to define but has certain properties such as being whole or harmonious, being alive, having self-maintenance capabilities, and having a certain simplicity. Buildings with this quality feel comfortable and pleasant to be in. They fit their context and evolve naturally over time. This quality, which he says is "nameless," is what he strives for in his design work.

Christopher Alexander's paradigm for quality is focused on user-centered design and the creation of spaces that feel good to be in. He emphasizes the importance of using local materials and methods, incorporating natural elements, and designing spaces that are adaptable and can evolve over time. His work has had a profound impact on the field of architecture and beyond, inspiring a new approach to designing spaces that are in harmony with their context and the people who use them.

Christopher Alexander's QWAN is an abstract and holistic concept that encapsulates certain esthetic and experiential properties. Christopher Alexander himself didn't provide a set list of buildings that embody this quality. Instead, he focused on the underlying principles that give rise to such quality.

That said, Christopher Alexander did refer to certain types of places that tend to have this quality. These include old, organically grown cities, handmade buildings using local materials and techniques, and natural landscapes. He noted that these places tend to have a sense of "rightness" and "life" that is often missing in modern, mass-produced buildings.

Following are the principles that can contribute to the QWAN:

- **Levels of Scale**: A building or place with QWAN often includes discernible elements at different scales, from the overall layout down to the smallest details.
- **Strong Centers**: The elements of the design create and reinforce "centers" or focal points that draw attention and provide structure.
- **Boundaries**: Elements are often defined by clear boundaries, which can be physical (such as a wall or a path) or perceptual (such as a change in color or material).
- **Local Symmetries**: Rather than strict, overall symmetry, buildings with QWAN often include local symmetries or echoes between different elements.
- **Positive Space**: All spaces, both buildings and the "negative space" around them, are shaped positively and intentionally.

- **Good Shape**: Each element of the design has a clear, distinct shape and contributes to the overall form.
- **Simplicity and Inner Calm**: The design avoids unnecessary complexity, leading to a sense of peace and order.
- **Non-separateness**: The design is in harmony with its surroundings and the wider context.
- **Alternating Repetition**: A pattern that consistently alternates, lending a rhythm to the design.
- **Deep Interlock and Ambiguity**: The design elements are so deeply interconnected that it is hard to distinguish one from the other, producing a complex, rich, and ambiguous design.
- **Contrast**: There is a play of opposites—light and dark, small, big, etc.
- **Gradients**: There is a gradual, gentle variation in some noticeable quality, such as size or density.
- **Roughness**: There is a certain welcomed imperfection and spontaneity that reflects the process of creation, acknowledging that exact precision isn't always ideal or human.
- **Echoes**: Elements within the design resonate with each other, creating a sense of unity and harmony.
- **The Void**: There is calm, emptiness in the mid of order.

Example 2.1 Traditional Mediterranean Villages

These villages, like those found on the Greek island city of Santorini (Figure 2.1), exhibit the principles of independence of scale, positive space, good shape, and deep interlock and ambiguity. Their simple yet distinct architecture and the harmony of the buildings with the landscape embody Christopher Alexander's vision of a living, timeless quality.

Figure 2.1 Village in Santorini. *Source:* gurgenb / Adobe Stock

---

The *Quality Without a Name* is not only about the physical aspects of a place or object but also the feelings and experiences these places or objects elicit. Therefore, it is the overall impression and the intangible sense of wholeness, harmony, and life that truly exemplify this quality.

Christopher Alexander applied the concept of "life" in a design as a way to describe the integration of multiple dimensions such as function, esthetics, human experience, fit to context, and QWAN. Christopher Alexander understood that holistic quality is "objective and precise, but it cannot be named" [7]. Nikos A. Salingaros describes QWAN as an "inner feeling, which we connect to unconsciously" [8]. He emphasized that design has become "focused on abstraction, formalism, and surface appearance to the exclusion of human adaptive factors. There was a little emphasis upon the people for whom the objects were being designed, no discussion about practicality or everyday usage." We see from this that every design is an "irreducible whole" which defies any attempt to quantify or analyze it within a reductionist framework so that reductionist approaches to science actually deter from an understanding of holistic quality.

2) **Human-centered design:** Christopher Alexander's approach to human-centered design places the user's experience, feelings, and comfort at the core of the design process. He stresses the importance of involving the people who will use the design in the process of creating it. This direct participation ensures that the design serves the needs and desires of its users. Christopher Alexander argues that the best designs evolve gradually and organically over time, responding to the changing needs of the users and the environment. This is in contrast to top–down, master-planned approaches. He believes that good design should resonate emotionally with its users, creating a sense of comfort, peace, or joy.

Many famous entrepreneurs benefit from this human centered design approach. Ward Cunningham, inventor of wiki (the technology behind Wikipedia), credits Christopher Alexander with directly inspiring that innovation, as well as the pattern languages of programming, and the even more widespread system of Agile (software development methodology). Will Wright—the creator of the popular games Sim City and The Sims—also credits Christopher Alexander as a major influence, as do the musicians Brian Eno and Peter Gabriel. Apple's Steve Jobs was also said to be a fan [9].

Christopher Alexander's human-centered design has been operationalized into a set of highly flexible design guidelines as follows: (i) Stepwise: Perform one adaptive step at a time; (ii) Reversible: Test design decisions using models; "trial and error"; if it does not work, undo it; (iii) Structure-preserving: Each step builds upon what is already there; (iv) Design from weakness: Each step improves coherence; (v) New from existing: Emergent structure combines what is already there into a new form. This is the way to produce deep value and meaning through a "Wholeness-Generating" design activity [10]. By following these steps, it becomes possible to achieve archetypal resonance, a blending of art and design. David Seamon has gone on to expand the articulation of Christopher Alexander's design guidelines into a list of 10 actions for working toward design quality as Table 2.1 below [11]:

Table 2.1 Christopher Alexander's Design Quality Practices

Christopher Alexander's 10 Structure Enhancing Actions
1) Step-by-step adaptation
2) Each step helping to enhance the whole
3) Always making centers
4) Following steps to unfold in the most fitting order
5) Creating uniqueness everywhere
6) Working to understand needs of clients and users
7) Evoking and being guided by a deep feeling of whole
8) Finding coherent geometric orders
9) Establishing a form language that rises from & shapes thing being made
10) Always striving for simplicity by which thing becomes more coherent and purer

3) **Timeless quality:** Christopher Alexander's concept of "timeless quality" refers to a profound, enduring, esthetic, and functional essence present in certain buildings, places, and objects, regardless of their cultural or historical context. This concept is the central theme of his book, *The Timeless Way of Building.*

   Following are some of the key points:
   - **Universality**: Timeless quality transcends culture, geography, and time. It represents an inherent sense of rightness and harmony that resonates with human beings universally. These are designs that—regardless of their cultural or historical context—remain beautiful and functional.
   - **Connotation**: Timeless quality connotes something enduring; tried, and, tested true; and not susceptible to trends or fads. It emphasizes durability and lasting appeal.
   - **Wholeness and harmony**: Achieving this timeless quality requires a sense of "wholeness," where all parts of a design fit together seamlessly and harmoniously, each contributing to the integrity and effectiveness of the whole.
   - **Emotional resonance**: Christopher Alexander believes that the timeless quality in a design should elicit a deep, emotional response—often a sense of peace, comfort, or joy.

In essence, Christopher Alexander's timeless quality is about creating designs that are deeply human-centered, adaptable, and emotionally resonant, leading to spaces and objects that feel intuitively "right" and enhance the quality of life for their users.

2.1.4.2 Possible Contributions for New Quality Paradigm

Christopher Alexander's views on quality, particularly as expressed in his seminal work on architecture and design, *A Pattern Language*, offer interesting insights that can contribute to new paradigms of quality. Christopher Alexander's views center around the concept of generating life within a design, with a focus on organic, adaptable, and human-centric structures. He underscores the importance of beauty, harmony, and comfort in creating an environment that resonates with human beings.

In this context, let us examine how his views could contribute to a new quality paradigm across various domains:

1) **User-Centric Design**: In the field of product and service design, Christopher Alexander's principles argue for a shift toward a more user-centric approach, focusing on the "fit" between the design and the user's needs and desires. This concept aligns with the new paradigm of quality, which emphasizes the role of the user's experience in determining the quality of a product or service.
2) **Adaptability and Evolution**: Christopher Alexander's concept of design underscores the idea that good design is not static; instead, it adapts and evolves according to the changing needs of the user and the environment. This idea can influence the new quality paradigm by emphasizing that quality is not a one-time achievement but an ongoing process of adaptation and improvement.
3) **Harmony and Balance**: Christopher Alexander argues that quality in design stems from achieving harmony and balance within a system. In the new quality paradigm, this can translate into the need for balancing various aspects of quality, such as efficiency, effectiveness, sustainability, and user satisfaction.
4) **Patterns and Wholeness**: Christopher Alexander's pattern language theory, which involves breaking down complex systems into repeatable "patterns," could inform the new quality paradigm by emphasizing that quality arises not just from individual components but from the interplay and integration of these components into a coherent whole.
5) **Esthetic Quality**: Christopher Alexander's work consistently stresses the importance of esthetic quality, arguing that environments that elicit a sense of beauty and joy contribute to human

wellbeing. This perspective could encourage a broader definition of quality that goes beyond functionality and efficiency to include elements of beauty, pleasure, and emotional resonance.

6) **Sustainability**: Christopher Alexander's views on design also incorporate elements of sustainability and respect for nature, suggesting that quality should be evaluated not only in terms of immediate results but also in terms of long-term impact on the environment and society.

Thus, while Christopher Alexander's views stem primarily from architecture and design, they have broader applicability and can significantly influence the way we understand and approach quality in various fields.

2.2 Changes Brought by Industry 4.0

2.2.1 Smart Manufacturing

The potential scope of Industry 4.0 is enormous, encompassing both a wide range of foundational technologies and various areas of application. These technologies span the spectrum from the Internet of Things (IoT) to extensive networks of interconnected smart machines and devices. The integration is facilitated by progressive tools such as 5G wireless technology, AI, scalable analytics through Big Data, and augmented reality (AR). The amalgamation of these technologies into sophisticated cyber–physical systems epitomizes the current stage of Industry 4.0 application. While still in the maturation phase, steady state implementation is being worked toward to demonstrate viable return on investment for businesses and society at large.

Industry 4.0 applications are not confined to manufacturing; rather, they reach into every corner of human activity. This includes sectors such as energy, agriculture, healthcare, business, and government.

A representation of the current state of Industry 4.0 application is provided by the National Institute of Standards and Technology (NIST) through its smart manufacturing ecosystem (as depicted in Figure 2.2). This showcases the application of 4.0 technologies and processes (outlined in the automation pyramid, Figure 2.3) across the product, production, and business dimensions of an enterprise. The smart manufacturing ecosystem [12], featuring both horizontal and vertical integration, is characterized by the following five key features: (i) Digitization; (ii) connected devices and distributed intelligence; (iii) collaborative supply chains; (iv) integrated Energy and resource management; (v) advanced sensors and big data analytics.

The smart manufacturing ecosystem typically integrates its applications with various capabilities across the enterprise, including

1) throughput or production capacity,
2) overall equipment effectiveness (OEE),
3) material and energy efficiency,
4) labor productivity,
5) flexibility or responsiveness to change,
6) on-time delivery,
7) fault tolerance,
8) product quality,
9) innovation at product or process levels,
10) customization or variety of products or services,
11) customer service,
12) product integrity (life cycle),
13) process integrity, and
14) logistics effectiveness

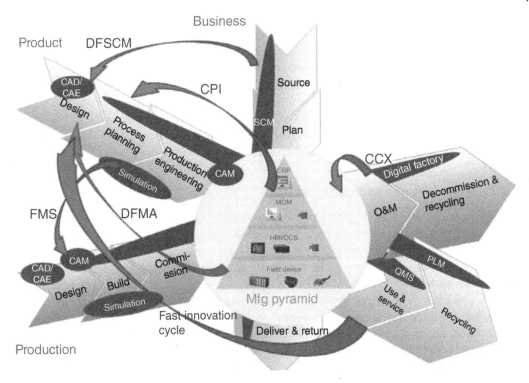

Figure 2.2 The NIST Smart Manufacturing Ecosystem

The technologies within the ecosystem are organized in a hierarchical structure, as depicted in Figure 2.3. This hierarchy ranges from level 1—smart field devices and sensors to level 2—smart machines, human machine interfaces (HMI), and distributed control systems (DCS). It continues up to level—3 with connected manufacturing operations management (MOM) or manufacturing execution systems (MES), and culminates at level 4—integrated enterprise and extended enterprise functionality [12].

Among the plethora of Industry 4.0 techniques, including cloud computing, Big Data, and additive manufacturing, the IoT stands as a pivotal enabler of the Fourth Industrial Revolution. The IoT technology hinges on two primary components, namely, IoT devices and IoT platforms [13]. The IoT devices are physical objects and everyday items imbued with internet connectivity through the integration of unconventional computing devices. Notable examples of IoT devices include smart TVs, wearable technology, and smart appliances. These devices differ from traditional ones in

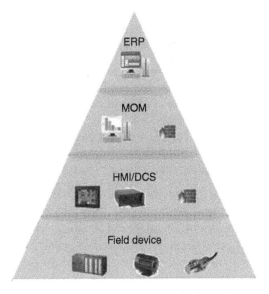

Figure 2.3 The Industry 4.0 Technology Pyramid.
Source: Lu et al. [12]/IEEE

their ability to transmit data, communicate, and interact over the internet. Moreover, they can be remotely monitored and controlled via IoT platforms.

As Industry 4.0 technologies continue to advance, an increasing number of products, along with their respective production equipment, infrastructure, and tools, are becoming embedded with IoT. This evolution is giving rise to smart products, smart factories, and smart suppliers. Furthermore, all stakeholders of products, such as customers, producers, and suppliers, are equipped with internet-enabled devices such as computers, smartphones, and tablets. In essence, the business world is evolving into a myriad of dynamic, internet-embedded ecosystems that thrive, adapt, and interact within various market environments.

2.2.2 Smart Enterprise by Superconnectivity

Leveraging the power of Industry 4.0 technology, particularly IoT platforms, enables a comprehensive integration of the extended enterprise. This includes research and development, engineering, sales and marketing, and suppliers, all converging to form a smart enterprise.

An IoT platform serves as a crucial facilitator. It functions as a software-based operating system that unifies all IoT components; it facilitates communication, streamlines data flow, manages devices, and empowers various applications using other Industry 4.0 techniques such as cloud computing, Big Data, and AI. In essence, the IoT platform orchestrates how numerous IoT devices, stakeholders, and smart systems within an ecosystem are connected, controlled, managed, and collaboratively functioning.

Various IoT platforms currently exist in the market, each at different stages of their evolution. Some are fundamental, capable of connecting and collecting data from IoT devices, providing device security, and offering data analytics services to users. Others, more advanced platforms such as PTC ThingWorx, AWS IoT Core, GE Predix, and Siemens MindSphere [14], provide additional capabilities. Beyond basic functions, these platforms monitor the condition and performance of smart devices and machines, plan condition-based maintenance, manage manufacturing operations, oversee wind farms, and track smart product usage, among other things.

A recent breakthrough in IoT platforms is the development of the CosmoPlat platform [15], which is the platform used by Haier Group, a multinational home appliance and consumer electronic company. Besides offering the regular functionalities of other IoT platforms, CosmoPlat presents several unique features:

1) **Superconnectivity**: It does not just link all the smart products produced and the devices in smart factories, but it also enables connection with customers, producers, suppliers, and extended business enterprises associated with those products through computers, smartphones, and network devices.

2) **Product Innovation and Production Management**: Utilizing its superconnectivity, the platform provides a fully interactive, transparent system to manage the entire process of product innovation, development, iterative enhancement, mass customization, and agile production.

Figure 2.4 illustrates how this platform interacts with, connects to, and manages a manufacturing ecosystem. This ecosystem encompasses all stakeholders of a product, including customers, wholesalers and retailers, product designers, suppliers, logistics, R&D providers, and even the "smart production system" made up of smart factories, modular subsystem providers, and parts factories. The platform can facilitate all activities related to the operation of this ecosystem, including business transactions, production management, virtual meetings, and more.

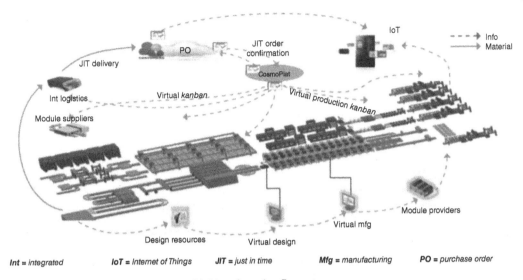

Figure 2.4 How CosmoPlat interact with Manufacturing Ecosystem

2.2.2.1 How Superconnectivity Affects Product Development and Production

Figure 2.5 demonstrates how a fully connected and interactive product development process operates. The initial step in this process involves generating a need for a new product. This need could be driven by customer input, new technology, the identification of a hidden market niche, or a combination of these factors. Once the needs are articulated on the platform, all connected designers and solution providers can submit their proposed designs and plans. If there are any technical challenges or obstacles, all connected researchers and solution providers can contribute their solutions.

Following this, the process moves into an interactive design improvement phase. During this phase, customers, designers, solution providers, smart plant engineers, and suppliers all communicate

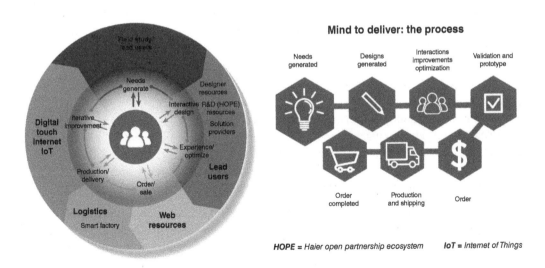

Figure 2.5 Fully connected and Interactive Product Development Process

interactively through the platform. At the end of this stage, several virtual prototypes will have been created and subjected to collaborative testing and iterative improvements.

After this phase, the first batch of products is produced and released into the market.

2.2.3 Other Changes Brought by Industry 4.0

In addition to the IoT technology already discussed, there are various other technologies currently in use:

- Pervasive sensing and connectivity ensure data can be generated, captured, aggregated, and shared anywhere, anytime.
- Internet of behaviors (IoB) systems, comprising wearable, mobile, and ambient devices provide advanced consumer insight and open feedback channels. Nelin and Winer posit that marketers can benefit from "almost unlimited" data [16].
- The AI utilizes limited input data to transform information into actionable insight.
- Hypertargeting advertising helps identify and reach niche markets.
- "Over-the-air" updates enable products to adapt to new and evolving use cases.
- Flexible manufacturing and adaptable architectures facilitate rapid, scalable customization of offerings.
- Cyclical design allows consumers' feedback to inform product offerings and vice versa.
- Emerging standards and tools are simplifying the integration of high technology into various systems.

Collectively, these technologies are driving Industry 4.0 closer to reality. If we consider historical patterns, we can observe that each significant industrial revolution precipitates substantial societal changes. The Second Industrial Revolution, for example, introduced the mass production model, effectively revolutionizing the traditional craftsman production model. This led to profound and enduring changes, such as division of labor, professional specialization, rigid hierarchical organizational forms, and standardized sequential processes for research, development, and production.

The Third Industrial Revolution, characterized by computing and automatic control technology, has already begun to alter some of these models. For instance, emerging industries such as software and electronics operate differently from traditional industries. As Industry 4.0 continues to evolve, we anticipate revolutionary changes, some of which are already occurring are listed in the following:

1) Hyperconnectivity for data and information.
2) Decentralization of R&D and production.
3) Autonomy and flexibility within organizational structures.
4) Revival of the modern craftsman model: small innovative companies.
5) Flexible and rapid collaboration of business elements.
6) Intelligent information systems capable of guiding and controlling many processes.
7) Standardized intelligent production processes becoming low cost and widely accessible.

More advanced devices, components, and technology are continually being integrated into smart manufacturing processes, bringing with them new capabilities and challenges. For example, a high-precision 3-D scanner can swiftly create solid 3-D computer models of a part, vastly outperforming coordinate measurement machines in quality monitoring. However, this also necessitates a "graphic-based quality control methodology."

2.2.4   Summary: Impact of Industry 4.0 on Quality

Industry 4.0 brings about transformative changes to the realm of quality management. At the core of this evolution is the shift from traditional methods of quality control to advanced, data-driven techniques empowered by emergent technologies. The key elements of this transformation include the following:

1) **Hyperconnectivity and Data Availability**: The advent of pervasive sensing and connectivity ensures the availability of data at an unprecedented scale. This enhances real-time monitoring and allows for proactive quality management.
2) **Advanced Analytics and Artificial Intelligence**: AI and other advanced analytics tools enable businesses to derive actionable insights from the vast amounts of data generated by interconnected devices and systems. This facilitates improved decision-making and predictive capabilities, which are vital for ensuring quality.
3) **Internet of Behaviors Systems**: These systems leverage a variety of devices to provide advanced consumer insights, opening up new channels for feedback and quality improvement.
4) **Flexible Manufacturing**: Adaptable architectures and flexible manufacturing processes facilitate rapid, scalable customization of offerings, allowing businesses to quickly adapt to quality demands.
5) **Over-the-Air Updates**: This capability enables products to adapt to new and evolving use cases, maintaining their quality over time.
6) **Cyclical Design Process**: A dynamic design process allows for continuous feedback and iterative improvements based on consumer and stakeholder input, leading to better quality products.
7) **Incorporation of Advanced Devices and Technologies**: New technologies such as high-precision 3-D scanners offer improved methods for quality control and monitoring.

Through these transformative changes, Industry 4.0 is enhancing the effectiveness and scope of quality management systems, leading to a new era of quality.

Since the introduction of the term "Industry 4.0," the quality community has been deliberating on its impact on quality management. Many believe that new technologies from Industry 4.0 will upgrade the current quality management system, leading to the evolution of "Quality 4.0." This concept will be discussed further in the following subsection.

2.3   Quality 4.0

2.3.1   What Is Quality 4.0

The term Quality 4.0 can be traced back to an article, *he Future of Quality*, [17] published by the ASQ in 2015. Since then, the concept has continued to gain traction, with numerous articles and lectures contributing to the dialogue [18, 19]. Beginning in 2017, ASQ has been hosting an annual Quality 4.0 Summit, drawing significant attention and participation. Many multinational corporations have similarly embraced this trend.

Some thought leaders have taken the initiative to align each stage of the industrial revolution with corresponding advancements in quality management. This parallel development from Quality 1.0 to Quality 4.0 corresponds with the evolution from Industry 1.0 to Industry 4.0, [20] as represented in Table 2.2:

Table 2.2 Industrial Revolutions and Quality (ASQ)

Period	Summary Description	Quality	Summary Description
Industry 1.0–Prior to 1890	• Humans harness water and steam power to build industrial infrastructure • Crude machines gain productivity over independent craft work • Increased output is achieved using mechanical advantages • Work focuses on performing tasks faster and more consistently • Transportation/moving goods occurs more frequently	Quality 1.0	• Quality is assured through measurement and Inspection • Production volume is emphasized rather than quality • Inspection does not focus on cost reduction, eliminating wastes, or loss and inefficiency • Work conditions are not important: maximizing worker productivity takes precedence
Industry 2.0–From 1890 to 1940	• Electricity powers industrial machines • Performance capability gains occur through application of new mechanisms • Scale of automation becomes broader as motor size can be varied to fit specific circumstances	Quality 2.0	• Maximizing productivity continues to be the primary focus • Adherence to standards that reflect the minimally acceptable quality level is prevalent • Financial quality is measured based on scrap and rework • Labor performance is used to measure productivity
Industry 3.0–1940 to 1995	• Computer power provided to workers to Increase productivity • Use of information and communication technology drives improvements • Human participation in workplaces declines • Stand-alone robotic systems replace manual work	Quality 3.0	• Quality is a business imperative • Meeting customer requirements (customer satisfaction) is emphasized • Continual Improvement Is applied • Gains in productivity occur by stabilizing highly efficient processes, standardizing work and involving all workers in the activities that create quality • Standardization activities (ISO 9001) and achieving business excellence through organization wide assessment (such as the *Baldrige Criteria tor Performance Excellence*) emerge
Anticipated changes that will occur during Industry 4.0–From 1995 to present	• Integrated cyber–physical interfaces automate working environments • Automated processes deal with end-to-end systems • Humans serve only in positions where human judgment cannot be automated and human Interactions cannot be simulated • Machines learn to learn (AI)	Quality 4.0	• Digitization is used to optimize signal feedback and process adjustment, and adaptive learning supports self-induced system corrections • Quality shifts its control-oriented focus from the process operators to the process designers • Machines learn how to self-regulate and manage their own productivity and quality • Human performance is essential; the emphasis shifts from production to system design and integration with the business system

The concept of mapping each stage of quality management to a corresponding industrial revolution is not precisely accurate. The first industrial revolution, which occurred in England, did not revolutionize the prevailing craftsman production model, and thus, no transformative quality system emerged from this era. There were some instances of industrial standardization, but these were sporadic. In fact, Juran observed that the ancient Chinese government had established standards and that the consistency of weapon processing during China's Qin dynasty approached modern industrial levels.

The modern quality system, which originated in the United States and was refined in Japan, particularly through total quality management (TQM), arose from the necessity of defect control within the mass production model of the Second Industrial Revolution. During the 1930s, pioneers such as Shewhart, Deming, and Juran developed statistical-based quality control systems. This foundation was later built upon by a series of innovations in Japan, forming the backbone of TQM as we discussed in Chapter 1.

The Third Industrial Revolution, marked by information technology and computers, had some influence on upgrading quality management. Examples include real-time fault detection and control in mass production, the application of computer-aided design (CAD) platforms for design quality, and the use of reliability simulation tests to improve product durability. Moreover, the data-driven Six Sigma movement helped popularize quality management systems and methodologies across all business sectors. However, these improvements were incremental and did not prompt "revolutionary" paradigm shifts.

Strictly speaking, if the quality assurance of the craftsman production system dating back to ancient times is classified as Quality 0.0, or the ancient quality system, then TQM could be deemed Quality 1.0, and our current quality management system could be referred to as Quality 1.0+.

However, it is apparent that the transformations brought about by Industry 4.0 are quite substantial. The traditional mass production system might undergo significant changes, or even become obsolete in many sectors of the economy. These shifts are likely to bring considerable, potentially revolutionary, alterations to our quality management system. Hence, we use "Quality 4.0" to denote "Quality Management under Industry 4.0," acknowledging that Quality 4.0 may not be the perfect name.

2.3.2 American Society of Quality's Descriptions on Quality 4.0

The American Society of Quality is the organization that promotes the course of Quality 4.0 [20], ASQ hosts Quality 4.0 conference annually since 2017, and publish many articles in its flagship magazine Quality Progress [13, 21–25]. Here I summarize some of the main points in ASQ's Quality 4.0 narratives.

2.3.2.1 American Society of Quality Definition of Quality 4.0
Quality 4.0 brings together Industry 4.0's advanced digital technologies with quality excellence to drive substantial performance and effectiveness improvements.

2.3.2.2 Key Features of Quality 4.0
1) Digitization is used to optimize signal feedback and process adjustment, and adaptive learning supports self-induced system corrections.
2) Quality shifts its control-oriented focus from the process operators to the process designers.
3) Machines learn how to self-regulate and manage their own productivity and quality.
4) Human performance is essential; the emphasis shifts from production to system design and integration with the business system.

2.3.2.3 Establishing and Implementing Quality 4.0 Principles

The ASQ believed that it is very important for quality professionals to help their organizations to help in the "digital transformation" and to upgrade their quality system by focusing on:

- **People**: Quality 4.0 is more than technology. It is a new way for quality professionals to manage quality with the digital tools available today and understand how to apply them and achieve excellence through quality. By speaking the digital language and making the case for quality in disruption, quality professionals can elevate their role from enforcers to navigators to successfully guide organizations through digital disruption and toward excellence.
- **Process**: As more work is automated the need for flawless processes remains the same, if not more important. Existing processes will be broken and the need to educate the next generation of workers to implement new processes and strategies will be vital to not only the quality professional but also the business operations. Quality is a vital link and should be included at the strategic level for sustainability during digital transformation.
- **Technology**: Technology is growing 10 times faster than it used to, and organizations' platforms, such as processes, systems, data, operations, and governance, must keep pace. Technology also is a great leveler because it gives any individual with the right idea and intent the capability that previously was available only to large organizations. Quality professionals must move from data analyst roles to data wrangler roles by engaging with new technologies, understanding these technological advancements and the potential outputs they create, and determining how and when to use them.

2.3.2.4 Quality 4.0 Tools

The ASQ believes that in the digital transformation process. In addition to the timeless and well-known quality tools and principles, the Quality 4.0 tools below should be leveraged to alleviate these challenges when implementing and deploying systems to support digital transformation.

- **AI**: Computer vision, language processing, chatbots, personal assistants, navigation, robotics, and making complex decisions.
- **Big Data**: Infrastructure (such as MapReduce, Hadoop, Hive, and NoSQL databases), easier access to data sources, and tools for managing and analyzing large datasets without having to use supercomputers.
- **Blockchain**: Increasing transparency and auditability of transactions (for assets and information), monitoring conditions so transactions do not occur unless quality objectives are met.
- **Deep Learning**: Image classification, complex pattern recognition, time series forecasting, text generation, creating sound and art, creating fictitious video from real video, adjusting images based on heuristics (make a frowning person in a photo appear to smile, for example).
- **Enabling Technologies**: Affordable sensors and actuators, cloud computing, open-source software, AR, mixed reality, virtual reality (VR), data streaming (such as Kafka and Storm), 5G networks, IPv6, and IoT.
- **Machine Learning**: Text analysis, recommendation systems, email spam filters, fraud detection, classifying objects into groups, forecasting.
- **Data Science**: The practice of bringing together heterogeneous datasets for making predictions, performing classifications, finding patterns in large datasets, reducing large sets of observations to the most significant predictors, applying sound traditional techniques (such as visualization, inference, and simulation) to generate viable models and solutions.

2.3.2.5 Quality 4.0 Value Propositions

The ASQ thinks that Quality 4.0 will deliver these values to quality communities, Value propositions for Quality 4.0 initiatives fall into six categories, listed in order of significance:

1) Augment (or improve upon) human intelligence.
2) Increase the speed and quality of decision making.
3) Improve transparency, traceability, and auditability.
4) Anticipate changes, reveal biases, and adapt to new circumstances and knowledge.
5) Evolve relationships, organizational boundaries, and concept of trust to reveal opportunities for continuous improvement and new business models.
6) Learn how to learn by cultivating self-awareness and other awareness as skills.

Quality professionals are perfectly positioned to propose and lead digital transformation initiatives because they have deep skills in

- systems thinking,
- data-driven decision making,
- leadership for organizational learning,
- establishing processes for continuous improvement, and
- understanding how decisions affect people in terms of lives, relationships, communities, wellbeing, health, and society in general.

2.3.3 Reflecting on ASQ's Quality 4.0 Narratives

1) The ASQ demonstrated an early recognition, starting in 2015, that Industry 4.0 technologies would substantially influence quality management and the quality community. They remain one of the leading voices in this discussion on the "new quality."
2) The ASQ's proposed Quality 4.0 framework includes many technology elements, such as Big Data, AI, and blockchain. This framework keeps pace with technological developments commendably well.

However, judging from the "Value Proposition of Quality 4.0" presented by ASQ, the focus remains predominantly on the traditional quality system and process management. The approach does not diverge significantly from the existing TQM system. It can be best described as "TQM enhanced with digital capabilities."

2.4 Hidden Gems: Lesser Known but Potent Ideas on Quality

In the dynamic sphere of Quality Assurance and Management, we have journeyed through a broad landscape of definitions, frameworks, and paradigms. After a thorough review of established models, from the traditional concepts of quality, Robert M. Pirsig and Christopher Alexander's intriguing thoughts, to the contemporary perspectives brought about by Industry 4.0 and the ASQ's Quality 4.0 framework, it is evident that the discourse of quality is ever evolving. The transformative impact of the Fourth Industrial Revolution on quality, marked by the integration of AI, cyber–physical systems, IoT, cloud computing, and cognitive computing, illuminates the necessity for a new, updated paradigm on quality.

However, the exploration does not stop at the juncture of these well-known definitions and paradigms. As we delve deeper into the labyrinth of quality discourse, there are numerous "hidden gems" awaiting discovery—potent, lesser-known ideas, and concepts that hold immense promise for our understanding and practice of quality in this new era. These hidden gems are not just theory; they are practical and applicable insights that can reshape our perspective on quality and help us navigate the complexities of the digital era.

In the forthcoming subsections, we will unearth these gems, focusing not only on their theoretical underpinnings but also on their practical implications and potential for shaping the future of quality. We aim to stimulate a discussion that goes beyond the established paradigms and ventures into new territories, fostering a culture of continuous learning, innovation, and improvement. After all, the essence of quality lies in the relentless pursuit of excellence—and that is the journey we are here to embark on together.

2.4.1 Quality as Customer Value

"Customer value" stands as a prominent business metric closely intertwined with quality. In Feigenbaum's 2004 paper [26], he stated that customer value is "The future of quality." In the same vein, several influential experts have proposed that customer value, viewed through the lens of value and expectation, should serve as the new definition of quality [27, 28].

Among the myriad of customer value models available, one particular model stands out for its comprehensive and straightforward approach [27, 28]:

$$\text{Customer value} = \text{Benefits} - \text{Liabilities}$$

The benefits comprise the following:

1) Functional benefits
2) Psychological benefits
3) Service and convenience benefits

Meanwhile, the liabilities are constituted by

1) economic liabilities (customer costs),
2) psychological liabilities, and
3) service and convenience liabilities.

To delve into these elements of customer value, the following features help:

1) "Functional benefits" encapsulate the direct capabilities and functionalities that products or services provide for customers. Consider a mobile phone serving as a multiplatform hub for communication, information, entertainment, and payments. The features, performance specifications (e.g., camera resolution, photo quality), safety, reliability, and durability of the product all fall into this category.
2) "Psychological benefits" refer to the emotional satisfaction derived from acquiring a particular product or service. This could be the pride associated with owning a luxury vehicle, the excitement of wearing a T-shirt adorned with a celebrity's image, or the cultural enrichment from visiting a renowned tourist destination.
3) "Service and convenience benefits" extend beyond the product itself to encompass the overall user experience, from the ease and convenience of delivery to the quality of after-sales service.
4) "Economic liabilities" incorporate the financial costs tied to a product, not only the initial purchase price but also the subsequent expenses such as loan interest, service fees, shipping and installation costs, and maintenance fees. Moreover, these liabilities also cover non-monetary costs such as time spent, spatial requirements, and potential lifestyle disruptions.

As such, this customer value model provides a comprehensive view of the balance between the benefits and liabilities of a product or service, aligning closely with the modern definition and expectations of quality.

Example 2.2   Sports Utility Vehicles

The concept behind the sports utility vehicle (SUV) ingeniously combines functionalities and features found in both trucks and passenger cars. When the SUV first emerged on the automotive scene, it boasted unique features that neither traditional cars nor trucks could singularly offer, such as exceptional comfort, ample cargo space, and generous seating capacity. This new breed of vehicle introduced customers to a fresh array of functional benefits and effectively addressed a number of previously unmet needs. Coupled with a reasonable pricing strategy, the SUV offered unparalleled customer value. Consequently, it rapidly expanded its market share, testifying to the strong appeal and utility it offered to a wide range of consumers (Figure 2.6).

| SUV | Car | Truck |

Figure 2.6   Sports Utility Vehicle's Customer Value. *Source:* Mike Mareen / Adobe Stock, nitinut380/ Adobe Stock, Ermell/Wikimedia Commons

---

Example 2.3   The Value of Brands

As can be seen from Figure 2.7, the physical functions of those T-shirts are the same, but after the brands are printed, the psychological benefits brought by the brand to customers greatly increase the value to some customers, so they can be sold at higher prices.

Figure 2.7   The Value of Brands

Recognizing, cultivating, and steering customer value effectively serves as a benchmark for business success [29–31]. Customer value emerges from a myriad of factors, including design choices, manufacturing quality, the outcomes of use, overall customer experience, and the relationship between the customer and the offering [32–34]. However, it is essential to note that perceived value fluctuates with the lifecycle stage of the offering and the context of the market [35, 36], cumulatively influencing the customer's lifetime value. Notably, this value is inextricably linked with quality, cementing the vital relationship between these two dimensions [28, 37].

2.4.2 Individualized Customer Value

The concept of customer value has evolved, leading to the recent proposition of "individualized customer value" [25, 38]. This term signifies that value propositions can vary significantly between individuals. For instance, when considering consumer goods like mobile phones or cars, individuals value functions and features differently, display varying levels of price acceptance, and hold distinct perspectives on brands.

Our previous analysis hinted at the quality perspectives of "value" and "expectation" aligning closely with the psychological process involved when "customers choose products or services." This process is inherently individualized, emphasizing that traditional quality, represented by a low defect rate, only encapsulates one aspect of quality. This aspect is a historical definition of quality arising from Industry 2.0, not entirely in line with the transcendent definition of quality as a "degree of excellence." Individualized customer value, in contrast, resonates more closely with the intrinsic meaning of quality, "value for money," and could hence be deemed inherent quality.

The quality concept of individualized customer value encapsulates both subjective and objective quality. Objective quality represents the value derived from "functional benefits," while subjective quality is tied to the "psychological benefits."

Robert M. Pirsig put forth the idea that quality is an emergent property rooted in the holistic integration between the product, the consumer, the producer, and most importantly, the context within which these entities interact. Hence, quality is more than a simple product metric; it evolves throughout the customer's usage experience and the ongoing relationship among all stakeholders.

Through this holistic perspective of quality, the fundamentally individualized nature of quality becomes evident. It transcends mere physical form and encapsulates unique subjective and spiritual aspects for everyone.

Building on Maslow's hierarchy of human needs, Yang [25, 38] put forth a model that accurately defines quality based on individualized customer value, as represented in Figure 2.8.

When considering the "quality of products," we must acknowledge the vast diversity among products. Broadly speaking, we can categorize products into the following five primary types in the business world:

1) **Consumer Products**: These are finished products available for sale to customers, including individuals, families, social groups, and small business units. Examples include cars, mobile phones, clothes, toys, appliances, and soft drinks.
2) **Industrial Products**: Businesses typically purchase these products to aid in their operations or to produce other goods. This category encompasses a wide range of products, from standard industrial equipment like power stations and container ships to components and parts of consumer goods or industrial equipment and industrial materials such as rolled steel and plastic films.
3) **Commodities**: Examples include crude oil, raw steel, copper, grains, etc.
4) **Service Products**: This vast category includes all types of services, such as healthcare, financial services, restaurants, entertainment, and more.
5) **Premier Products**: This category encompasses artworks, luxury products, premium services, and landmark architecture.

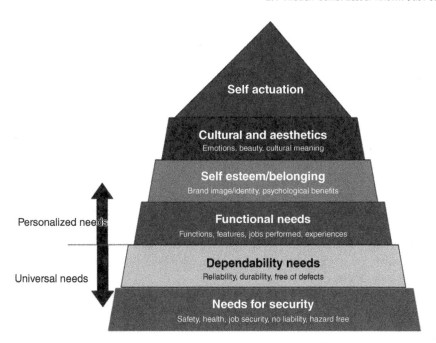

Figure 2.8 Quality Pyramid: Individualized Customer Value

Referring to the "quality pyramid" depicted in Figure 2.8, quality is conceptualized as meeting needs. The pyramid's foundational level, "safety requirements," encompasses product safety, legal compliance, authenticity, food safety, and more. These requirements must be fulfilled 100%. The pyramid's second level, "dependability," signifies durability, reliability, and the absence of failures. These two levels represent basic needs that must be satisfied universally, without exceptions for all products and all customers. These needs align well with the scope of traditional quality management.

The pyramid's third level pertains to the product's functionalities. For industrial products, the functional needs are likely standard and well defined. For commodities, both providers and users typically cannot modify their functionalities or properties, and their price might correspond to their grade level. For consumer products and service products, the functional needs may be varied, dynamic, or individualized according to different customers. For instance, customers' functional requirements for cameras can be highly diverse, with different performance, indicators, and form factors satisfying different customers. Fulfilling functional requirements depends on the alignment between product design and customer needs.

For all levels above the pyramid's third, including self-esteem, culture, and esthetics, customer experience, and emotional needs, these might be crucial for premier products and require meticulous attention. Most higher level needs derive from psychological aspects and tend to be more personalized. Hence, we usually cannot satisfy them with a "one size fits all" approach. This fact underscores why the design aspect of quality is so challenging.

2.4.3 Peter Drucker's View: Good Quality and Poor Quality

In 2001, Gregory Watson, then-chairman of ASQ's Board of Directors, interviewed Peter Drucker, often considered the "founder of modern management." Peter Drucker articulated that "a business is an organization that adds value and creates wealth. Value is created for customers, and wealth is generated for owners." He proposed a quantitative approach to assessing quality, stating that "One is the cost of poor quality; another is the profit of good quality as shown in additional purchases, and a third is the long-term gain in brand image, consumer loyalty, and repeat purchases" [39].

Here, "poor quality" refers to issues such as defective products, failures or safety problems, sub-par performance, and anything that falls short of expectations. This aspect of quality has been the focus of the quality community for over a century, since the Second Industrial Revolution. The community is quite familiar with it.

In contrast, "good quality" relates to the value a product delivers to customers. As Peter Drucker stated, "a business is an organization that adds value and creates wealth. Value is created for customers, and wealth is generated for owners." The extent of the value a product can provide to customers is predominantly determined at the product development stage. Regrettably, a consensus on this notion of "good quality" is still elusive in our quality community. According to Bradley Gale's description of the process of establishing standards for the Malcolm Baldrige National Quality Award [28], some members are yet to accept that "customer value creation" should be a significant aspect of quality. They argue against this notion, citing the difficulty in developing objective, quantitative metrics for its assessment.

The reluctance of some members to accept "customer value creation" as a crucial aspect of quality during the process of establishing standards for the Malcolm Baldrige National Quality Award [28] is reflective of a traditional mindset. This resistance to change often stems from a deep-rooted belief that quality can only be assessed through quantitative metrics, such as process capability indices (Cpk), (defective) parts per million (ppm), and others. These metrics, while vital for defect prevention and process improvement, only offer a partial view of quality.

However, quality and value creation are indeed "different animals," each requiring its unique set of metrics. The success of value creation is not only judged by conventional quality control metrics but also by the product's market success. While measuring customer value creation might seem challenging, it can be achieved through other methods, which I will discuss in chapter 3. It is critical to remember that the value creation is a customer-centered metric, rather than a process-focused one.

In the evolving landscape of quality management, it is paramount to embrace an expanded view of quality—one that appreciates the significance of value creation alongside defect prevention. While traditional metrics are essential for maintaining product standards, they must be complemented by a deeper understanding of the customer's perception of value to fully grasp the product's quality.

Peter Drucker's articulation of "good quality" and "poor quality" presents a profoundly insightful perspective on the nature of quality, akin to observing both sides of a coin. For an extended period, this dualistic understanding of quality had been overlooked, with the focus primarily placed on mitigating "poor quality." Peter Drucker's conceptualization broadens this view by introducing the essential counterpart: the creation of "good quality."

Peter Drucker's framework underscores that "good quality" is about value creation for customers. This aspect goes beyond just rectifying defects and avoiding failures; it involves contributing tangible and intangible benefits to the customer's experience and is intrinsically linked to the product development phase.

By suggesting a quantitative approach to assessing quality, Peter Drucker provides a foundation for comprehensive quality strategy in the 21st century. He proposes assessing quality based on three significant facets: the cost of poor quality, the profit arising from good quality (evidenced through additional purchases), and the long-term gains in brand image, customer loyalty, and repeat purchases.

This holistic understanding of quality invites us to consider not only the prevention of subpar quality but also the creation of high-value products that enhance the consumer's experience, foster loyalty, and drive profitability. In this way, Peter Drucker's perspective both deepens our understanding of quality and broadens its scope, inviting us to consider how quality contributes to value creation and wealth generation. His vision thereby continues to be instrumental in shaping quality management strategies and practices in the modern business landscape.

2.5 Evolving Paradigm for Quality in the Era of Industry 4.0

Dating back to the end of the last century, a long-term business research project, Profit Impact of Market Strategy) (PIMS), which incorporated more than 3000 companies and several prestigious universities, concluded, "In the long run, a company's competitive quality advantage in its products and services is the most significant determinant of business success" [40]. Nevertheless, the project also acknowledged the lack of a single definition for "quality." Indeed, quality has always been highly subjective and personalized, despite its fundamental linguistic definition as a degree of excellence. Various perspectives of quality, such as from the product, producer, or customer viewpoints, hold importance and relevance within their own contexts.

However, under the Taylorism-based mass production system that emerged after the Second Industrial Revolution, featuring linear process flow and a focus on single tasks, it was unfeasible to manage and coordinate all these aspects simultaneously. Consequently, the defining feature of quality since the Second Industrial Revolution has been the ability to maintain an extremely low defect rate. However, with the Fourth Industrial Revolution, revolutionary advancements in communication, information, and AI have transformed the impossible into possible:

- Highly developed information and communication technology enables stakeholders like innovators, research and development teams, product designers, users, producers, and suppliers, dispersed globally, to communicate, transfer information, and coordinate their efforts seamlessly. Digital twin technology, which simulates a real-world product or service, can significantly enhance coordination, sharing, and cocreation throughout all stages of development. This represents a significant shift from the Taylorism system's rigid, compartmentalized product development tasks and stage-gate processes, resulting in a substantial reduction in product development time.
- Smart manufacturing systems, incorporating superconnectivity, the IoT, and AI, can manufacture a high variety of personalized, small-batch products with remarkable flexibility and efficiency.
- Flexible, decentralized manufacturing has become a reality for many products.
- Enhanced customer interaction with product and service providers is made possible due to superconnectivity and continuous advancements in information technology. This interaction and cocreation between customers and producers can be facilitated through numerous information channels, where opinions, feedback, and suggestions on existing products or services can be shared, interacted with, and continuously improved.

With the transition from Industry 2.0 to Industry 4.0, we are witnessing a revolutionary shift in productivity and flexibility within production systems. Consequently, a similar revolutionary change should transpire in quality management. We propose expanding the traditional quality concept, centered on an "extremely low defective rate" since Industry 2.0, into a comprehensive and holistic quality concept. This concept would return to the original definition of "excellence" and incorporate all perspectives, including product, customer, producer, and value. It would consider qualities such as subjective, objective, and spiritual.

The reasoning is simple. Before the Second Industrial Revolution, the concept of quality under the craftsman model was holistic and well rounded. Communication between producers and customers was abundant; quality encompassed design, craftsmanship, esthetics, and meeting specific customer requirements. This model was in line with the inherent literal definition of quality as "excellent and decent," catering to all quality perspectives, since the designers, producers, and users were closely connected. The craftsman model's main drawback was its high cost and low productivity. The Second Industrial Revolution supplanted the craftsman model with mass production, drastically reducing costs and increasing productivity. However, this transition lost the craftsman model's advantages such as personalization, diversity, meticulous craftsmanship, and high esthetics, along with seamless coordination between consumer, design, and production.

We believe that Industry 4.0's progress can offset most of these shortcomings introduced by the mass production system and reintroduce the advantages of the craftsman model while incorporating the conveniences of new technologies.

As we travel in the era of Industry 4.0, with its characteristic interconnectedness, digitization, and smart technologies, the paradigm of quality is undergoing a profound transformation. Quality, in this brave new world, is no longer confined to the traditional view of defect prevention and conformance to specifications; it is evolving into a broader, more holistic concept that encapsulates customer value creation and user experience. This shift necessitates a reframing of our understanding and approach to quality management.

This section, titled "Evolving Paradigm for Quality in the Era of Industry 4.0," seeks to explore this paradigm shift in detail, shedding light on the changing facets of quality and the increasing importance of value creation. We will delve into the "dual facets of quality," where we re-envision "good quality" and "poor quality (prevention)" as complementary elements.

We will also journey into the landscape of "value creation in the new era," scrutinizing the role of digitization and customization in enhancing the customer's experience. We will examine the "expanded roles of quality assurance," studying how modern quality practices have evolved to accommodate the complexities of the Industry 4.0 environment.

Finally, we will anticipate the "evolving trends," making sense of the future of quality management in an ever-changing technological landscape. Through this exploration, we aim to equip you with a comprehensive understanding of the changing tides of quality management as we sail into the era of Industry 4.0.

2.5.1 Dual Facets of Quality

Quality, in its broadest sense, presents itself in two facets: *customer value creation* and *quality assurance*. Both are inseparable elements that constitute the cornerstone of effective quality management, each playing a crucial role in maintaining the overall quality equilibrium. Peter Drucker's concept of "good quality" and "poor quality" paints an insightful picture of this duality, wherein "good quality" is synonymous with the creation of customer value and "poor quality" represents the risks that need to be mitigated through quality assurance measures.

In the context of Industry 4.0, characterized by AI, superconnectivity, and other advanced technologies, adopting this dual-faceted approach to quality becomes more feasible and critical. These technologies provide us with enhanced capabilities to better understand and meet customer needs, thus amplifying the potential for customer value creation. Simultaneously, they equip us with sophisticated tools and techniques to assure quality and control risks more effectively and efficiently.

However, as we stride forward into this new era of quality management, it is essential to understand that these two facets are not to be addressed sequentially or in isolation. Instead, they are interdependent, needing to be integrated and addressed concurrently. For instance, when we introduce an innovation that promises to deliver high customer value, we need to immediately engage our quality assurance mechanisms to manage any associated risks and test and eliminate bugs in the innovation. It is about striking a complementary, making sure that we push a win–win strategy both to create value and do a good job in quality assurance.

In conclusion, this dual-faceted approach to quality—one that harmonizes the creation of customer value with the right quality assurance—is not about prioritizing one over the other or executing one at the expense of the other. Rather, it is about fostering a win–win relationship between them, harnessing their combined power to drive superior quality outcomes. This holistic approach forms the foundation for quality management in the era of Industry 4.0, underpinning our journey toward greater value creation and superior quality assurance.

2.5.2 Customer Value Creation in the New Era

As we venture into the transformative landscape of Industry 4.0, the concept of customer value creation has gained a renewed emphasis. Driven by digitization, hyperconnectivity, and advanced analytics, value creation in this new era transcends traditional boundaries, sparking a fundamental shift in the way we understand and deliver value to customers.

The central tenet of customer value creation in the new era is personalization. Today's consumers are discerning, informed, and expect products and services to be tailored to their unique needs and preferences. Advances in technology, particularly AI and data analytics, provide unprecedented opportunities to understand customer behavior and preferences at a granular level, enabling businesses to deliver personalized value propositions.

In the context of Industry 4.0, the concept of "individualized customer value," has become feasible. Historically, the large-scale industrial model, informed by Taylor's system, found it challenging to cater to individual customer needs. The siloed, sequential, and granular research and product development process, reminiscent of a waterfall approach, struggled to integrate efficiently with the rigid, standardized, and inflexible production process. However, Industry 4.0 is revolutionizing these traditional paradigms.

We are now in a prime position to nurture and develop this intrinsic quality, founded on individualized customer value. As we have previously noted, traditional quality, exemplified by extremely low failure rates, only represents a portion of the total quality spectrum. There is significant room for enhancing quality through the development of customer value. For instance, Mikel Harry, a pioneer of the Six Sigma movement, proposed that "the third wave of Six Sigma should focus on creating value." While he acknowledged the necessity at that time, there was neither a robust theoretical framework nor mature technical conditions to support this shift. Today, however, both the theoretical underpinnings and the technology have advanced considerably, making it entirely feasible to transition toward this value-centric approach.

Moreover, the evolving digital ecosystem facilitates a broader, more dynamic view of value creation. Businesses are now able to extend their value propositions beyond the point of sale, delivering continuous value throughout the customer journey. From prepurchase interactions and user experiences to postpurchase services and support, every touchpoint provides an opportunity to create and enhance customer value.

In addition, the concept of value has broadened to encompass not only functional benefits but also emotional, social, and environmental aspects. Consumers increasingly value brands that align with their values and contribute positively to society. Consequently, businesses that integrate sustainable practices and social responsibility into their value propositions often enjoy a competitive edge.

However, the new era also presents challenges. Heightened customer expectations, rapidly evolving technologies, and increased competition necessitate continuous innovation and agility in value-creation strategies. Furthermore, with increased data utilization comes the need for robust data management and privacy practices to uphold customer trust.

In conclusion, customer value creation in the new era requires a shift from product-centric to customer-centric strategies, from transactional to relational engagements, and from static to dynamic value propositions. It calls for a holistic understanding of customer needs, an innovative application of technology, and a commitment to delivering sustained, personalized, and meaningful value throughout the customer journey. This approach to value creation, underpinned by ethical and responsible practices, holds the key to thriving in the era of Industry 4.0.

2.5.3 Expanded Role of Quality Assurance

In the era of Industry 4.0, the roles of quality assurance have evolved beyond their traditional scopes, necessitating a shift in focus that simultaneously emphasizes defect prevention and customer value creation. This evolution is driven by the massive technological advancements and superconnectivity that hallmark Industry 4.0, which allows for real-time data analysis, predictive modeling, and highly efficient, automated processes.

1) With the widespread adoption of AI and machine learning technologies, quality assurance roles now incorporate real-time monitoring of production processes, immediate defect detection, and swift remedial actions. The AI algorithms can learn from historical data, detect patterns, and even predict potential issues before they occur, allowing organizations to prevent defects rather than merely react to them.

2) The rise of the IoT and advanced sensor technologies have resulted in unprecedented data availability. This explosion of data enhances the quality assurance process by providing a richer understanding of both production processes and product use. Real-time data streams from connected devices and sensors make it possible to track product performance in detail, from production to the point of use, thus improving the capability to assure quality throughout the product's lifecycle.

3) A key aspect of this expanded role includes risk management during the early stages of innovation and production concept definition. With the rapid pace of technological evolution, innovations bring not only novel opportunities but also new uncertainties. Quality assurance now involves proactive risk identification, assessment, and mitigation strategies during these initial stages to ensure successful, risk-balanced innovation outcomes.

4) The integration of quality assurance into the product development process is more critical than ever. With rapid prototyping and digital twin technologies, quality assurance teams can identify and address potential quality issues during the design stage itself, minimizing costly postproduction modifications.

5) Further, as systems grow more complex and interconnected, the risk management responsibilities of quality assurance extend to ensuring the robustness of these intricate systems themselves. Quality assurance now has a crucial role in guaranteeing the reliability, safety, and security of these systems, and in preventing or minimizing the impact of any systemic failures.

6) With the explosion of data availability in the era of Industry 4.0, ensuring data quality has emerged as another pivotal aspect of quality assurance. With decision-making processes becoming increasingly data-driven, the quality of data gathered, processed, and interpreted directly

impacts the quality of these decisions. Quality assurance, in this context, involves implementing stringent checks and balances to maintain data integrity, accuracy, and relevance.

7) Moreover, in this new era, quality assurance extends beyond the product itself to the entire customer experience. Quality assurance professionals are increasingly involved in ensuring seamless user interfaces, personalized customer interactions, and responsive service—all of which contribute significantly to perceived value and customer satisfaction.

8) Finally, the role of quality assurance in risk management has expanded. As products and services become more complex and interconnected, the potential risks and consequences of quality failures increase. Quality assurance teams are now integral to identifying, assessing, and mitigating these risks.

In summary, the era of Industry 4.0 has expanded the roles of quality assurance, making it a strategic, integral part of organizations' operations and value-creation process. The fusion of technology with traditional quality assurance practices is driving proactive defect prevention, comprehensive product lifecycle oversight, enhanced customer experience, and robust risk management, ensuring that organizations remain competitive in this rapidly evolving landscape.

2.5.4 Evolving Trends

In the era of Industry 4.0, quality management is undergoing a transformative journey, marked by continuous evolution and innovation. The intersection of emerging technologies, rapidly changing customer expectations, and shifting market dynamics is reshaping the landscape of quality, and guiding its future trends.

One such trend is the shift from reactive to proactive quality management. Instead of dealing with issues after they have occurred, companies are now leveraging predictive analytics and AI-based tools to identify potential problems before they arise, promoting a more proactive and preventative approach.

Simultaneously, there is a growing emphasis on personalized, customer-centric quality. Companies are leveraging Big Data and advanced analytics to understand individual customer preferences and tailor their offerings accordingly, creating enhanced, personalized value.

Meanwhile, digital twin technology is becoming a cornerstone in quality assurance. By creating a virtual representation of a product, process, or system, companies can test and optimize their designs before moving to production, improving product quality and reducing time to market.

However, as these trends evolve, uncertainty and ambiguity become inherent components of the process. It is challenging to precisely predict the trajectory of these advancements and their impacts on quality management. Thus, to navigate this rapidly changing landscape, continuous learning, adaptability, and improvement become essential.

Organizations need to adopt a culture of learning and innovation, where teams are encouraged to experiment, fail fast, and learn faster. By remaining agile and adaptable, organizations can swiftly respond to new developments and incorporate them into their quality strategies.

Furthermore, continuous improvement, a long-standing principle of quality management, remains as relevant as ever in this era. Organizations need to constantly scrutinize their processes, seek feedback, and implement iterative improvements.

The evolution of quality in the era of Industry 4.0 represents an exciting journey into uncharted territory. By fostering a culture of continuous learning, improvement, and innovation, organizations can ride the wave of change and transform their quality paradigms to thrive in this new era.

References

1 Garvin, D.A. (1984). What does product quality really mean. *Sloan Management Review* 25: 25–43.
2 Reeves, C.A. and Bednar, D.A. (1994). Defining quality: alternatives and implications. *Academy of Management Review* 19 (3): 419–445.
3 Pirsig, R.M. (1992). *Lila: An Inquiry into Morals*. Bantam.
4 Pirsig, R.M. (1999). *Zen and the Art of Motorcycle Maintenance: An Inquiry into Values*. Random House.
5 Alexander, C. (1977). *A Pattern Language: Towns, Buildings, Construction*. Oxford University Press.
6 Alexander, C. (1979). *The Timeless Way of Building*, vol. 1. New York: Oxford University Press.
7 The Quality Without a Name Blog Article by Victoria Scribens. Victoria Scribens, Blog (December 2011). https://roseandphoenix.wordpress.com/2011/12/16/the-quality-without-a-name (accessed 29 June 2023).
8 Salingaros, N.A. (2020). Connecting to the world: Christopher Alexander's tool for human-centered design. *She Ji: The Journal of Design, Economics, and Innovation* 4: 455–480. https://doi.org/10.1016/j.sheji.2020.08.005.
9 Mehaffy, M. (2022). Why Christopher Alexander still matters. *Planetzen* (22 March) https://www.planetizen.com/features/116600-why-christopher-alexander-still-matters
10 Salingaros, N.A. (2008). "Algorithmic Sustainable Design: The Future of Architectural Theory", a Series of 12 Lectures, 2008. *International Journal of Architectural Research* 2 (2): 231–233.
11 Seamon, D. (2007). Christopher Alexander and a phenomenology of wholeness. *Annual meeting of the Environmental Design Research Association (EDRA)*, Sacramento, CA (May 2007).
12 Lu, Y., Morris, K.C., and Frechette, S. (2015). Standards landscape and directions for smart manufacturing systems. In: *2015 IEEE International Conference on Automation Science and Engineering (CASE)*, Gothenburg, Sweden, 998–1005. IEEE https://doi.org/10.1109/CoASE.2015.7294229.
13 Yang, K. (2020). Hyper linked: how an ecosystem including the Internet of Things will change product innovation and quality. *Quality Progress* 53 (8): 14–21.
14 Simonds, D. (2016). Siemens and General Electric gear up for the internet of things. *The Economist* (03 December). ISSN 0013–0613, accessed April 26, 2019.
15 Marianna Glynska (2017). Haier COSMOPlat provides world-class originality for global intelligent manufacturing. *The Huffpost* (01 June), accessed April 26, 2019.
16 Neslin, S.A. and Winer, R.S. (2014). The history of marketing science: beginnings. *The History of Marketing Science* 3: 1–6.
17 ASQ (2015). 2015 Future of quality report. Quality throughout (report from ASQ, American Society for Quality).
18 Broday, E.E. (2022). The evolution of quality: from inspection to quality 4.0. *International Journal of Quality and Service Sciences* 14 (3): 368–382.
19 Zonnenshain, A. and Kenett, R.S. (2020). Quality 4.0—the challenging future of quality engineering. *Quality Engineering* 32 (4): 614–626.
20 Quality Resources (2023). American Society of Quality "Quality 4.0". https://asq.org/quality-resources/quality-4-0 (accessed 11 January 2023).
21 Watson, G.H. (2019). The ascent of quality 4.0. *Quality Progress* 52 (3): 24–30.
22 Watson, G.H. (2020). Constant evolution toward quality 4.0. *Quality Progress* 53 (8): 32–37.
23 Carvalho, A.V., Enrique, D.V., Chouchene, A., and Charrua-Santos, F. (2021). Quality 4.0: an overview. *Procedia Computer Science* 181: 341–346.

24 Siegel, J. and Yang, K. (2021). Going deep. *Quality Progress* 54 (10): 14–19.

25 Yang, K., Mejabi, O., and Rogers, J. (2022). Paradigm shift: the new model for quality in industry 4.0 is predicated on individualized experiential and relationship-based quality. *Quality Progress* 55 (10): 28–35.

26 Feigenbaum, A.V. and Feigenbaum, D.S. (2004). The future of quality: customer value. *Quality Progress* 37 (11): 24–29.

27 Sherden, W.A. (1994). *Market Ownership: The Art & Science of Becoming# 1*. American Management Association.

28 Gale, B., Gale, B.T., and Wood, R.C. (1994). *Managing Customer Value: Creating Quality and Service that Customers Can See*. Simon and Schuster.

29 Porter, M.E. (1985). Competitive Advantage – Creating and sustaining superior performance. *Competitive Advantage* 167: 167–206.

30 Porter, M. (1996). What is strategy? *Harvard Business Review* 74 (6): 61–78.

31 Woodruff, R.B. (1997). Customer value: the next source for competitive advantage. *Journal of the Academy of Marketing Science* 25 (2): 139–153.

32 Haemoon, O. and Kim, K. (2017). Customer satisfaction, service quality, and customer value: years 2000–2015. *International Journal of Contemporary Hospitality Management 29* (1): 2–29.

33 McColl-Kennedy, J.R., Hogan, S.J., Witell, L., and Snyder, H. (2017). Cocreative customer practices: effects of health care customer value cocreation practices on well-being. *Journal of Business Research* 70: 55–66.

34 Leroi-Werelds, S. (2019). An update on customer value: state of the art, revised typology, and research agenda. *Journal of Service Management*.

35 Eric Boyd, D. and Koles, B. (2019). Virtual reality and its impact on B2B marketing: a value-in-use perspective. *Journal of Business Research* 100: 590–598.

36 Rantala, T., Ukko, J., Saunila, M. et al. (2019). Creating sustainable customer value through digitality. *World Journal of Entrepreneurship, Management and Sustainable Development* 15 (4): 325–340.

37 Bolton, R.N. and Drew, J.H. (1991). A multistage model of customers' assessments of service quality and value. *Journal of Consumer Research* 17 (4): 375–384.

38 Yang, K. (2020). The quality vision under Industry 4.0. *ASQ Quality 4.0 Virtual Summit* (28–30 September 28–30).

39 Watson, G.H. (2002). Peter F. Drucker: delivering value to customers. *Quality Progress* 35 (5): 55–61.

40 Buzzell, R.D., Buzzell, R.D., Gale, B.T., and Gale, B.T. (1987). *The PIMS Principles: Linking Strategy to Performance*. Simon and Schuster.

3

Quality by Design and Innovation

3.1 The Trend of Quality: Going Upstream

In first two chapters, we delved into the historical and linguistic interpretations of the term "quality," tracing its evolution over time. The concept of quality invariably pertains to specific products or services. In a conventional business setting, all products or services undergo a cycle of development, design, and delivery, referred to as the product lifecycle. In order to assure superior quality, it is crucial to infuse every stage of this product lifecycle with rigorous quality work.

Table 3.1 serves to highlight how diverse quality assurance techniques and tools are seamlessly integrated and executed at every phase of a typical product lifecycle. This traditional approach to product development and manufacturing is sequential, unfolding gradually from stage 0 through stage 7. Here, the term "traditional" signifies a hardware-dominated product development paradigm. However, it is important to note that the process might not follow a completely sequential pattern for software.

Modern quality methods took root following the Second Industrial Revolution in the early 20th century. At the onset of this revolution, manufacturers endeavored to refine tolerances for component interchangeability, enhance finishes to elevate esthetics and feel and establish material uniformity to ensure performance consistency. Notable figures such as Shewhart, Deming, and Juran led the way in the quality control of materials, parts, and products in the early 20th century. By the 1940s, frameworks such as Plan-Do-Check-Act and Statistical Process Control facilitated process monitoring and control to identify, rectify, and eliminate defects. Quality assurance tasks were primarily focused on the final stage of the product development cycle, namely production, or the inspection of incoming suppliers' materials or parts. During this period, the foremost priority of quality assurance was defect and failure reduction.

The 1960s saw a shift toward proactive quality assurance in the design phase, ushered in by the advent of Failure Mode and Effect Analysis. Subsequently, the use of manufacturing-line and equipment sensors, coupled with numerical controllers, allowed for closed-loop process supervision. Concurrently, the Toyota Production System worked toward minimizing waste and expediting process changeover, while Total Quality Control and Poke Yoke enabled employees to identify and rectify issues and make processes error-proof. The outcome was enhanced product specification conformance and defect reduction. In the 1980s, quality management transformed into an organization-wide effort, acknowledging the significance of human capabilities and employee engagement, leading to the concept of Total Quality Management. Although defect reduction remained the predominant focus of quality assurance, efforts gradually moved toward

Quality in the Era of Industry 4.0: Integrating Tradition and Innovation in the Age of Data and AI,
First Edition. Kai Yang.
© 2024 John Wiley & Sons, Inc. Published 2024 by John Wiley & Sons, Inc.

Table 3.1 Traditional Product Life Cycle and Quality Assurance

Product/Service Life Cycle Stages	Quality Assurance Tasks	Quality Methods
0) Impetus/ Ideation	• Ensure new technology/Ideas to be robust for downstream development	• Robust Technology Development
1) Customer and business requirements study	• Ensure new product/service concept to come up with right functional requirements which satisfy customer needs	• Quality Function Deployment
2) Concept development	• Ensure the new concept can lead to sound design and free of design vulnerabilities • Ensure the new concept to be robust for downstream development	• Taguchi Method/Robust Design • TRIZ (Theory of Inventive Problem Solving) • Axiomatic Design • DOE (Design of Experiment) • Simulation/Optimization • Reliability based Design
3) Product/service design/ prototyping	• Ensure designed product (design parameters) to deliver desired product functions over its useful life • Ensure the product design to be robust for variations from manufacturing, consumption, and disposal stages	• DFMEA (Design Failure Mode and Effect Analysis) • Taguchi Method/Robust Design • DOE • Simulation/Optimization • Reliability-based design/testing and estimation
4) Manufacturing process design/ preparation/ product launch	• Ensure the manufacturing process is able to deliver the designed product consistently	• PFMEA (Process Failure Mode and Effect Analysis) • DOE • Taguchi Method/Robust Design • Trouble shooting and diagnosis
5) Production	• Produce designed product with high degree of consistency and free of defect	• SPC (Statistical Process Control) • Trouble shooting and diagnosis • Inspection • Poka Yoke
6) Product/service consumption	• Ensure customer have a satisfactory experience in consumption	• Quality in after-sale service
7) Disposal	• Ensure customers have trouble free in disposing of used products/service	• Service Quality

the earlier stages of product development, such as product design and manufacturing preparation, to preempt failure and defects through well-considered product and process design. Moreover, particularly for Japanese manufacturers, greater emphasis was placed on understanding customer needs to design products that customers desire.

Post-1980s, with the remarkable progress of computer-aided design technologies boasting advanced modeling, analysis, and simulation capabilities, many tasks, such as functional and validation testing, could be conducted virtually during the early product design stage. Focus began to shift from enhancing product performance in the later stages of the design life cycle to the initial phases, where product development occurs at a more abstract level—essentially, prevention versus

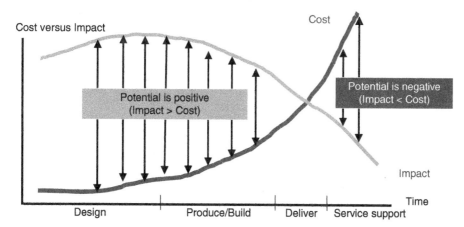

Figure 3.1   Effect of Design Phases on Life Cycle

problem-solving. This shift was driven by the realization that decisions made in the early stages of the design life cycle have the most substantial impact on the total cost and quality of the system. It is often argued that up to 80% of the total cost is committed during the concept development phase [1]. The authors of this document concur, believing that at least 80% of design quality is also determined during the early phases, as depicted in Figure 3.1.

The "potential" is defined as the difference between the influence of the design activity at a particular design phase and the total development cost up to that phase. This potential is positive but decreases as the design progresses, indicating reduced design freedom over time. As financial commitments are made (e.g., purchasing production machines and facilities, hiring staff), the potential begins to transition from positive to negative. When the product reaches the consumer, the potential turns negative, with the cost outweighing the impact significantly. At this stage, design changes for corrective actions can only be made at high costs, which may include customer dissatisfaction, warranty issues, marketing promotions, and in many instances, under the watchful eye of the government (e.g., recall costs).

In recent years, software has claimed an increasingly significant portion of all products. It is even projected that by 2030, 30% of a car's cost will be attributed to software development. Similar to trends seen in hardware, software development is also experiencing a "shift to the left," a strategy referring to an earlier intervention in the development lifecycle timeline. This shift translates into integrating testing and quality assurance activities earlier in the development process, specifically during the requirements and design phases, rather than deferring these activities until the development process concludes. By implementing this "shift to the left," software development teams can detect and resolve defects sooner, thereby reducing time and costs associated with testing and bug-fixing. Moreover, it helps ensure that the software aligns with desired quality standards, as testing and quality assurance activities are woven into the entire development process.

The tendency to "go upstream" or "shift to the left" is observed in both hardware and software developments and extends to quality assurance (QA) activities as well. Traditionally, QA tasks were executed later in the development process, closer to the cycle's end. However, this approach often led to a plethora of issues, including long development lead time, excessive engineering hours, and costs. By advancing QA activities upstream, product development teams can identify and rectify defects earlier in the development process, thereby helping to cut costs and enhance product quality.

Some of the primary advantages of shifting QA upstream in the development process include:

1) **Reduced Defects**: By eliminating faulty or inadequate designs and correcting design bugs earlier in the development process, product development teams can ensure that the final product will have good quality.
2) **Lowered Costs**: Identifying problems early in the development process can help teams evade the expenses associated with later-stage faulty design resolution.
3) **Quicker Time-to-Market**: Reducing the engineering hours required for product development to accelerate the product's journey to the market.
4) **Elevated Customer Satisfaction**: Faster delivery of higher-quality products can increase customer satisfaction and loyalty.

Overall, the trend of "going upstream" in QA is spurred by the desire to enhance product quality, decrease costs, and expedite product delivery.

3.2 The Journey into Quality by Design

From the discussion in the previous section, it is evident that the early stages of the product life cycle, which include market research, customer needs identification, product definition, product concept, system innovation, and product design, are instrumental in determining a product's success or failure. The notion of *quality by design* has gained traction in business, engineering, and quality communities since the 1970s [2, 3], following the Japanese automotive industry's rising market share in the United States, partly attributed to their focus on quality and continuous improvement. Quality by design refers to a product development approach that stresses quality integration at every process stage, moving away from the reliance on quality control in manufacturing and post-production inspections and corrective actions.

In the initial stages of this quality by design initiative, as early as the late 1970s, partially in response to Japanese successes in the automotive and electronics industries, the main priorities typically included:

1) Designing products tailored to customer needs, such as deriving customer insights from market research and deploying customer needs into the design using tools like QFD (Quality Function Deployment).
2) Erasing design weaknesses and reliability risks early in the design process using methods like failure modes and effects analysis, computer-aided engineering, simulation, and virtual testing, rather than expensively correcting them later after the hardware or structure design is finalized. This approach also saves total engineering hours in the product development process.
3) Reducing the product development lead time. In the early 1980s, Japanese companies were quicker at developing new cars compared to their Western counterparts. They were found to be early adopters of concurrent engineering, modular design approach, and many other principles and tools [4]. The West began to adopt these methods aggressively, integrating digital technology such as computer-aided design (CAD) and CAE. Integrated Product Development (IPD) was the Western response to the Japanese product development process. The concept of IPD, first proposed in the early 1990s by a team of researchers at the Massachusetts Institute of Technology (MIT) [2], embraced best practices from successful companies, emphasizing the importance of cross-functional teams, shared goals and metrics, open communication, and collaboration throughout the product development process. They also identified several key practices

associated with successful IPD, including early involvement of manufacturing and suppliers, rapid prototyping and testing, and continuous improvement and iteration. IPD has been widely adopted in several industries, including automotive, aerospace, and consumer goods. Many companies have discovered that IPD can help reduce development lead times, improve product quality, and increase customer satisfaction.

4) Making product performance robust against manufacturing variations and usage conditions. This strategy, known as robust design, was first proposed by Japanese quality expert Genichi Taguchi [5, 6]. Taguchi's approach to robust design aimed to reduce variability in a product or process by optimizing the settings of design parameters. This involved using a specialized experimental design statistical method to identify the most important factors and their optimal settings and designing the product or process to be tolerant of variations in these factors. The Taguchi Method gained popularity in the 1970s and 1980s, particularly in the automotive industry, to improve product quality and reduce costs. By the 1990s, the concept of robust design had become a widely accepted approach to product development in other industries, such as electronics and semiconductor manufacturing industries.

In the late 1990s, several leading Western companies, including General Electric and Ford Motor Company, initiated a comprehensive quality by design roadmap, Design for Six Sigma (DFSS), an approach to product development that employs statistical and quality tools to design products that meet customer needs and are defect-free. One of the main drivers of DFSS was the need to improve product quality and reduce development costs. DFSS provided companies with a means to design products that were less prone to failure or requiring rework, thus reducing warranty claim costs and enhancing customer satisfaction. The increasing complexity of products and manufacturing processes also drove the adoption of DFSS. As products became more complex, ensuring they were defect-free and met customer needs became more challenging. DFSS offered a way to manage this complexity and design more reliable and easier manufacturing products.

At this juncture, it becomes apparent that the definition and scope of "quality" in "quality by design" still align with the traditional quality community's definition, focusing primarily on defect prevention and compliance with standards.

3.3 Design for Six Sigma, A Serious Attempt for Quality by Design

DFSS [7, 8] is a methodology dedicated to creating products or processes that meet customer demands and are devoid of defects. This approach extends the Six Sigma methodology, originally developed by Motorola in the mid-1980s, to enhance the quality of its products and services during the design stage.

Conceived in the late 1990s, DFSS sought to address the constraints of traditional Six Sigma methods, which primarily concentrated on improving existing products and processes rather than designing new ones. DFSS strives to ensure new products or processes are designed with high quality, reliability, and customer satisfaction from the start rather than depending on subsequent improvement efforts.

Typically, DFSS employs a five-step process known as DMADV (Define, Measure, Analyze, Design, and Verify), akin to the Define, Measure, Analyze, Improve, and Control process used in traditional Six Sigma. However, DMADV concentrates on the design of new products or processes rather than enhancing downstream production or service processes. DFSS also incorporates various effective methods and tools suitable for the early design stage, such as QFD, Simulation and Modeling, Design for X, and the Taguchi Method.

Several companies have successfully implemented DFSS and have significantly enhanced their processes, products, and services. These companies include General Electric, Motorola, Caterpillar, Ford, and Samsung.

Yet, there are criticisms about DFSS [9]. Concurrently, another initiative, Lean Six Sigma, seems to have gained much stronger momentum and widespread acceptance and is still active today.

Why did not DFSS, a theoretically solid, well-disciplined Quality by Design approach, yield results comparable to Lean Six Sigma? In my opinion, the answers are as follows:

1) Lean Six Sigma merges Lean operation principles, or the Toyota Production System, with Six Sigma processes and systems. Lean operation principles, based on a few highly effective, simple principles such as waste elimination and pull-based production, can be applied to all kinds of businesses, governmental operations, and services. They can quickly improve operating efficiency and speed and are relatively easy to learn. Thus, they provide many readily available benefits that people can seize. It is an easy win.
2) Conversely, DFSS is far more complex and sophisticated, making it more challenging to learn and implement. It is most suitable for manufacturing and high-tech industries.
3) However, I believe the most crucial reason lies in how to define DFSS's goals and objectives. The term "DFSS" suggests that achieving Six Sigma, or Six Sigma Quality—a rate of 3.4 defects per million—is the design goal, as some earlier literature advocated [10]. In my opinion, this is the problem. When designing a product, our aim should extend well beyond achieving a low defect ratio; it should be to deliver superior customer value than all competitors.

There is unquestionably nothing wrong with *Quality by Design*, but we must first understand the appropriate definition of quality in this context.

Before I formally discuss the definition of design quality, I want to recount the most successful DFSS story so far—the story of Samsung. This narrative will offer us the inspiration for the correct role model of quality by design.

3.3.1 Samsung's Journey for DFSS and Innovation

Samsung Group [10], a multinational manufacturing conglomerate, comprises numerous affiliated businesses, most notably Samsung Electronics, is one of the world's leading information technology companies and consumer electronics manufacturers. However, since its establishment in 1937, Samsung has maintained a relatively low profile, often following market trends rather than leading them, for a long time. This trajectory changed drastically around the year 2000, following Samsung's commitment to the Six Sigma movement [11, 12].

In the early 2000s, Samsung embarked on its journey to implement Six Sigma [13], aiming to enhance quality and efficiency across all business units. The company realized that Six Sigma could facilitate defect reduction, improve customer satisfaction, and boost profitability.

In order to implement Six Sigma, Samsung initially launched a training program to familiarize its employees with the methodology and tools. Tailored to the requirements of each business unit, this program enlightened employees about the core principles and techniques of Six Sigma.

In tandem, Samsung established a robust Six Sigma infrastructure, encompassing a dedicated Six Sigma team, a steering committee, and a suite of metrics and goals. The Six Sigma team took the lead on specific projects, while the steering committee supervised the program, providing valuable guidance and support.

As Samsung's commitment to Six Sigma deepened, the company began to witness significant improvements in quality and efficiency. Reports indicated a reduction in defects, heightened customer satisfaction, and a boost in profitability, clearly demonstrating the success of their Six Sigma implementation.

3.3.1.1 DFSS and TRIZ Greatly Helped Samsung's Innovation Initiatives

Samsung, particularly Samsung Electronics, operates within the rapidly evolving electronic and information technology sector. Consequently, superior research and product development is crucial for the company's competitiveness. Only a couple of years into its Six Sigma implementation, Samsung recognized the potential of DFSS and promptly launched its own DFSS initiative. The company has then integrated DFSS into its product development processes to guarantee the production of high-quality products that meet or surpass customer expectations [14, 15].

While the regular DFSS approach focuses on traditional quality metrics like defect reduction, reliability, and robustness, Samsung enhanced its DFSS strategy by incorporating additional methods and tools. These include TRIZ (Theory of Inventive Problem Solving) [16], technology roadmap [17], and technology tree methodologies [18]. These have played a pivotal role in propelling Samsung's technological innovation.

TRIZ, the Russian acronym for "Theory of Inventive Problem Solving," is a theory about creativity developed in the former Soviet Union between 1946 and 1985 by engineer and scientist Genrich S. Al'tshuller and his team. With over 1500 person-years of research into the world's most successful problem solutions across science, mathematics, and engineering, systematic analysis of globally successful patents, and studies on the psychological facets of human creativity, TRIZ represents a rich intellectual resource [7, 16].

Following the dissolution of the Soviet Union, TRIZ became globally recognized, and many TRIZ experts from former Soviet Republics migrated West. Despite high expectations from the business community, the results have been mixed [19]. However, in 2000, Samsung Electronics, recognizing a dire need for innovation capabilities to succeed in the marketplace, embraced TRIZ as a key tool for fostering innovation capabilities.

As part of its DFSS initiative, Samsung facilitated extensive TRIZ training for a large number of engineers and R&D professionals. Given that TRIZ originated in Russia, there were cultural and language barriers to overcome. To address this, Samsung hired numerous TRIZ experts from former Soviet Republics to form cross-functional teams with local engineers and researchers [20, 21]. These teams collaboratively undertook many challenging technical innovation projects, which led to significant advancements in Samsung Electronics' core technologies and key products [22, 23].

3.3.1.2 A Dual-Track Innovation Strategy: Technology Push and Market Pull

In the early 2000s, Samsung Electronics adopted a dual-track innovation strategy that leverages both technology-driven and market-pull mechanisms to foster innovation. This strategy is illustrated in Figure 3.2.

Technology-driven innovation is an approach focusing on the development and introduction of new technologies into products. Examples of notable technology-driven innovations include Microsoft's operating systems and Intel's CPU. Owing to their "first in market" positioning and technological superiority, they became "dominant designs" or industrial standards. In this approach, corporations heavily invest in research and development to create new products or technologies capable of addressing emerging customer needs or existing problems. Samsung has adopted this "technology-driven" strategy because, as used to a quick learner and adopter, it recognizes the importance of staying ahead of the competition in the rapidly developing technological

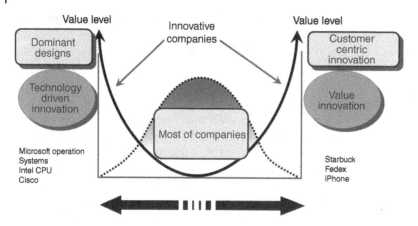

Figure 3.2 Innovation Road Map

landscape of the 21st century. The goal is to lead in technology and offer innovative solutions to customers' needs.

In a technology-driven innovation strategy, the momentum for innovation emanates from within the organization rather than external factors such as market demand or customer feedback. This proactiveness entails identifying potential innovation areas and investing in the necessary research and development to bring new technologies to market. In Samsung Electronics' technology-driven strategy, tools like the technology roadmap and technology tree [17, 18] are extensively used to align emerging technologies with market trends and minimize resource wastage. TRIZ is employed as a tool to enhance the ability of innovators to create breakthrough inventions.

On the other hand, market-pull innovation is driven by customer needs or market demand. Companies using this approach focus on identifying customers' needs and developing products or technologies to meet those needs. Identifying unmet or hidden needs is particularly crucial. Not being product experts, customers may not know what options are available. For instance, early in the 20th century, when Henry Ford was asked whether he had conducted market research or customer feedback surveys to understand what customers wanted in a car, he famously said, "If I had asked people what they wanted, they would have said faster horses." This anecdote emphasizes that customers may not always articulate their product desires effectively. Consequently, innovation often involves going beyond stated customer desires to identify underlying needs and drivers of behavior. Prominent examples of market-pull innovations include Starbucks and Federal Express; they successfully identified and met hidden market needs, thereby becoming success stories. In Samsung's perspective, successful technology-driven or market-pull innovation can significantly benefit both customers and the company. If both are present, the value can be exceptional, as illustrated by the touchscreen mobile phone initiated by Apple Inc.

Samsung acknowledged the power and complementarity of both technology-driven and market-pull mechanisms in enhancing the customer value of a product. To bolster Samsung Electronics' market-pull innovation strategy, the company promoted several systematic innovation methodologies, providing extensive training for R&D and engineering professionals. These methodologies include the Blue Ocean Strategy [24], the Medici Effects [25], and design thinking [26, 27]. A notable example of Samsung's use of design thinking was the development of the Galaxy smartphone series, launched in 2010. The Galaxy series incorporated a number of user-centered design features, such as a large touchscreen interface and an intuitive user interface, setting Samsung apart from competitors and establishing it as a leader in the early 2010s smartphone market.

3.3.1.3 Summary of Samsung Experiences

In conclusion, Samsung's journey started with DFSS and TRIZ. However, from the onset, Samsung's aim with DFSS was much more comprehensive and far-reaching than other DFSS practitioners. Samsung's goal extended beyond achieving Six Sigma Quality; it sought to employ design and innovation to create exceptional customer value and leapfrog its competitors by fundamentally enhancing its innovation capabilities. Thus, Samsung's definition of quality in its DFSS practice included customer value creation, not just defect reduction. This perspective made a substantial difference.

Several years post the initiation of DFSS, Samsung ceased referring to its strategy as DFSS. It transformed into a comprehensive strategy, intertwining innovation, Six Sigma, and Lean methodologies, which has contributed to the company's sustained success and competitive edge in the global marketplace.

Here are some of the ways Samsung has amalgamated these methodologies:

1) **Innovation**: Samsung has fostered a robust culture of innovation, promoting experimentation, risk-taking, and collaboration. The company encourages technology-driven and market-pull innovation through solid strategies, roadmaps, and training. It has heavily invested in research and development, establishing several innovation centers worldwide to inspire creativity and novel ideas.
2) **Six Sigma**: Samsung has applied Six Sigma, a data-driven process improvement method, across its operations. The company utilizes Six Sigma methodologies to identify and eradicate defects and inefficiencies in its processes, resulting in higher-quality products and services, reduced costs, and elevated customer satisfaction.
3) **Lean**: Samsung has also embraced Lean methodologies, focusing on waste elimination and process streamlining throughout its operations. The company leverages Lean principles to enhance efficiency, cut costs, and optimize its supply chain.

3.3.2 Apple Inc.'s Innovation Journey Under Steve Jobs

Having delineated Samsung's success story, it is instructive to draw a parallel to another exemplary organization in the consumer electronics industry: Apple Inc. [28]. Apple's noteworthy breakthroughs also materialized in the early 2000s, under the leadership of Steve Jobs [29, 30]. This period was punctuated by several key milestones, including the launch of the iPod in 2001, which disrupted the music industry and positioned Apple as a consumer electronics heavyweight. Further, the debut of the iPhone in 2007 revolutionized the smartphone industry, bolstering Apple's reputation as a groundbreaking innovator and market leader.

Steve Jobs' achievements at Apple in the early 2000s can be attributed to several fundamental factors:

1) **Innovation**: Jobs was celebrated for his penchant for innovation and his knack for crafting products that were well ahead of their time. Apple introduced many products and features that were unprecedented and did not necessarily align with existing customer preferences [31, 32]. Realizing the limitations of market surveys and the voice of the customer approach, Jobs and Apple devised an alternative methodology. They pushed technological boundaries to create entirely novel functionalities, grasped and extended the inherent capabilities of products, and designed offerings optimized around these breakthroughs. For instance, the iPhone was designed with a focus on simplicity, ease of use, and esthetics. It was the first smartphone to blend a touchscreen interface with Internet connectivity, multimedia capabilities, and a broad range of apps. Apple's cohesive approach to hardware and software development ensured

a seamless user experience and distinguished the iPhone from its competitors. Under Jobs' leadership, Apple unveiled a series of pioneering products, including the iPod, iPhone, and iPad, which revolutionized the consumer electronics industry.

2) **Design-Centric**: Jobs underscored the importance of design in product development, emphasizing the creation of products that were both visually appealing and user-friendly. This design ethos helped differentiate Apple's products from its competitors and fostered a strong brand identity.

3) **Hardware and Software Integration**: Jobs advocated for the simultaneous design of hardware and software components, allowing for superior control over the user experience and enabling seamless integration between the two. This holistic approach helped create an unmatched user experience and distinguished Apple's products from its competitors.

4) **Robust Marketing and Branding**: Jobs' marketing acumen and his ability to foster a robust brand identity and customer loyalty through Apple's advertising and promotional efforts were widely recognized.

5) **Leadership and Vision**: Jobs was a charismatic, visionary leader who inspired his employees to push boundaries and think creatively. His leadership fostered a culture of innovation and excellence at Apple.

Comparing the trajectories of Samsung and Apple in the early 2000s, it is evident that both organizations employed comprehensive innovation strategies, leveraging technology-driven and customer-centric approaches to develop superior products. Apple, with its longstanding culture of innovation and leadership under Steve Jobs, has consistently introduced groundbreaking products. Conversely, Samsung embarked on its commitment to robust innovation in the early 2000s, initially leveraging DFSS, TRIZ, and other innovative tools to establish a formidable innovation system. This system has been continually refined with the inclusion of contemporary innovation methodologies.

Most importantly, both companies' strategies aim to deliver superior customer value through innovation and product development processes, demonstrating effective Quality by Design. The approaches of both Apple and Samsung set an important milestone in this respect.

3.4 Quality by Design in the Era of Industry 4.0

3.4.1 Overviews of Design Quality and Quality by Design

For the rest of the chapter, we will delve into the concept of "Quality by Design" in the context of Industry 4.0, a topic of significant relevance. Upon delving into various literatures, it becomes apparent that there is not a universally accepted definition of "Quality." For the purpose of this book, I have meticulously explored the most suitable definition of "Quality" as it pertains to Industry 4.0 in Chapter 2, and will be using the definition and paradigm detailed in Figure 2.8, the *quality pyramid*, throughout this book. Similarly, the term "Design" also engenders different interpretations across various disciplines and industries. To ensure clarity before we proceed, I have outlined below the definitions of key terminologies that will be used in this chapter.

Design: It refers to the systematic process of creating solutions to problems or fulfilling needs by employing a blend of creative and analytical thinking. Design involves recognizing user needs, establishing design requirements, ideating, and developing concepts, evaluating and refining design alternatives, and finally, crafting a product or solution that satisfies the needs of the intended users. As seen in Table 3.2, the scope of design encompasses all phases from 0 to 4 in

Table 3.2 Survey Form for Customer Values

Quality (Nonprice) Attributes	Importance Weights (Add up to 100)	Performance Scores 1–10, 1 = lowest, 10 = highest			
		Company A (Our company)	Company B (Competitor 1)	Company C (Competitor 2)	Company D (Competitor 3)
1					
2					
3					
4					
5					
6					
7					
8					
9					
10					
Sum of Importance Weights =	**100**				
Price (Perceived Transaction Price)					
Price Competitiveness					

the traditional product life cycle. The concept of design extends across various disciplines such as product design, graphic design, web design, architecture, and more, each with a slightly unique design process.

Design Quality: This denotes the value that a design offers to customers. As the output of design spans products, services, and user experiences, design quality or the quality of the design corresponds to the degree of excellence or value of these outputs. Design is the origin of all qualities, extending to the quality of products, services, and experiences. Design quality encapsulates all dimensions of quality including functionality, usability, reliability, performance, safety, esthetics, and more. Factors such as customer satisfaction, brand perception, sales volume, market share, reliability, and performance often serve as metrics to assess design quality.

Quality by Design: This term signifies a dedicated approach to product development and manufacturing that emphasizes embedding quality at each stage, with a particular focus on the initial stages of the product life cycle. Quality by design incorporates innovation and QA into the design and development process instead of relying on quality control in manufacturing and post-production inspections and rectifications.

The role of design in the product life cycle has been underscored by scholars across disciplines like innovation/R&D management, operations management, and marketing management. There is substantial agreement across these fields on the substantial contribution of design to the competitive performance of a product and a company [3, 33, 34]. The field of quality management and quality engineering is also witnessing a growing recognition of "Quality by Design" as an effective and significant approach to achieve superior quality. The success stories of Apple Inc. and Samsung offer prime examples of this.

A pressing question in the current scenario, as we transition into the era of Industry 4.0, is how will we implement "Quality by Design"? Before delving into this, let us first understand the changes brought about by Industry 4.0.

3.4.2 Some Significant Changes in Business Ecosystem in Digital Revolution

Just as humans and animals exist within ecosystems in nature, businesses operate within their own ecosystems. A business ecosystem comprises a multifaceted network of individuals, organizations, and technologies, all interconnecting to create, deliver, and capture value. It encapsulates every actor involved in the conception and provision of a specific product or service, as well as the supporting technologies and infrastructure. While the constituents of a business ecosystem vary depending on the specific industry and market, they typically include the following:

1) **Customers**: The end-users whose needs and preferences drive the creation and delivery of value.
2) **Producers**: The organizations and individuals responsible for creating and manufacturing products and services, thereby delivering value.
3) **Suppliers**: Those providing the necessary raw materials, components, and other inputs for product creation and manufacturing.
4) **Distributors**: Entities responsible for transporting and delivering products to customers, including wholesalers, retailers, and other intermediaries.
5) **Service Providers**: They offer supportive and complementary services necessary for product creation and delivery.
6) **Competitors**: Other organizations or individuals offering similar products or services and competing for market share.
7) **Regulators**: Government agencies or other organizations responsible for setting and enforcing industry or market standards and regulations.
8) **Investors**: Those who provide the financial resources necessary for growth and expansion.

In the predigital era (the later stages of the 3rd Industrial Revolution and ongoing in the 4th Industrial Revolution), business ecosystems were more localized and less interconnected. Businesses operate within specific geographic regions, with limited competition and a more pronounced division of labor.

The automobile industry, for instance, featured vertical integration, with a single company handling vehicle design, manufacturing, and sales. Similarly, in the retail industry, most transactions take place in physical stores, with minimal competition from online retailers or other channels. The digital revolution, which began before the 4th Industrial Revolution, has significantly impacted business ecosystems, introducing fundamental changes to business interactions and stakeholder engagement. Key changes include the following:

- **Product Types**: Prior to the digital revolution, products were predominantly hardware-based, with minimal electronic and digital elements. Now, we see an influx of products embedded with electronics and software, Internet of Things (IoT) features, and artificial intelligence. Digital products and services such as computer applications, social media, and social networks have emerged, bringing with them unique business ecosystems.

- **Increased Connectivity**: Digital platforms and tools have allowed businesses to connect and collaborate in unprecedented ways, spawning new innovation and growth opportunities.
- **Heightened Competition**: The digital revolution has fostered increased competition as new entrants disrupt established markets and challenge existing business models.
- **Customer Empowerment**: Customers now have enhanced access to information, granting them more control over their purchasing decisions. This shift toward customer-centric business models means businesses are increasingly focused on meeting customer needs and preferences.
- **Data Emphasis**: With an increased focus on data and analytics, businesses can inform decisions and improve operations. New business models based on data and insights, such as predictive analytics and artificial intelligence, have emerged.

The digital age has brought about significant changes to the roles of ecosystem actors due to emerging technologies and the ubiquity of digital platforms. These changes encompass three major groups—producers, customers, and stakeholders:

Producers: Product manufacturers are increasingly integrating digital technologies to enhance product functionality and user experiences. Digital product and service providers, such as software developers and app creators, have vastly different business models. The rise of digital platforms enables them to reach global audiences and distribute products more efficiently.

Stakeholders: Stakeholders in industries are increasingly focused on sustainability, environmental impact, data privacy, and security. The data-driven digital economy has introduced new stakeholders, such as data providers and analytics firms.

Customers: Customers are demanding more personalized and connected experiences, focusing on convenience and ease of use. The rise of e-commerce and on-demand services has allowed customers to make purchases from anywhere at any time, with personalized recommendations based on their browsing and purchasing history. The increased emphasis on data privacy and security has also raised awareness among customers of their rights and responsibilities in the digital economy.

3.4.3 More Changes Expected by Industry 4.0

The evolution of Industry 4.0 continues to unfold, and as we move forward, certain technologies and methodologies are expected to profoundly impact product life cycle processes and the business ecosystem. Here are a few key elements, many of which are already in play:

1) **Pervasive Sensing and Connectivity**: This technology involves the widespread application of sensors and interconnected devices designed to gather and exchange data within various contexts such as homes, buildings, cities, and industries. By enabling real-time monitoring, tracking, and analysis of diverse physical and environmental parameters, we can optimize resource use, improve quality of life, and enhance the efficiency of different processes. Take, for instance, modern electric vehicles, which transmit vast amounts of data about vehicle usage and performance. This data not only enriches the product experience but also informs future developments.

2) **Internet of Behaviors (IoB)**: IoB is a recent concept that uses technology to track, collect, and analyze data on individual behaviors, habits, preferences, and emotions. This data can influence these behaviors through personalized recommendations, targeted advertising, or feedback loops, allowing businesses to tailor their products and services to user behaviors.

3) **Hyper-Targeting Advertising**: This modern marketing strategy leverages data analysis to identify and engage highly specific consumer groups with personalized ads. By generating a

detailed consumer profile from data sources such as browsing history, search queries, social media activity, and demographic information, marketers can enhance their campaign effectiveness and efficiency.

4) **Over-the-Air (OTA) Updates**: These wireless updates for software and firmware on electronic devices, such as smartphones and IoT devices, allow manufacturers to deliver new features, security patches, and bug fixes without requiring a physical connection. For example, Tesla has utilized OTA updates to address issues and enhance its electric vehicles' functionality even post-sale.

Example 3.1   Over-the-Air (OTA) Updates by Tesla

Tesla is well-known for its use of OTA updates to enhance and improve the functionality of its electric vehicles (EVs) even after they have been sold to customers. For example, in 2018, Consumer Reports conducted a review of the Tesla Model 3 and found that the vehicle's stopping distance from 60 miles per hour was longer than expected. This issue raised concerns about the safety of the vehicle and prompted Tesla to release an OTA update to address the issue [35, 36].

Tesla's CEO Elon Musk quickly acknowledged the issue on Twitter and stated that the company was working on a fix. He also emphasized the importance of OTA updates in addressing issues quickly and efficiently. He said, "Tesla will be rolling out a software update to fix braking issue that Consumer Reports says it uncovered in its Model 3 testing. With further refinement, we can improve braking distance beyond initial specs. Tesla won't stop until Model 3 has better braking than any remotely comparable car." The OTA update, which was released in May 2018, included improvements to the Model 3's anti-lock braking system that reduced the vehicle's stopping distance by about 20 ft. According to Tesla, the update was based on data collected from the vehicle's sensors and from real-world driving. Following the OTA update, Consumer Reports retested the Model 3 and found that the vehicle's stopping distance had indeed improved significantly. The organization praised Tesla for its prompt response to the issue and for using OTA updates to address safety concerns.

---

5) **Flexible Manufacturing and Adaptable Architectures**: These systems can quickly and efficiently adjust to changes in production demands and product designs. By leveraging modular and reconfigurable systems, businesses can respond to evolving customer demands, market trends, and product design requirements, resulting in faster time-to-market, improved quality, and cost reductions.

6) **Iterative Product Design**: This design approach involves testing, refining, and improving product designs through iterative cycles. By collecting user feedback and incorporating it into the design process, products can be more functional, user-friendly, and esthetically pleasing. Apple's iPhone development is a great example of this iterative product design.

Example 3.2   Iterative Product Design of Apple iPhone

A good example of iterative product design is the development of the Apple iPhone. Apple has always been known for its focus on user experience and design, and the development of the iPhone was no exception.

The development of the iPhone involved multiple iterations and prototypes, each of which was tested and refined based on user feedback. For example, the original iPhone prototype was a simple rectangular shape with a plastic screen, but after testing with users, Apple decided to switch to a glass screen and a curved design that was more comfortable to hold.

Similarly, the development of the iPhone's user interface involved multiple iterations and refinements based on user feedback. Apple's designers tested different layouts, icons, and animations until they arrived at the final design that is now familiar to millions of users around the world.

Even after the initial launch of the iPhone, Apple continued to use iterative product design to improve the device's performance and user experience. Each new version of the iPhone has introduced new features and refinements based on user feedback, such as improved cameras, faster processors, and new user interface elements.

7) **Human in the Loop (HITL) Artificial Intelligence (AI)**: This design approach integrates human decision-making into AI systems. By combining AI algorithms with human oversight and intervention, HITL AI can create more accurate, reliable, and trustworthy systems. For instance, the application of AI-powered chatbots for customer service can be enhanced through human intervention when necessary.

As we continue our journey into Industry 4.0, these technologies and methodologies promise to transform not only how products are designed and used but also how businesses operate within an increasingly interconnected ecosystem.

3.4.3.1 Summary: Benefits of Industry 4.0 Technologies for Quality by Design

Industry 4.0 technologies and methodologies confer substantial benefits, specifically in two key areas:

1) **Enhanced Customer Understanding**: These technologies substantially boost a business's ability to recognize, comprehend, and react to customer needs. This is especially crucial given the dynamic, individualized, and diverse nature of these needs.
2) **Tailored Business Solutions**: Industry 4.0 technologies empower businesses to address customer needs through accelerated innovation and design cycles, mass customization of product offerings, highly agile and flexible production methods, and intelligent services.

3.4.4 The Objective of Quality by Design in Industry 4.0: Cultivating Customer Value

In the opening chapters, we retraced the historical and linguistic roots of the term "quality" and its evolution over time. The concept of quality is inextricably tied to specific products or services. From antiquity, quality has symbolized "excellence." However, given the diverse needs, tastes, and purchasing powers of consumers, "relative excellence," or "customer value," is a more apt definition of quality. This interpretation gained popularity among academics post-1980s [37]. Despite the natural fit of customer value as a definition of quality, it proves challenging for practitioners to quantify, manage, and control, leading them to prioritize defect control, reliability enhancement, and standard maintenance. Such "defect-free" notions are so deeply embedded within the quality sector that they have become the working definition of traditional quality.

Nonetheless, in the wake of the third and fourth industrial revolutions, complete implementation of "quality as a means to create customer value" is now within reach.

To excel at quality by design, we must first unravel the essence of "quality." Based on the discussions in the initial chapters of this book, the proposed "individualized customer value" (Figure 2.8) serves as a comprehensive and fitting definition.

Adopting individualized customer value as the quality definition, revisiting Figure 2.8 seems prudent. Traditional quality forms the two foundational layers in this diagram, representing the universal prerequisites for all customers. In its absence, quality ceases to exist. Any discrepancies in these layers, namely safety and dependability, can tarnish a company's reputation. In the era of Industry 4.0, traditional quality retains its significance, but its scope is not the only one defining quality.

Next, the third layer of the pyramid focuses on fulfilling functional requirements, i.e., designing and developing products catering to the functional needs of customers. This often entails unique features or superior performance not offered by competitors, essentially an exceptional product at a fantastic price. Pre-Industry 4.0 era witnessed the formulation of many theories and methodologies like QFD, Kansi engineering from Japan, and Western market research. These methods have had numerous triumphs but also encounter significant limitations. It is challenging to align a rigid, stage-gate-based product development process with the diverse, evolving, and subjective needs of customers. The information revolution has facilitated effective strategies to deliver customer value through improved design quality. We anticipate that Industry 4.0 will further popularize and make feasible the notion of "customer value creation by design," which we will discuss more in this chapter.

In the previous chapter, we classified products into five categories: consumer products; industrial products; commodities; service products; and premier products. For many consumer products, service products, and premier products, the maximization of individualized customer value should be the quality by design goal. For other categories like industrial products, commodities, and some consumer products, like toilet paper, the products are not individualized. Hence, the traditional quality aspects, safety and dependability layers in Figure 2.8, and functional quality form the scope of our design objectives. Of course, after-sales service may still be somewhat individualized.

Items above the third level in the pyramid cater to the spiritual, psychological, and cultural needs of customers. These elements are vital for establishing a world-class company and brand, a topic we will delve into within this chapter.

3.4.5 Identifying Customer Needs in the Era of Industry 4.0

The initial step in comprehending customer value is the identification of customers' needs. Several conventional methods exist for this purpose:

1) **Conduct Market Research**: This involves the collection of customer preferences and needs data using methods like surveys, focus groups, and interviews. Such information assists businesses in understanding their customers' requirements and pinpointing potential opportunities for new products or services.
2) **Analyze Customer Feedback**: Reviews on social media, online platforms, and customer service interactions provide valuable insight into customer needs and potential areas for improvement. They also reveal trends and patterns in customer behavior.
3) **Utilize Analytics**: Tools for analytics offer significant insights into customer preferences and behaviors. Analyzing data regarding customer interactions with a company's website, social media platforms, and other digital channels allows businesses to discern the most sought-after products or services.

4) **Monitor Competitors**: Competitor monitoring provides insights into customer expectations for similar products or services. Analyzing competitors' marketing strategies, product offerings, and customer feedback uncovers potential market gaps that a business can fill with its offerings.

5) **Engage with Customers**: Customer engagement through social media, email marketing, and other channels helps foster relationships with customers and provides insights into their needs and preferences. Active listening and responsive feedback can enhance a business's understanding of their customers' needs.

While these traditional methods will continue to be relevant, new approaches are gaining momentum.

3.4.5.1 Voice of Customer (VoC) 4.0

This approach gathers and analyzes customer feedback using digital tools to improve the feedback process's accuracy, speed, and efficiency. Key aspects of VoC 4.0 [38, 39] include:

- **Social Listening**: This involves monitoring and analyzing social media channels to identify customer feedback and sentiment related to a brand, product, or service.
- **Customer Feedback Surveys**: This involves using online surveys to collect feedback from customers, often through email or other digital channels.
- **Customer Experience Analytics** [40, 41]: This involves analyzing customer interactions with digital channels, such as websites and mobile apps, to identify areas for improvement and track customer sentiment.
- **Voice Analytics**: This involves using natural language processing and machine learning algorithms to analyze customer voice interactions, such as phone calls or chatbot conversations, to identify patterns and insights [42].
- **Journey Mapping**: This involves mapping out the customer journey and identifying key touchpoints where customers may have feedback or pain points [43].
- **Personalization**: This involves using customer data and analytics to personalize the customer experience and tailor products and services to individual customer needs and preferences.

VoC 4.0 aims to provide real-time, comprehensive views of customer feedback and sentiment, driving improvements in product development, customer service, and overall customer experience.

3.4.5.2 Mining Customer Needs with IoT (Internet of Things)

The IoT can serve as a formidable tool in mining customer needs by gathering and analyzing data from interconnected devices. Here are a few ways businesses can employ IoT to extract customer needs [40, 44]:

1) **Data Collection from Connected Devices**: Utilizing IoT sensors enables businesses to accumulate data from connected devices, including smart appliances, wearables, and home automation systems. This data can shed light on how customers interact with these devices and the features they find most beneficial.

2) **Analysis of Customer Usage Patterns**: By scrutinizing data derived from IoT devices, businesses can discern customer usage patterns and trends, such as the frequency and timing of certain products or service usage. Such information aids businesses in comprehending their customers' needs and inclinations.

3) **Monitoring of Product Performance**: IoT sensors facilitate real-time monitoring of product and service performance. This real-time data aids businesses in identifying potential issues or areas for enhancement, allowing for timely adjustments.

4) **Personalization of Customer Experiences**: By leveraging data collected from IoT devices, businesses can personalize customer experiences and tailor their products and services to meet individual customer preferences. For instance, a smart thermostat could learn a customer's temperature preferences and adjust the home temperature automatically to suit their comfort.

5) **Real-Time Customer Feedback**: IoT devices can be utilized to gather real-time customer feedback. A customer could use a voice-activated device to provide feedback on a product or service, which can then be analyzed to uncover areas for improvement and guide product development.

In summary, IoT can be a precious tool for unearthing customer needs by offering real-time data and insights into customer behavior and preferences. By harnessing IoT technology and analytics, businesses can better understand their customers and create products and services that align with their needs and expectations.

3.4.5.3 Mining Customer Needs with IoB (Internet of Behaviors)

The IoB represents the aggregation and analysis of customer behavior data to derive insights into their needs and preferences. Here are several ways businesses can harness IoB to extract customer needs [45, 46]:

1) **Customer Data Collection**: IoB can be employed to amass data on customer behavior from various sources, such as social media, mobile apps, and website interactions. This data can illuminate customer predilections, interests, and behavioral patterns.

2) **Customer Behavior Analysis**: Through the analysis of customer behavior data, businesses can discern patterns and trends in how customers interact with their products and services. This information aids businesses in identifying areas for enhancement and developing products and services that cater more effectively to customer needs.

3) **Customer Sentiment Monitoring**: IoB can be used to track customer sentiment in real-time. This can be achieved via social media monitoring and sentiment analysis tools, which can assist businesses in understanding how customers perceive their products and services.

4) **Personalization of Customer Experiences**: By evaluating customer behavior data, businesses can personalize the customer experience and tailor their products and services to meet individual customer needs and preferences. For instance, a mobile app could utilize customer behavior data to provide personalized product or service recommendations.

5) **Prediction of Customer Needs**: By assessing customer behavior data, businesses can forecast customer needs and anticipate future trends. This can empower businesses to stay ahead of evolving customer needs and preferences and create products and services that fulfill these requirements.

Overall, IoB provides businesses with a powerful tool to delve into customer behaviors and preferences, facilitating the development of products and services that effectively meet and anticipate customer needs.

3.4.5.4 Social Listening

This refers to the practice of tracking social media platforms [47], online forums, and other digital channels to glean insights into customer needs and preferences. Here is the procedure:

1) **Define Your Goals**: The initiation of social listening involves clarifying your objectives. What do you aspire to achieve by monitoring social media channels? Are you seeking feedback on a specific product or service, or attempting to spot trends in customer behavior?

2) **Select Your Tools**: An array of social listening tools exists, spanning from complimentary options like Google Alerts to premium platforms like Hootsuite and Sprout Social. These utilities enable you to scan social media platforms for distinct keywords, hashtags, or brand mentions.

3) **Monitor Social Media Platforms**: Upon selecting your tools, you can commence the monitoring of social media platforms for pertinent conversations. This might encompass searching for specific keywords or hashtags, or tracking mentions of your brand or products.

4) **Data Analysis**: As you accumulate data from social media platforms, it is crucial to analyze it to procure insights into customer needs and preferences. This could entail recognizing common themes or trends in customer dialogs or employing sentiment analysis tools to comprehend how customers perceive your brand or products.

5) **Action Initiation**: Once you have scrutinized the data, you can take action grounded in your insights. For instance, you might devise new products or services that more effectively fulfill customer needs, or modify your marketing strategy to better engage your audience.

Overall, social listening serves as a valuable method for understanding and catering to customer needs, enabling businesses to enhance their offerings and marketing strategies accordingly.

Example 3.3   Application of Social Listening to Improve Customer Relations in Academic Libraries

A recent case study [48] documented how social listening can be used to improve customer relations in academic libraries.

The case study focuses on the University of Arizona Libraries, which used social listening tools to monitor conversations on social media platforms and other digital channels. The project aimed to identify opportunities to improve customer service and enhance the customer experience.

The study found that social listening was an effective way to identify customer needs and concerns, as well as to identify opportunities to improve customer service. By monitoring social media platforms, the library was able to identify issues that customers were experiencing with library services, such as long wait times for books and difficulty accessing online resources.

The library was also able to use social listening to improve customer relations by responding to customer inquiries and feedback in a timely and professional manner. This helped to build trust and loyalty with customers and improve the overall customer experience.

In summary, these novel methods employ technological advancements to adapt to the dynamic nature of customer needs and preferences. By utilizing these techniques, businesses can stay ahead of the curve, promptly respond to market trends and competitive pressures, and ultimately enhance their customer value proposition.

3.4.6   Evaluating Customer Value and Analyzing Value Proposition

After recognizing customer needs, the subsequent step is to ascertain and elucidate the customer value [37] for a specific product. This process involves comprehending the advantages the product offers to the customers and how these benefits align with their needs and expectations. Below are some steps businesses can employ to ascertain and elucidate customer value for a product:

1) **Understanding Customer Needs**: The initial step in ascertaining customer value is to comprehend the needs and expectations of your target audience. This could entail conducting customer surveys, analyzing customer feedback, observing customer behavior and trends, and using contemporary methods.

2) **Identifying Product Benefits**: Once you have a firm grasp of customer needs, you can pinpoint the benefits that your product provides. These benefits may include aspects such as convenience, quality, reliability, or affordability.

3) **Evaluating Product Benefits**: After determining the benefits of your product, you need to assess them in the context of customer needs and expectations. This might entail conducting market research or user testing to understand how customers perceive and utilize your product.

4) **Defining Customer Value Proposition**: Based on your evaluation of product benefits, you can define your customer value proposition. This should succinctly articulate the benefits of your product and how they fulfill customer needs and expectations.

5) **Communicating Value Proposition**: Once you have delineated your customer value proposition, you need to effectively communicate it to your target audience. This could involve designing marketing materials, developing messaging strategies, and leveraging social media and other digital channels to reach your audience.

Ultimately, this process helps businesses not only understand their customers' perception of the value of their products but also communicate this value effectively, thus aiding in their overall business and marketing strategy.

Example 3.4   Identifying and Evaluating Customer Value in Gallbladder Surgery

Endoscopic surgery, also referred to as laparoscopic surgery or minimally invasive surgery, is a surgical technique utilized for gallbladder removal in patients suffering from gallbladder disease. During endoscopic surgery, a surgeon creates small incisions in the abdomen and introduces a laparoscope, a slender tube equipped with a camera and light, to visualize the internal organs. The surgeon then employs specialized surgical instruments to remove the gallbladder through these small incisions.

In comparison to conventional open surgery, which requires a large incision in the abdomen for gallbladder removal, endoscopic surgery offers numerous benefits:

1) **Accelerated Recovery Time**: Given that endoscopic surgery involves smaller incisions and results in less tissue damage, patients typically benefit from a quicker recovery time and experience less pain and discomfort.

2) **Diminished Risk of Complications**: Endoscopic surgery is linked with a lower risk of complications, such as infections and bleeding, compared to conventional open surgery.

3) **Smaller Scars**: As endoscopic surgery involves smaller incisions, patients are left with smaller scars, thus offering a more aesthetic outcome.

4) **Shorter Hospital Stay**: Patients who undergo endoscopic surgery usually have a shorter hospital stay compared to those who undergo traditional open surgery.

The above four points highlight the product benefits of endoscopic surgery.

In this context, the following survey form, as illustrated by Table 3.2, recommended by Gale [37], is utilized to evaluate and compare the customer values of traditional surgery and endoscopic surgery.

In this form, the "quality attribute" column represents the product benefits, while the second column lists the relative importance of each attribute from the customer's perspective. Typically, this table is used to compare the customer values of our product with those of our competitors. Since customer value is a balance of benefits versus cost, the last two rows detail the cost.

Table 3.3 Customer Value Studies of Gallbladder Operations Endo-Surgery Versus Traditional Surgery

Quality Attributes	Customer's Weight of Attributes (Total = 100)	Industry Comparison (Real metric, *performance, score*)		Quality scores Comparison	
		Endo	Traditional	Ratio	Customer Weight × Ratio
At home recovery period	40	1–2 wk *Score = 9*	6–8 wk *Score = 3*	3.0	120
Hospital stay	30	1–2 d *8*	3–7 d *4*	2.0	60
Complications rate	10	0–5% *9*	1–10% *6*	1.5	15
Postoperative scar	5	0.5–1 in *7*	3–5 in. *5*	1.4	7
Operation time	15	0.5–1 h *10*	1–2 h *5*	2.0	30
Total	100				232 = **Market perceived quality score**

In Table 3.3, the relative importance of each customer value attribute, or quantifiable product benefit, is provided.

For instance, the duration of the home recovery period, measured in days, is a vital customer value attribute. In this case, the patient's weight is 40, and the recovery period for traditional surgery spans six to eight weeks, earning a performance score of 3. Endoscopic surgery has a much quicker recovery period of one to two weeks, thus, it garners a higher score of 9 out of 10.

The aforementioned table merely provides a benefits comparison of the two surgical methods.

However, since customer value represents benefits against cost, we also need to evaluate the cost comparison. Generally speaking, endoscopic surgery is linked with lower costs compared to traditional open surgery, due to a shorter hospital stay, reduced risk of complications, and smaller incisions.

Therefore, the customer value of endoscopic surgery significantly surpasses that of traditional surgery. However, it is critical to remember that the choice to undergo a specific type of surgery should be informed by numerous factors, including medical necessity and patient preference, and should be made in consultation with a certified healthcare professional.

Let us now discuss the concept of a "customer value proposition." This is a statement that clearly outlines the unique benefits that a product or service provides to customers and how these benefits fulfill their needs and expectations. It is a critical component of a company's overall value proposition, used to set their product or service apart from competitors.

Take the previous example of the two surgical methods. Both surgeries offer the same health benefits to patients, but one is notably superior in terms of the benefits, or attributes. From a customer value standpoint, they share very similar value attributes, albeit at differing performance levels. Endoscopic surgery outperforms traditional surgery across all attributes, and at a lower cost as well. Hence, these two surgeries have different customer value propositions, with the difference lying in performance levels, not different customer value attributes. If you consider two hospitals both offering endoscopic surgery, they essentially provide the same customer value proposition. In such a case, they would have to compete on factors like price or service—a classic "red ocean" competition scenario. Therefore, it is advisable to differentiate your customer value proposition from those of your competitors.

Example 3.5   Differentiating Customer Value Propositions for Fitness Trackers

A fitness tracker, typically worn on the wrist, is a device capable of recording a variety of physical activities such as walking steps, running distance, heart rate, sleep patterns, and swimming laps. These devices interact with a mobile app via Bluetooth, allowing for device configuration and the download of user activity data. While most smartwatches incorporate some of these features, dedicated fitness trackers usually offer a more comprehensive range of functions. Figure 3.3 showcases a few typical fitness trackers.

Here are three distinct customer value propositions that target different consumer segments:

1) **For the Health-Conscious**: "Maximize your potential with our state-of-the-art fitness tracker. Monitor your steps, calories burned, and heart rate while receiving personalized insights and coaching to support your fitness aspirations." The Fitbit Charge 4 exemplifies this category well.

2) **For the Tech-Savvy**: "Experience the pinnacle of fitness tracking technology. Our tracker is equipped with the latest sensors and algorithms for delivering precise data and insightful analytics. Connect with our app to explore innovative ways to enhance your workouts and overall health." The Garmin Venu 2 aptly represents this category.

Figure 3.3   Typical Fitness Trackers. *Source:* theverge.com

3) **For the Budget-Conscious**: "Maintain your fitness without straining your wallet. Our cost-effective fitness tracker provides all the essential features for a healthy, active lifestyle without the premium price tag of competing devices. Benefit from personalized coaching and support at a price that accommodates your budget." The Xiaomi Mi Band 6 is a fitting example of this category.

3.4.6.1 Willingness to Pay (WTP) as a Customer Value Indicator

WTP is a key metric indicating the maximum price a customer is prepared to pay for a product or service. It serves as a valuable tool in evaluating the perceived value of a product from the customer's perspective. This simple yet powerful metric assists businesses in aligning a product's value proposition with customer needs and preferences [49, 50].

Here is a step-by-step guide on how to leverage WTP in assessing a product's customer value:

1) **Segment the Customer Base**: The initial step entails identifying the product's customer segments, which may be defined by demographics, psychographics, or other factors influencing customer behavior.
2) **Ascertain WTP for Each Segment**: Subsequently, businesses should employ market research methodologies to determine the WTP for each customer segment. This can involve executing surveys or hosting focus groups to collect data on customer preferences and price sensitivity.
3) **Compare WTP with Cost**: Once the WTP is ascertained for each segment, businesses can then compare it to the product's production and delivery costs. If the WTP outweighs the cost, it signifies that the product is not only valuable to customers but also profitable for the business.
4) **Pinpoint Areas for Enhancement**: If the WTP is less than the cost, it suggests potential gaps in the product's value proposition. Businesses can use the WTP data to discern the features or benefits most valued by customers, and consequently identify areas where the product can be enhanced to increase its value proposition and subsequently foster a higher WTP.

3.4.6.2 Survey-Based Customer Value Evaluation Methods

Customer value evaluation is the process of assessing the value that a product or service provides to its customers. It involves understanding the needs, preferences, and pain points of customers and identifying the ways in which the product or service satisfies those needs and provides value. The goal of customer value evaluation is to ensure that the product or service is aligned with customer needs and preferences and that it provides value to customers that justifies the price and supports long-term customer loyalty.

There are several popular survey-based methods for customer value evaluation, including:

1) **Customer Surveys**: Surveys are a common method for gathering feedback from customers about their perceptions of the value provided by a product or service. Surveys can include questions about customer satisfaction, WTP, and loyalty, among others.
2) **Customer Interviews**: Interviews with customers can provide more in-depth insights into their needs, preferences, and pain points. These can be conducted in-person or over the phone and can provide valuable qualitative data that can be used to inform product development and marketing strategies.
3) **Net Promoter Score (NPS)**: NPS [51, 52] is a metric that measures customer loyalty and satisfaction by asking customers how likely they are to recommend a product or service to others. A higher NPS indicates greater customer loyalty and satisfaction.

4) **Customer Lifetime Value (CLV)**: CLV [53, 54] measures the total value of a customer over the entire lifetime of their relationship with a business. It takes into account factors such as customer acquisition cost, average purchase value, and customer retention rate.

5) **Value Proposition Score (VPS)**: VPS [55, 56] measures how well a company's value proposition resonates with customers. It is typically based on a survey question that asks customers to rate the clarity and effectiveness of the company's value proposition.

Each of these methods provides a different perspective on customer value and can be used to identify areas for improvement and make strategic decisions that align with customer needs and preferences. The most effective approach will depend on the specific business and product or service being evaluated.

3.5 Customer Value Creation by Innovation

In the highly competitive landscape of modern business, the key to a company's success lies in its ability to consistently create value for its customers. To accomplish this, companies must ceaselessly innovate, developing new products and services that align with the evolving needs and preferences of their customers. Customer value creation extends beyond the mere production of new offerings—it involves refining existing products and services and enriching the overall customer experience. The power of innovation allows businesses to distinguish themselves from their competitors, foster customer loyalty, and boost revenue growth.

It is very important to have a multidisciplinary approach; innovation is often the result of the collective efforts of professionals from diverse fields and responsibilities. Successful innovation necessitates a collaborative mindset, leveraging the unique skills, knowledge, and perspectives each professional brings to the table. Engineers, designers, marketers, operations managers, and quality professionals, among others, all have vital roles to play. By working together, these professionals can create a synergy that enables the development of holistic solutions that are not only innovative but also perfectly aligned with customers' needs and preferences. This collaborative approach not only drives innovation but also helps foster a culture of continuous improvement, ensuring that customer value creation remains at the forefront of every business initiative.

Quality professionals are uniquely equipped to facilitate the creation of customer value through innovation. They can provide invaluable insights and expertise, assisting businesses in identifying areas for potential enhancement, creating innovative products and services, and elevating the customer experience.

Beyond conceptual development, quality professionals are instrumental in ensuring that these innovative products and services not only cater to customer needs and preferences but are also reliable, safe, and consistently delivered.

The following sections delve into the idea of customer value creation through innovation, introducing a range of strategies and methodologies used in its achievement. It will also present several contemporary methods aimed at bolstering personalized customer value.

3.5.1 Blue Ocean Strategy

In the contemporary marketplace, we are deluged with various products and services. The rapid advancement of information technology and the global economy have considerably accelerated the tempo of product development and cost reduction. Today, even a superior product offering substantial customer value and consequent appreciation could quickly be rivaled by an influx of competitors, who emerge rapidly with offerings of decent quality, similar features, and aggressive pricing. In such a scenario, businesses are cornered into ruthless competition, contending on every aspect of current customer values while simultaneously being pressed to improve them and reduce prices. This ferocious rivalry often culminates in the shrinking of profit margins.

This scenario, prevalent in industries like airlines, automotive, and several others, is often referred to as the "red sea." The analogy likens businesses to fishermen crowded along a coast, fiercely competing for a dwindling fish population. The red sea denotes a saturated market packed with companies offering comparable goods or services.

Contrasting the "red sea" is the "blue ocean," an analogy referring to fishermen in an expansive, open blue ocean with a plethora of fish and no competition. The blue ocean symbolizes an untapped and uncontested market, with minimal or no competition due to its uncrowded nature. To chart a blue ocean, businesses must deviate from the norm, crafting something unprecedented. The Blue Ocean Strategy is a business approach designed to capture uncontested market space, thereby rendering competition irrelevant [24].

In particular, the Blue Ocean Strategy equips a company to independently modify the conventional key customer value items, establishing a new set of customer values. Consequently, a new product or service category emerges, devoid of competition.

The Blue Ocean Strategy framework is composed of four key elements:

1) **Eliminate**: This step entails identifying and eliminating the factors that the industry or market typically associates with customer value but are taken for granted. It may involve diminishing or discarding certain features or services that do not hold significant value for customers.
2) **Reduce**: This involves scaling back on specific features or services associated with some customer value factors that customers deem essential but are not critical to their needs. It can include cost reduction or process streamlining to enhance product or service efficiency.
3) **Raise**: This requires identifying and augmenting the performance levels of certain customer value factors that are crucial to customers and distinguish the company from its competitors. This could mean enhancing the quality or performance of the product or service.
4) **Create**: This necessitates devising new customer value factors that have never been introduced in the market before. It could involve the development of new products or services that cater to unmet customer needs or preferences.

The subsequent Example 3.6 will demonstrate how the Blue Ocean Strategy functions.

Example 3.6   Cirque du Soleil Cirque du Soleil

Cirque du Soleil, French for "Circus of the Sun," is a prosperous entertainment entity based in Montreal, Quebec, Canada. Founded in Quebec in 1984, this venture, while referred to as a circus, differs significantly from traditional circuses. Rather than relying heavily on animal acts, Cirque du Soleil is predicated largely on human performances. This venture not only preserves the legacy of the circus but also incorporates elements from street performances, opera, ballet, and rock music. Unlike the traditional circus that predominantly emphasized stunts and skills without much regard for narrative, Cirque du Soleil's performances are story-centric. The company is also known for its vibrant colors, delightful live music, and a medley of global talents.

As depicted in Figure 3.4 through the "Strategy Canvas" tool by Kim and Mauborgne [24], part of the Blue Ocean Strategy, there are two axes: the horizontal axis represents the key

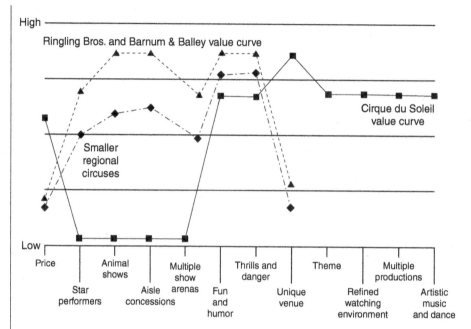

High

Ringling Bros. and Barnum & Balley value curve

Cirque du Soleil value curve

Smaller regional circuses

Low

| Price | | Animal shows | | Multiple show arenas | | Thrills and danger | | Theme | | Multiple productions |
| Star performers | | | Aisle concessions | | Fun and humor | | Unique venue | | Refined watching environment | | Artistic music and dance |

Figure 3.4   Customer Value Curves of Cirque du Soleil

customer factors considered during a purchasing decision, while the vertical axis represents the level of offering or performance level a company provides for each factor.

Examining from the customer value factors perspective, Cirque du Soleil amalgamated value factors from various alternative industries such as regular circus shows, opera, Broadway-style musical shows, Las Vegas-style shows, and Chinese acrobats. This amalgamation forged a unique entertainment style, leading to substantial success. Cirque du Soleil managed to attract not just the traditional circus audience—children—but also those typically uninterested in circuses—corporate clients and adults.

Here is how Cirque du Soleil reimagined customer value:

- **Eliminate**: It discarded four customary customer value factors associated with a regular circus—star performers, animal shows, aisle concessions, and multiple show arenas.
- **Reduce**: It retained two customer value factors—fun and humor, thrills and danger, albeit at a slightly reduced level.
- **Raise**: It maintained a unique venue as a customer value factor but offered it at a significantly higher level. Cirque du Soleil upgraded its venue to the "Cirque du Soleil Theater," specifically designed to accommodate its unique shows featuring acrobatics, dance, and music. The modular design of the theater allows for easy reconfiguration to cater to different show formats, complemented by state-of-the-art sound and lighting systems.
- **Create**: It introduced four new customer value factors—theme, refined watching environment, multiple productions, and artistic music and dance. These final four customer value factors are typically associated with musicals, Las Vegas shows, among others.

This unique customer value curve of Cirque du Soleil not only sets it apart as a one-of-a-kind circus but has also rendered it so successful that it is now the largest cultural export of Canada.

It is evident that the fundamental objective of the Blue Ocean Strategy is to enhance the customer value proposition, thereby leading to a significantly improved product or service offering. The Strategy Canvas, an integral part of this approach, serves as an excellent planning template for the reconfiguration of the customer value proposition.

3.5.2 Medici Effect

The term "Medici Effect" is attributed to Frans Johansson from his book "The Medici Effect: Breakthrough Insights at the Intersection of Ideas, Concepts, and Cultures" [25]. This phrase encapsulates the burgeoning of creativity and innovation at the convergence of ideas, concepts, and knowledge from distinct disciplines, cultures, and domains.

The "Medici Effect" owes its name to the influential Medici family of Renaissance Italy. Celebrated patrons of arts and sciences, the Medici family catalyzed the exchange of ideas and knowledge across diverse fields and cultures. Their contribution to the Renaissance, stimulating innovation and creativity, embodies a historical instance of the Medici Effect.

According to Johansson, the Medici Effect transpires when ideas from disparate fields, cultures, and disciplines intersect, giving rise to novel breakthroughs and innovations. This phenomenon can occur in various contexts, including the arts, business, science, and technology. By amalgamating diverse perspectives and knowledge, the Medici Effect can lead to the inception of unprecedented products, services, and solutions.

Johansson contends that the key to igniting the Medici Effect is fostering an environment that encourages diversity, cross-disciplinary collaboration, and innovation. This might involve assembling individuals from various backgrounds and cultures and encouraging them to exchange ideas and perspectives. It could also include creating spaces and platforms that facilitate knowledge exchange across different disciplines.

The Medici Effect has been witnessed in multiple scenarios, from the evolution of new technologies to the birth of unique art forms. By merging ideas and concepts from different fields, cultures, and disciplines, the Medici Effect holds the potential to foster innovation and creativity, capable of transforming industries and revolutionizing the world.

Here, I present two examples of innovations that align with the Medici Effect.

Example 3.7   East Gate Centre of Harare, Zimbabwe

The Eastgate Centre in Harare, Zimbabwe [57], is hailed for its groundbreaking design and construction (Figure 3.5). This emblem of architectural innovation in Africa was conceived by architect Mick Pearce in collaboration with engineers at Arup Associates. Upon completion in 1996, it stood as the country's most expansive commercial and shopping complex. Interestingly, its heating and cooling system is inspired by the principles employed within local termite mounds.

Inside these mounds, termites cultivate a fungus, which serves as their primary food source. This fungus necessitates a steady temperature of 87 °F, regardless of the fluctuating external temperatures on the African veld that range from 35 °F at night to 104 °F during the day. Termites maintain this stable environment by creating ventilation at the mound's base, directing air into chambers cooled by damp mud fetched from deep water tables, and then upward through a flue to the peak. To regulate the temperature, termites continuously craft new vents and seal off the old ones.

Figure 3.5 East Gate Center. *Source:* David Brazier/Wikimedia Commons

The Eastgate Centre's design draws its inspiration from these termite mounds. The building is specifically designed to harness natural airflows and cooling systems, thereby reducing reliance on artificial air conditioning. The structure's shape, along with judiciously positioned vents, allows cool air to circulate while warm air is expelled, resulting in a comfortable and sustainable indoor environment. Remarkably, the building maintains a consistent temperature of 73–77 °F using less than 10% of the energy consumed by comparably sized buildings. Thus, the Eastgate Centre is an exemplary manifestation of the Medici Effect, where a blend of ideas from disparate fields, cultures, and disciplines results in innovative breakthroughs.

Example 3.8   Airbnb

Airbnb [58] is a platform that allows people to rent out their homes or apartments to travelers. It was founded in 2008 by Brian Chesky and Joe Gebbia, who were struggling to pay rent on their San Francisco apartment. They noticed that a large design conference was coming to town, and all of the hotels were sold out. They decided to rent out air mattresses in their living room to conference attendees, and the idea for Airbnb was born.

The founders of Airbnb drew on their background in design to create a platform that not only solved a practical problem (finding affordable accommodation) but also offered a unique and personalized experience for travelers. By combining the concept of home sharing with a user-friendly platform and a focus on design, Airbnb was able to create a new market that disrupted the traditional hotel industry. This is an example where ideas from different fields (home sharing, design, and technology) intersected to create a breakthrough innovation, which is a key characteristic of the Medici Effect.

3.5.3 Design Thinking

Design thinking is a problem-solving approach rooted in design principles. This methodology took shape in the 1990s, thanks to the concerted efforts of David Kelley and Tim Brown [59, 60] at IDEO, a leading global design and innovation consultancy. The pair is recognized for mainstreaming the concept of human-centered design, which forms the cornerstone of design thinking. Their firm underscores the significance of grasping the needs, behaviors, and experiences of users and stakeholders in order to devise efficacious solutions. To this end, IDEO has formulated a broad array of tools and methods encompassing brainstorming, prototyping, and user testing.

Another significant contributor to design thinking is Peter G. Rowe [61], a renowned architect, urban planner, and distinguished professor at Harvard University. Rowe's insights into design thinking pivot around the notion that design transcends the mere creation of objects or structures; it involves formulating solutions to intricate problems. He advocates for a profound understanding of a problem's context and constraints and promotes a multidisciplinary perspective toward design.

The design thinking model generally comprises five phases:

1) **Empathize**: Gain insights into the needs and experiences of users and stakeholders related to the problem at hand.
2) **Define**: Elucidate the problem and express the associated challenges and opportunities.
3) **Ideate**: Stimulate the generation of a diverse range of ideas and solutions, fostering creative thinking.
4) **Prototype**: Develop and assess various prototypes and solutions, refining them based on user feedback.
5) **Test**: Appraise the efficacy of the prototypes and solutions, making requisite adjustments prior to implementation.

Design thinking is often an iterative process, with each stage interacting with and informing the others. The objective is to generate innovative, effective resolutions to complex issues by tackling them from a human-centric viewpoint.

Design thinking has been embraced and utilized by many leading companies to drive success and innovation. Enterprises such as Google, Apple, and Procter & Gamble have integrated design thinking into their strategic approaches, recognizing its potential in solving complex business challenges and cultivating a culture of innovation. The methodology's human-centered focus helps these companies better understand their customers' needs, leading to the development of products and services that truly resonate with their target market. These practical applications of design thinking have resulted in tangible improvements in customer satisfaction, operational efficiency, and overall business performance. This widespread corporate adoption and the resultant successes underscore the effectiveness and transformative power of design thinking in today's business landscape.

Example 3.9   Apple's First Mouse

In the early 1980s, Apple was developing its first personal computer, the Macintosh. One of the biggest challenges the company faced was how users would interact with the computer. At the time, most computers used a command-line interface that was difficult for the average person to understand and use (Figure 3.6).

To solve this problem, in 1980, Apple asked IDEO, led by Tom Kelley, to develop a mouse for their radical new computer, the Lisa [62]. Previous attempts at mouse design, by Xerox, yielded results that were too expensive and hard to make. The Apple mouse needed to be more

Figure 3.6 Apple's First Mouse. *Source:* Fotoproff/Adobe Stock

reliable and less than 10% of the cost of the earlier versions. The team started by empathizing with the user to understand how people used computers and their pain points.

They brainstormed ideas and created many prototypes, testing them with users to get feedback on what worked and what did not. Then design team created a cheaper, much-improved mechanism that would operate the mouse and found that a well-designed, plastic "ribcage" would hold the pieces together. The team similarly tested and refined the mouse's other key components, from the audible and tactile click of the button to the rubberized coating on the ball. They also repeatedly click and move the mouse for days, logging "mouse miles" to check the reliability of the electromechanical assembly.

The mouse was easy to use and could be operated with just one hand, making it a revolutionary innovation in the world of personal computing.

Apple's first mouse became an icon of design thinking, and it helped to establish the company as a leader in user-centered design. Today, the mouse is still a widely used input device.

Example 3.10 Procter & Gamble's Swiffer

In 1994, Procter & Gamble was looking to grow through innovation. The company established a New Ventures group for innovative design. The group assembled a team with experts in design and marketing, together with some experts from the hard-surface cleaning and paper divisions at P&G. The mission for this diverse group was simple: find a better way to clean a floor; out floor cleaning afresh, through the eyes of consumers [63] (Figure 3.7).

Figure 3.7 Procter & Gamble's Swiffer. *Source:* EPAM services

The team set out as design anthropologists to watch how people cleaned their kitchen floor. They visited 18 homes in Cincinnati and Boston, spending about an hour and a half in each.

The team listed each of the steps required to clean a floor: move furniture, sweep loose dirt, locate components, mix solution, prepare mop, and so on. They noticed something strange: about half the steps were for cleaning the floor; the other half were for cleaning the mop. They looked more closely at the dirty mop. Under a microscope, it was clear that most of the dirt was stuck to the outside surface of the mop. We concluded that mops also worked by the adhesion of dirt to a cleaning surface. And then we recognized the second contradiction: the better a mop was at grabbing dirt off the floor, the more difficult it was to clean the mop itself.

From this process of ethnographic fieldwork and studio-based analysis, a clear set of design requirements emerged. They were looking for a solution that:

- Works for both sweeping and mopping;
- Does not require you to clean the broom or mop after cleaning the floor;
- Is clean to use and does not make you want to wear old clothes;
- Is quick and fun, so people want to clean more and can have cleaner homes.

The team came up with a proposed solution, called "a diaper wipe on a stick," inspired by disposable diapers. Finally, a disposable sheet is attached to a handle; dirt sticks to the sheet, and when it is full, you do not clean it—you just throw it away. This is the creation of Procter & Gamble Swifter.

3.5.4 Co-creation with Customers and Stakeholders

Co-creation is a process of involving customers or users, and possibly other stakeholders in the creation or design of products, services, or experiences. It is a collaborative approach to innovation that seeks to understand and meet the needs of customers more effectively by engaging them in the design process [64].

The concept of co-creation has its roots in the field of service design and was first popularized in the mid-2000s. The idea was that by involving customers in the design of services, companies could create more personalized and effective experiences that better meet the needs of customers.

With the development of information technology and digital platforms, it is easier to create a collaboration infrastructure [65]; subsequently, other stakeholders can participate in co-creation if needed [66]. Co-creation can involve any stakeholder who has an interest in the creation or design of a product or service. This could include suppliers, distributors, employees, and innovators.

The co-creation framework typically involves four main stages:

1) **Discover**: This involves understanding the needs and desires of customers through research, surveys, and interviews.
2) **Ideate**: This involves generating ideas for products or services in collaboration with customers through workshops, brainstorming sessions, and other forms of collaboration.
3) **Prototype**: This involves creating prototypes of products or services based on the ideas generated in the ideation stage and testing them with customers.
4) **Deliver**: This involves refining and finalizing the product or service based on feedback from customers and launching it to the market.

The co-creation framework emphasizes the importance of involving customers in the design process and creating a collaborative environment that fosters creativity and innovation. By involving customers in the design process, companies can create products and services that better meet the needs of their customers and build stronger relationships with their customers based on trust and collaboration.

Here are a few excellent cases of co-creation:

- **LEGO Ideas**: LEGO Ideas is a platform that allows LEGO fans to submit their own ideas for new LEGO sets. Users can submit their designs, and if they receive enough votes from the LEGO community, they may be produced and sold by LEGO. This is a great example of co-creation because it allows customers to be involved in the design and development of new LEGO products.
- **Starbucks My Starbucks Idea**: Starbucks created the My Starbucks Idea platform, which allows customers to submit their own ideas for new products or improvements to existing products. Customers can vote on each other's ideas, and Starbucks reviews the most popular ideas and may implement them in their stores. This is another great example of co-creation because it allows customers to be involved in the design and development of new Starbucks products.
- **Nike Flyknit**: Nike used a co-creation approach to develop their Flyknit line of shoes. The company worked with athletes and designers to create a shoe that was both lightweight and supportive. This is an excellent example of co-creation because it allowed Nike to work with external partners to create a product that better met the needs of their customers.

Now, I give a detailed case study involving multi-stakeholders and customers.

Case Example 3.1   Self-Cleaning Air-Conditioning Unit Development (Figure 3.8)

The development of a self-cleaning air-conditioning unit emerged from the specific needs of households with newborns and mothers, where primary considerations include low noise, gentle airflow, and clean air; this real case study was carried out in the Haier Group [67].

This project began with "Concept Generation," incorporating over 150,000 social network users, including customers and subject matter experts, who contributed over five million pieces of feedback concerning the product's requirements.

In the "Design" stage, the IoT platform attracted 35 solution providers, including designers, R&D specialists, and suppliers, to participate in this groundbreaking design project, undergoing five significant cycles of iterative design enhancements. Three universities, several equipment suppliers, and the company's in-house design engineers made considerable improvements in control, ventilation, and self-cleaning using advanced techniques such as intelligent sensors and active noise control.

During the "Experience" phase, many customers assessed a "virtual product" on the platform and engaged with designers and resource providers, which led to additional enhancements. The first production batch of 11,000 units was manufactured and sold. Since the new products are IoT-enabled, a wealth of product usage data, such as power consumption and noise profiles, was collected.

The company also received further customer feedback through various digital touchpoints, leading to the "Feedback" stage. With all this information, the company commenced the next round of the design process to further refine the product; the second batch sold 180,000 units.

This product development and production process, enabled by super connectivity, diverges significantly from traditional methods with several distinct features:

1) Customers can participate throughout all stages of the product development process by contributing their ideas, testing product prototypes, and providing feedback, from Concept Generation to design improvement, and even individualization. In traditional product

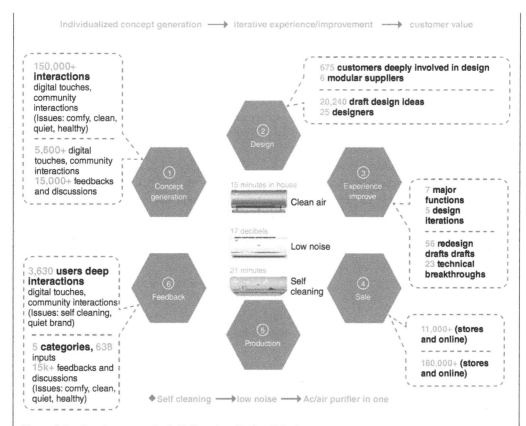

Figure 3.8   Development of a Self-Cleaning Air Conditioning

development, manufacturers usually extract customer needs before design and receive customer feedback after the product hits the market.

2) Super connectivity allows a product innovation and development project to quickly engage and secure a more extensive, higher-quality team of designers, solution providers, and external resources to devise superior product solutions.

3) The IoT platform facilitates concurrent engineering practices by providing a massive, easily accessible venue that surpasses traditional product development processes. The limitations of physical workspace are no longer an issue, and all product development work is transparent and traceable on the platform.

4) The IoT platform also enables mass customization, as any customer can co-design and order personalized products. For instance, on CosmoPlat, 16% of products sold were individualized, 32% were co-developed, and 52% were modular customized products.

5) With an increasing number of products being IoT-embedded, many product usage data and customer feedback are continually sent back to manufacturers for rigorous analysis. The insights derived can be promptly applied to product design enhancement, software upgrades, and manufacturing process improvements.

6) In the case of CosmoPlat, the IoT platform has greatly enhanced the efficiency and speed of product development and production.

7) The embedding of more IoT devices into traditional hardware products and equipment significantly enhances their capabilities. However, this complexity also poses considerable challenges to quality, reliability, and risk assurance.

3.5.5 Design for Individualized Customer Value

Design for Individualized Customer Value [68–70] is a framework that focuses on creating products and services that meet the specific needs and preferences of individual customers. The framework consists of the following steps:

1) **Customer Segmentation**: The first step is to segment customers based on their needs, preferences, and behaviors. This can be done using a variety of methods, such as surveys, social network mining, online footprints, and data analytics. With Industry 4.0, there could be many segments of few or one customers.

2) **Customer Value Analysis**: Once customers are segmented, the next step is to analyze the value that each segment places on different product features and attributes. This can be done using methods such as machine learning, AI, and conjoint analysis, which allows customers to trade off different features and attributes to determine their value.

3) **Product Customization**: Based on the customer segmentation and value analysis, the next step is to design products and services that can be customized to meet the specific needs and preferences of individual customers. This can involve offering a range of options and configurations, allowing customers to choose the features and attributes that are most important to them.

4) **Real-Time Customer Feedback**: To ensure that products and services continue to meet the evolving needs and preferences of individual customers, it is important to collect real-time feedback and data. This can be done through methods such as online reviews, social media monitoring, and customer surveys.

5) **Continuous Improvement**: Based on customer feedback and data, the final step is to continuously improve products and services to better meet the needs and preferences of individual customers. This may involve tweaking product features and attributes, offering new customization options, or developing entirely new products and services.

Overall, the "Design for Individualized Customer Value" framework is focused on creating products and services that are tailored to the specific needs and preferences of individual customers. By using customer segmentation, value analysis, and real-time feedback, companies can create products that better meet the needs of their customers and improve customer satisfaction and loyalty.

Example 3.11 Social Network

Social Networks are intricate business ecosystems that primarily generate revenue through advertising and data collection. These networks encompass several categories of stakeholders, each with distinct objectives and values:

- **Social Network Platform Providers**: These are the entities that own and operate the platforms. Some well-known networks include Facebook, Twitter, TikTok, and YouTube. Their core business objectives typically involve maximizing Internet traffic volume on their network and maximizing income from advertisers and vendors.
- **Users**: The general populace that frequents the network, which, for larger social networks, can range from millions to hundreds of millions. Their primary value propositions lie in connecting with friends and like-minded individuals, discovering new and desirable content and information, and participating in and forming online communities. Since users have highly individualistic tastes and desires, their core customer values are also highly personalized.

- **Content Creators**: A segment of users that generate content, including short videos, articles, and web posts. They seek to increase the reach and influence of their content, attract fans and traffic, form online communities, and earn income from the network and advertisers.
- **Advertisers**: These stakeholders place advertisements on the network to gain exposure. Their core objective is to increase engagement from potential customers on their advertisements and drive sales.
- **Online Producers or Vendors**: These entities leverage the network to seek business opportunities and sales. Their primary value proposition is to connect with likely customers and increase sales.

Each stakeholder category has unique core customer values, and every individual within each category also possesses distinct desires, offerings, and preferences. However, a win–win solution is achievable if the network platform can intelligently route and distribute online traffic.

Central to this is the social network's recommendation system, a machine learning algorithm that utilizes user data and behavior to suggest content and connections to users. It analyzes user data such as likes, comments, shares, and search history to discern patterns and preferences. This information is then used to recommend content and connections that align with the user's interests. A well-designed recommendation system also takes into account contextual information, such as the user's location, device, and time of day, to deliver personalized suggestions.

This recommendation system employs a collaborative filtering approach, looking for similarities between users and their behavior to make recommendations. For instance, if User A and User B have liked similar content in the past, the system may recommend content to User A that User B has engaged with.

Additionally, the system uses a content-based filtering approach, analyzing the content itself to make recommendations. For example, if a user frequently interacts with travel-related posts, the system may recommend similar content.

The system may also use a hybrid approach, merging both collaborative filtering and content-based filtering methods to make recommendations.

In essence, the success of a social network hinges on its ability to accurately deliver individualized customer values to stakeholders. This is achieved by offering an engaging user experience, curated content, and targeted marketing, all customized to stakeholder preferences.

3.5.6 Emotional, Psychological, and Culture Value Creation for Stakeholders

In our discussion in Chapter 2, we discussed that quality is a holistic concept; quality pyramid illustrated by Figure 2.8 was adopted as our contemporary framework for this holistic quality.

For some products or services, especially for some premier products and service products, emotional, psychological, and cultural values are important parts of the overall customer value.

In product and service development, there are several strategies for emotional, psychological, and cultural value creation for stakeholders:

1) **Emotional Value Creation**: Emotional value creation is about creating an emotional connection between the product or service and the stakeholders. Some strategies for emotional value creation include creating a brand personality that resonates with the stakeholders, using

storytelling to create an emotional connection, and creating experiences that evoke positive emotions. For example, Apple is a company that has successfully created an emotional connection with its stakeholders through its brand personality and product design. Apple's products are known for their sleek and elegant design, which creates a sense of sophistication and luxury. The company's marketing campaigns also focus on the emotional benefits of using its products, such as feeling creative or connected.

2) **Psychological Value Creation**: Psychological value creation is about creating a sense of well-being and satisfaction for the stakeholders. Some strategies for psychological value creation include creating products or services that help to reduce stress or anxiety, using gamification to create a sense of achievement or progress, and creating products or services that align with the stakeholders' values or beliefs. For example, Headspace [71] is a meditation app that has successfully created a sense of well-being and satisfaction for its users. The app provides guided meditations and mindfulness exercises that help users to reduce stress and anxiety, improve focus and productivity, and improve overall well-being. Headspace also uses gamification techniques to create a sense of achievement and progress, such as awarding badges for completing daily meditation goals.

3) **Cultural Value Creation**: Cultural value creation is about creating products or services that reflect and celebrate the culture of the stakeholders. Some strategies for cultural value creation include incorporating cultural symbols or traditions into the product or service, partnering with local communities or organizations to create products or services that are culturally relevant, and creating products or services that celebrate diversity and inclusion. For example, Louis Vuitton is known for its iconic monogram canvas, which has become synonymous with the brand. The monogram canvas is not only a symbol of luxury and exclusivity, but it also represents the brand's heritage and history. Louis Vuitton has also created limited edition collections that celebrate different cultures and traditions, such as the "Tribute to Africa" collection.

Overall, the key to creating emotional, psychological, and cultural value for stakeholders is to deeply understand their needs, values, and beliefs and to design products and services that align with these. This requires a user-centered approach to design and development, as well as ongoing engagement and feedback from the stakeholders to ensure that their needs and desires are being met.

Example 3.12   Li Ziqi's Video Channel

Li Ziqi is a Chinese content creator who produces videos showcasing her daily life in rural Sichuan province. Her videos feature traditional Chinese recipes, handicrafts, and rural life and have gained a huge following both in China and internationally. Her YouTube channel has produced only 128 videos as of May 2023 since its inception in 2017. But it has 17.5 Million subscribers, more than that of CNN, BBC.

Her videos got more than 2.9 billion views. As verified on August 21, 2021, the number of subscribers of her channel is a Guinness World Record for "The most subscribers for a Chinese language channel on YouTube" [72]. She rarely speaks in her videos, and the sounds of nature, cooking, and calm music are most prominent. There is barely any international language, such as English was used in the video, but her channel draws the audience from all over the world.

Li Ziqi is a great example of an entrepreneur who has successfully created emotional, psychological, and cultural value for her stakeholders.

One way in which Li Ziqi creates emotional value for her viewers is through her soothing and calming videos, which provide a sense of relaxation and escape from the stresses of daily life. Her videos also showcase the beauty of nature and traditional Chinese culture, creating a sense of cultural pride and appreciation.

Li Ziqi also creates psychological value for her viewers by providing a sense of inspiration and motivation. Her videos showcase her incredible skills in cooking, handicrafts, and farming, inspiring viewers to try new things and pursue their passions.

Finally, Li Ziqi creates cultural value by showcasing and preserving traditional Chinese culture. Her videos feature traditional Chinese recipes and handicrafts that are in danger of being lost in modern society. By sharing these traditions with a global audience, Li Ziqi is helping to keep them alive and celebrate the rich cultural heritage of China.

3.5.7 Design for Quality of Experience

Quality of experience (QoE) [73, 74] is a term used to describe a user's overall experience when interacting with a product or service. It refers to the user's subjective assessment of how well the product or service meets their expectations and satisfies their needs. QoE encompasses various factors, including usability, ease of use, accessibility, responsiveness, reliability, esthetics, and emotional satisfaction.

QoE is important in today's marketplace because customers expect products and services that not only meet their functional requirements but also provide a positive experience. A product that performs its intended function but is difficult to use, unattractive, or frustrating to operate will likely result in a negative customer experience, leading to lower satisfaction and potentially lost sales.

Design for QoE [75, 76] refers to designing products or services with the goal of creating positive experiences for users. It involves a holistic approach that takes into consideration various factors such as usability, esthetics, emotional appeal, and brand values.

To implement design for QoE, the following steps can be taken:

1) **Understand the User**: Conduct research to understand the user's needs, behaviors, and pain points. This can be done through methods such as user interviews, surveys, and usability testing.
2) **Understand the Experience**: Based on user research, find the current typical experience of an existing product and define the ideal experience that the product or service should deliver. This includes considering factors such as usability, esthetics, emotional appeal, and brand values.
3) **Design the Product/Service**: Use the insights gained from the research to design the product or service with the desired experience in mind. This may involve creating prototypes, wireframes, or mockups.
4) **Test and Refine**: Conduct usability testing to evaluate how well the product or service delivers the desired experience. Make necessary adjustments and refinements to improve the experience.
5) **Monitor and Update**: Once the product or service is launched, continuously monitor user feedback and update the product or service to maintain or improve the quality of the experience.

I am giving one example, and one case study to show how this design for QoE works.

Example 3.13 Design of Nest Learning Thermostat

This "smart thermostat" [77] was invented by Nest Labs, the company was founded by Apple alums Tony Fadell and Matt Rogers, who played major roles in designing iPod and iPhone. At a time when terms like smart device, connected home, and IoT were not well known to consumers. The original product was a smart thermostat for the home that could be controlled through an app, and the design team focused on creating a device that was not only functional but also intuitive and esthetically pleasing to use. They adopted a new round shape (Figure 3.9), in contrast of rectangular shapes of most thermostat at that time. The interface is designed to be simple and easy to use, allowing users to adjust the temperature with a simple turn of the dial or through the accompanying mobile app.

The device is designed to learn each user's behavior and preferences, and record the temperature profiles during the days when the user is at home. Nest thermostat has an intelligent Auto-Away feature; whenever the Nest knows you are out of the house, it will adjust the home temperature to save energy. The result is a thermostat that provides a seamless and enjoyable personalized user experience while also helping to reduce energy costs. The original Nest Thermostat was a hit with reviewers on release in November of 2011, it sold out on the first day.

Figure 3.9 Nest Learning Thermostat. *Source:* Cody/Adobe Stock

Case Example 3.2 Quality of Experience Mining and Deployment in an Electrical Vehicle Producer

The company at which this study was conducted is SAIC-GM-Wuling automobile company (SGMW), a joint venture between SAIC Motor, General Motors, and Liuzhou Wuling Motors Co. Ltd. based in Liuzhou, Guangxi Province, China. SGMW is a major mass-volume producer of mini electrical vehicles in China. Initially, after SGMW was incorporated, General Motors' *Global Customer Audit (GCA)* was adopted as the SGMW's product quality assessment methodology; in GCA, a group of quality engineers will spend a total of up to 25 hours on an eight-mile specialized-course road test and quality control reviews on product metrics such as dimensional quality, quality of paint, door opening and closing effort, turning radius, breaking distance, etc., and use the resulting measurement data to evaluate the quality. Clearly, GCA's focus is on cars, but after adopting GCA for several years, SGMW found that it is not adequate to ensure customer satisfaction. For example, even if the fit and finish satisfied the specification, some customers still thought the car looked ugly, and even if the odor inside the new car met the requirements of health regulations, customers still reported disliking the smells.

Step 1: Understand the Customers

SGMW realized that Generation Z had become the most important customer group for the company. Generation Z [78] refers to young people born between 1995 and 2010. They are digitally savvy and heavily influenced by the Internet, instant messaging, and social media. This generation values experiences over possessions and is more likely to spend money on experiences such as travel, events, and activities [79]. This feature of Generation Z will make QoE an important factor in making buying decisions. SGMW wanted to develop a hybrid electrical vehicle. According to the user behavior (see Figure 3.10) data mined by telematics of a similar vehicle fleet, the daily travel mileage is concentrated within 60 km, and urban driving accounts for 90%. The average speed is mainly distributed between 10 and 35 km/h, accounting for 54.9%, and the working conditions below 80 km/h account for 95%. It can be concluded that most of these vehicles are used for urban travel and daily commuting.

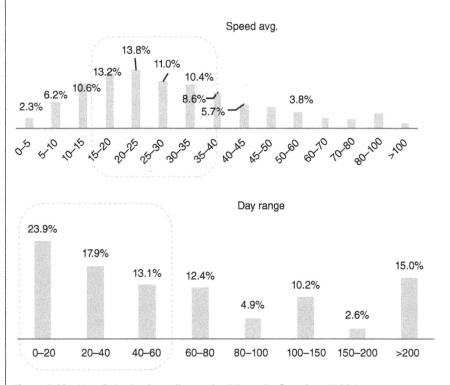

Figure 3.10 User Behavior According to the Telematics Data from Vehicle

Step 2: Understand the Experience by Conducting Large Scale Usability Testing

QoE refers to the degree of satisfaction in feelings that a user has before, during, and after using a product or system; these feelings include the user's emotions, beliefs, preferences, cognitive impressions, physiological and psychological reactions, behaviors, and achievements (ISO 9241-210). QoE has three key factors: user, system (product/service), and scenario. The scenario refers to the circumstance/environment that a car is used; it is a very important factor that influences the QoE.

Scenarios and Scenario Chains
Car usage experience is a very complicated matter. People drive cars on different roads and use cars for different purposes.

SGMW developed its own QoE mining and deployment process, and it is believed to be the first company in the automotive industry to practice design for QoE [80].

SGMW divided the whole overall car ownership and usage experience into segments of well-defined usage scenarios. A scenario is an independent usage segment under certain circumstances, such as entering a car, loading the trunk, starting the car, driving, urban travel, highway driving, parking, etc. These individual scenarios can be logically connected to form scenario sequences in line with user habits, so the overall journey QoE and key scenarios can be assessed completely, as shown in Figure 3.11.

Scenario Refining and QoE Rating
For each scenario, detailed features of each scenario were identified by the quality engineer; for example, typical features of "urban travel" scenario were illustrated in Figure 3.12, such as type of roads, type of urban parking places, etc., based on the circumstance of each scenario. This "urban travel" scenario can be further refined into a tree structure as described in Figure 3.13, which will be convenient to itemize users' QoE inputs as the basic for merit-rating scores.

The measurement of QoE needs a good scale, since QoE reflects people's attitude and sentiment. SGMW defined a 10-point scale, with each point scale paired with a precise definition of its meaning. For example, scale level 10 means that the user experience "greatly exceeds user expectations, leads user demand, establishes industry benchmark and redefines market segment standards." This 10-point QoE assessment scale with paired definitions is shown in Table 3.4.

Figure 3.11 Typical Car Usage Experience Scenarios and Scenario Chain

Figure 3.12 Urban Traveling Refining to QoE Merit Scores

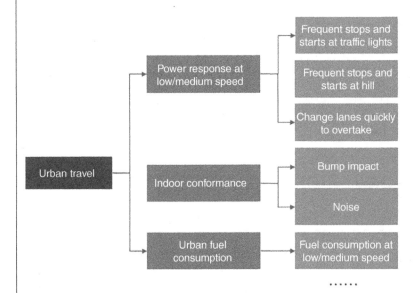

Figure 3.13 Refined Scenario Tree to Obtain QoE Merits Rating Scores

Scenario QoE Mining, Rating
In the SGMW's design for the QoE process, for each scenario, such as urban travel, the company recruits customer testers to try out new car prototype models, fill out the survey template, and give their evaluations and feedback on QoE. Quality engineers also conduct their own scenario tests and give evaluations as well. For one of the new car models, the company used 3000 prototype testing cars and did more than 20,000 user scenario tests and QoE assessments. These assessments provide massive and first-hand customer input on their user experiences. SGMW subsequently uses this valuable input for car redesigns and improvements.

Table 3.4 QoE Assessment Scale

Scale explanation		Remark
Far more than users expected Redefine benchmarks and set industry and market segment standards	10	This product can realize the industry monopoly
Exceed user expectations Become the most competitive product in the industry and market segment, without major defects	9	Bring users surprise, have the expectation of buying again
Fulfill user expectations Lead the market segment, product existence of slight defects can be ignored	8	Parts without error and leakage, basically no design and manufacturing defects, minor defects can be ignored
Meet most of expectations Be competitive in the product and know that users expect the highlights to make them happy	7	There are no errors or omissions in parts, and all parts can reach the marketable quality level, without major defects and defects
Only partial user expectations are achieved Execution quality is average at the border, and achieved at an entry price	6	There are no errors or omissions. The main parts and all functional parts can be used normally without major defects
Most of the customer's expectations were not fulfilled Product defects and weaknesses lead to the decline of product competitiveness	5	There are no errors and misses in the functions and safety regulations, and the safety parts can be used normally. There are major defects and some users do not buy this car
User expectations are not met There are major defects which seriously affect the competitiveness of products	4	There are no errors or omissions in the safety regulations, and some users cannot accept the defects. The vehicle cannot be sold
User requirements are not met No obvious competitiveness	3	The vehicle shall not be sold if there are defects unacceptable to all users
Failure Most parts need to be redesigned	2	The user observes and then walks away
Total failure The problem cannot be solved without a redesign	1	

Step 3: Redesign the Product/Service

In this SGMW's QoE case, as in many typical automotive product developments, the company is not designing a completely new product without prior user experiences; it is redesigning and upgrading a current model with hundreds of prototype car models. So after the large-scale usability tests in Step 2, These assessments provide massive and first-hand customer input on their user experiences. SGMW subsequently uses this valuable input for car redesigns and improvements. By comparing with traditional quality testing, verification, and validation process, this QoE process identified many more action items for the car. Because SGMW has a very flat and flexible organization structure, this rich and massive trove of QoE assessment information can be quickly digested and transformed into redesign action items and deployed into the final redesign of the vehicle, so the final vehicle in mass production has superior QoE. The QoE system greatly helped SGMW, as its Wuling EV was one of the top global EV sellers in 2021 and, by August 2022, had sold 1 million units [81].

3.6 Quality Management and Assurance in Early Product Life Cycle

3.6.1 Quality in Product Development: Crafting Customer Value and Controlling Quality Loss

Let us first revisit the insights from Peter Drucker, recognized as the "father of modern management," who suggested a quantitative approach to evaluating quality [82]. He proposed three core factors: the cost of *poor quality*, the profit from *good quality* as reflected in additional purchases, and the long-term gains in brand image, consumer loyalty, and repeat business. By "poor quality," Drucker was referring to defective products, failures or safety issues, poor performance, or anything falling below expectations. These elements have been the focal point of quality management since the Second Industrial Revolution, and our quality community is well-versed in dealing with them.

Conversely, "good quality" denotes the value a product offers to its customers. As Drucker put it, "a business is an organization that adds value and creates wealth. Value is created for customers, and wealth is generated for owners." The value a product can bring to customers is mostly determined at the product development stage.

Regrettably, the quality community still lacks a consensus on this "good quality." As Bradley Gale's description [37] on the process of establishing standards for the Malcolm Baldrige National Quality Award indicates, some in the community do not even acknowledge that "customer value creation" should be a crucial aspect of quality, because "it is difficult to establish an objective, quantitative metric to assess them."

This stance is not surprising if we consider the evolution of modern quality management. During the early phase of the Second Industrial Revolution, there was little variety in product design. For instance, the Ford Model T was a single design and once that design was finalized, there were no alternatives. Consequently, all the work on QA was unrelated to design; it was solely about controlling the quality loss due to deviations from the design, defects, faults in the production process, supplier issues, and so on. (Note: here, "quality loss" means both monetary and nonmonetary losses caused by "substandard quality." It is different from the concept used by Dr. Taguchi in his "quality loss function.") This focus has persisted in most quality practices since Japan's quality revolution in the 1970s.

However, the significance of "good quality" has been escalating since the 1980s, particularly since the dawn of the 21st century, with innovation becoming the most critical value creation mechanism. In my view, if our quality community wants to remain relevant in the 21st century, we cannot afford to neglect this "good quality" and customer value creation. Crucially, the majority of this customer value is established during the product development phase.

3.6.1.1 Dual Responsibilities in Quality Management

Hence, we identify two key responsibilities in quality management at the product development stage: QA and customer value creation, as previously outlined. However, my intention is not to introduce a new buzzword into our profession. Given that the purpose of "quality loss control" closely aligns with our understanding of "QA," I will use the term "QA" in subsequent discussions.

QA and customer value creation represent two different yet complementary tasks in quality management at the product development stage. QA aims to identify and eradicate defects, errors, or risks that compromise a product or service's reliability, thereby enhancing dependability and curbing costs. In contrast, customer value creation concentrates on comprehending and satisfying customer needs and expectations by developing products or services that deliver unrivaled value in comparison to competitors.

While both methods strive to enhance a product or service's overall quality, QA typically emphasizes safety and dependability, such as defect rates, reliability, and durability. On the other hand, customer value creation targets aspects of "good quality," including features, design, user experience, and even emotional, psychological, and cultural aspects. Ultimately, both tasks are crucial for achieving elevated customer satisfaction and loyalty levels, as they cater to different quality aspects that matter to customers.

The Scope of the Two Tasks These two tasks, QA and customer value creation at the product development stage, encompass the following responsibilities:

1) **QA in Product Development**:
   - Identifying and addressing design-stage quality issues to prevent propagation into production.
   - Implementing design reviews and testing to detect and resolve quality issues before they escalate in terms of cost.
   - Ensuring design integrity, for example, making sure that new features, new technologies, new parts will not adversely affect the other parts of product.
   - Assisting in establishing quality standards.
   - Designing and deploying quality control plans to ensure compliance with quality standards throughout the development process.
   - Instituting change management processes to ensure that design changes undergo thorough review and testing, avoiding unintended quality problems.
   - Providing feedback and data to design engineers to enhance product quality and prevent future quality losses.
2) **Customer Value Creation in Product Development**:
   - Implementing market research to comprehend customer needs and preferences.
   - Involving ground breaking innovation and making sure that it creates good customer value.
   - Developing product features and benefits that cater to customer needs and preferences.
   - Undertaking user testing to ensure that the product satisfies customer expectations and delivers value.
   - Ensuring that product packaging, labeling, and marketing materials effectively articulate the product's value to customers.
   - Offering customer support and service to ensure a positive customer experience with the product.
   - Constantly improving product features and benefits based on customer feedback and evolving market trends.

In product development, QA and customer value creation are instrumental in ensuring that products meet customer expectations while adhering to time and budget constraints. By identifying and addressing quality issues early in the development process, companies can avoid expensive rework and boost product dependability, reliability, and durability. By concentrating on customer needs and preferences, companies can develop products that deliver value to customers and foster long-term relationships.

Necessary Skills for These Two Tasks Based on the scope of these two tasks, the required skillsets and knowledge modules for quality professionals can be easily identified. Some examples are:

For Quality Assurance

- Knowledge of statistical process control and data analysis.
- Familiarity with quality management systems such as ISO 9001 and Six Sigma.
- Understanding of product and process design.
- Knowledge of root cause analysis and problem-solving techniques.
- Ability to manage and interpret customer feedback and complaints.
- Awareness of regulatory and industry standards.

For Customer Value Creation

- Understanding of customer needs and market trends.
- Familiarity with design thinking and co-creation processes.
- Ability to conduct market research and customer analysis.
- Knowledge of innovation and product development processes.
- Understanding of branding and marketing strategies.
- Ability to collaborate with cross-functional teams including marketing, engineering, IT, and design.

From the above analysis, it is clear that these two tasks differ significantly in terms of objectives, scope, required knowledge bases, and qualifications. For QA, the task scope and the qualifications of capable professionals align closely with those of traditional quality experts. However, when it comes to customer value creation, it is a whole different scenario.

Now we must address the following question: "who should undertake these tasks?"

3.6.2 Whose Responsibilities for Quality?

3.6.2.1 Emergence of the Quality Department

In the aftermath of the Second Industrial Revolution, the importance of quality management began to be acknowledged, leading to the establishment of dedicated quality departments within manufacturing organizations. One of the earliest examples of a formal quality department dates back to the 1920s at the Western Electric Company, a part of the Bell System [83].

The 1950s saw the emergence of the Total Quality Management concept, which emphasized the importance of quality across the entire organization, not just within the confines of the quality control department. This approach promoted the participation of all employees in quality management, along with the utilization of statistical tools to enhance processes.

Between the 1960s and 1970s, quality control started to extend beyond manufacturing into the service industries. In the 1980s, the International Organization for Standardization (ISO) devised a series of quality management standards, including the ISO 9000 series. This provided organizations with a framework to design and implement quality management systems.

As previously discussed, from the early 1920s to 1980, the primary task for quality was QA, predominantly in the production stage. Over this 60-year period, substantial knowledge bases and know-how were developed and refined. Quality departments evolved to become a central function within numerous companies. These departments were tasked with developing, leading, and implementing quality management systems throughout the company, monitoring and analyzing quality data, and coordinating continuous improvement activities across various business units.

Post the 1980s, "Quality by Design" started gaining importance, and by 2000, innovation had become the most critical driver for business success. Creating customer value through innovation was an entirely new task. Now, as the fourth industrial revolution is well underway, the question arises—is it still practical to centralize all quality responsibilities within a single quality department?

People have recognized several disadvantages of assigning a single quality department with the responsibility for all quality matters within an organization:

1) **Insufficient Capability and Expertise**: Quality is collectively created by numerous individuals from various sectors and disciplines, such as innovators, engineers, developers, etc. All specific quality issues are invariably linked to particular aspects of products or processes; there are no isolated, generic quality problems. This is especially true for design quality at the product development stage. A solitary quality department typically lacks the capability, expertise, experience, and resources needed in all relevant areas to tackle quality problems, often leading to suboptimal outcomes or even failures in the quality process.

2) **Inefficient Communication and Processes**: When a single department is accountable for all quality issues, communication between that department and others may become slow and inefficient, causing delays in resolving quality issues. For instance, if a company employs a rigid stage-gate process in product development, with the quality department being part of the gate review, communication could be hectic, slow, error-prone, and highly inefficient, potentially causing substantial development lead time delays.

3) **Overburdening**: A solitary quality department may become overwhelmed with the task of managing quality issues for the entire organization, leading to burnout and reduced effectiveness.

4) **Restricted Perspective**: A single quality department might possess a narrow perspective on quality issues as they primarily focus on their own processes and systems. This can result in a lack of understanding of the broader context of quality issues within the organization.

5) **Risk of Conflicts of Interest**: In certain cases, a single quality department might be more concerned with meeting their own internal targets, audit requirements, and metrics rather than addressing real quality issues. This can generate conflicts of interest and undermine the efficacy of the quality process.

3.6.2.2 Realignment of Quality Management Functions: Integration and Deep Collaboration

In recent years, there has been a growing trend toward integrating quality management with other business functions, such as supply chain management, product development, marketing, and risk management. This approach recognizes that quality is interconnected with other aspects of business operations and that managing quality effectively requires collaboration across departments and functions.

Example 3.14   Siemens' Integrated Platform for Product Life Cycle Management

Siemens has been developing its very comprehensive, sophisticated integrated digital platform since 1990s. In 1999, Siemens introduced its initial version of Teamcenter, as a product data management (PDM) solution, and it has since evolved into a comprehensive product lifecycle management (PLM) software solution. In 2018, Siemens launched its Xcelerator Platform, in which Teamcenter is one of the major components, bringing together several other existing solutions under a single brand. In 2021, Teamcenter, which is a part of Xcelerator, added Teamcenter Quality as a module inside the Teamcenter portfolio.

Now I briefly describe the structure and functions of this Xcelerator digital platform.

On the top level, Xcelerator is a portfolio of software and services offered by Siemens that includes several solutions for product design, engineering, and manufacturing. It includes software tools for CAD, simulation, PDM, which is the Teamcenter, and manufacturing execution systems.

Other solutions included in the Xcelerator portfolio include:

- **NX**: A CAD tool that enables designers and engineers to create and edit 3D models of products.
- **Simcenter**: A simulation software that allows engineers to simulate and test product performance in a virtual environment.
- **Capital**: A software solution for electrical systems and wire harness design.
- **Opcenter**: A manufacturing execution system that enables manufacturers to optimize production processes and improve quality.

Teamcenter is a key component of the Xcelerator portfolio, serving as the PLM software solution that enables organizations to manage the entire lifecycle of a product, from initial concept and design to manufacturing, service, and disposal. Teamcenter provides a centralized repository for product data and supports collaboration among all stakeholders, including designers, engineers, manufacturing personnel, and suppliers.

Some key functions of Siemens Teamcenter include:

1) **Product Data Management**: Teamcenter provides a centralized repository for managing all product-related data, including 3D CAD models, bills of materials, specifications, and documentation.
2) **Workflow and Process Management**: Teamcenter enables organizations to define and manage workflows and processes for product development, ensuring that tasks are completed in the correct order and that approvals and reviews are properly documented.
3) **Change Management**: Teamcenter supports change management processes, allowing organizations to track changes to product data and ensuring that all stakeholders are notified of changes and their impact.
4) **Supplier Collaboration**: Teamcenter enables collaboration with suppliers, allowing organizations to manage supplier data and documents and to track supplier performance.
5) **Quality Management**: Teamcenter supports quality management processes, including inspection planning and execution, nonconformance management, and corrective and preventive actions.
6) **Requirements Management**: Teamcenter enables organizations to manage product requirements, ensuring that all requirements are properly documented and traced throughout the product lifecycle.

The 5th item in the above Teamcenter Quality. Siemens Teamcenter Quality is a software platform designed to help organizations manage and improve their quality management processes. It is part of the larger Teamcenter suite of PLM software from Siemens.

As a fully connected platform, Teamcenter Quality provides users with a comprehensive set of tools and capabilities for managing quality throughout the product lifecycle. This includes functions such as:

- **Quality Planning**: Teamcenter Quality allows users to create and manage quality plans, which outline the specific quality requirements for a product or project.
- **Nonconformance Management**: The platform includes tools for managing nonconformances and quality incidents, allowing users to track issues and take corrective actions as needed.

- **Document Control**: Teamcenter Quality provides a centralized repository for quality-related documents, ensuring all stakeholders can access the most up-to-date information.
- **Audit Management**: The platform includes capabilities for managing audits and assessments, including scheduling, tracking findings, and reporting.
- **Supplier Quality Management**: Teamcenter Quality provides tools for managing supplier quality, including supplier performance monitoring and supplier corrective action management.

About this integration of quality with Xcelerator Platform, Dr. Raffaello Lepratti, Siemens vice president of Business Development and Marketing for Manufacturing Operations Management, said in an interview [84], "Creating this single source of truth for product, process and quality data can help ensure each step of the process is synchronized and compliant while also offering traceability and a high level of transparency within the process," He added that the main goal of the Teamcenter Quality is to create "aligned workflows between quality, manufacturing and engineering teams to help improve collaboration and reduce the need for coordination." By centralizing around the end product and building communication channels, Teamcenter Quality will be able to do away with much of the day-to-day need for coordination and reduce the number of meetings that need to take place to ensure everyone is on the same page.

3.6.3 Quality Assurance in the Early Stage of Product Life Cycle

3.6.3.1 Is the Separation of Value Creation and Quality Assurance a Good Idea?

Earlier in this chapter, I explored the importance of both QA and customer value creation during the product development phase. These tasks are inherently different and require distinct knowledge sets and skills. Traditional quality professionals are well-versed in QA tasks, efficiently managing the "cost of poor quality." However, for "creating good quality" or customer value creation, a different knowledge base and skill set may be necessary. This likely requires collaboration from professionals across multiple disciplines, such as scientists, engineers, marketing professionals, outside contractors, etc. For traditional quality professionals, some may argue, "I know my work, but that product innovation task is not my responsibility. It isn't a part of quality, so I will focus on my part, while you focus on yours."

Regrettably, in the 21st century, this is an outdated approach. Firstly, the tasks of QA and customer value creation are complementary and interconnected as they address different aspects of delivering a superior product or service to customers. If customers do not value the product, it will not achieve good sales. Conversely, a well-designed product prone to defects and errors will also fail. Therefore, both tasks are essential. Secondly, not only do we need to execute both tasks, but we also need to do them concurrently and collaboratively. Here is why:

1) **Alignment of Objectives**: Concurrent operation allows the teams to align their objectives, ensuring coordinated efforts. Both teams share the common goal of delivering a high-quality product. Through collaborative work, they can align their strategies, actions, and resources to achieve that goal effectively.

2) **Early Detection and Prevention of Issues**: Concurrent working enables early detection and resolution of issues. For instance, when a product development team introduces an innovative product that involves new technology, components, materials, etc., the QA team must assess, contain, and mitigate risks due to a lack of technical maturity. They must also establish

standards for production and suppliers. This proactive collaboration prevents quality issues from reaching customers, saving time and cost, and potentially protecting the brand's reputation.

3) **Efficiency and Speed to Market**: Working concurrently minimizes delays and expedites the product's time-to-market. By resolving quality issues early and aligning efforts, the teams can streamline processes, reduce rework, and avoid costly delays.

4) **Cross-Functional Collaboration**: Concurrent work promotes collaboration and knowledge sharing among the teams.

As we can see, customer value or "good quality" (as referred to by Peter Drucker) is created through innovation and design. Setting up and enforcing quality standards is a crucial strategy to contain "poor quality." Here, I would like to discuss the relationship between quality and standards, and why quality standards need to be designed and established from the early stages of product development.

3.6.3.2 Quality and Standards: An Interconnected Relationship

Quality and standards are interrelated concepts within the sphere of product or service development. Standards act as a set of guidelines or benchmarks that outline a specific level of quality that a product or service is expected to meet. They offer a uniform language and framework for evaluating and comparing quality, even across diverse industries and geographical locations. On the other hand, quality denotes the level of excellence or superiority of a product or service as perceived by its users or customers.

Standards can facilitate the achievement of a minimum level of quality in products or services, but they do not assure superior quality or customer satisfaction. Quality is a more comprehensive concept that encompasses not only the attributes of a product or service, but also the total experience and value it provides to the customer. While adherence to standards can serve as a valuable tool for quality control and assurance, true quality can only be realized by understanding and satisfying the needs and expectations of the customer.

Quality professionals should be engaged in designing and establishing standards right from the inception of the product development stage. This is because addressing quality issues early on is typically less expensive than rectifying them later. By incorporating quality professionals during the design stage, companies can prevent quality issues from seeping into the production phase, thereby avoiding costly rework or product recalls.

Example 3.15   Agile Development and Agile Quality Assurance

Agile development found its roots in concurrent engineering practices, as discussed in a 1986 Harvard Business Review article [85]. This article showcased the innovative practices and results achieved by companies like Fuji, Honda, and 3M. The practices involved empowering product development teams to make decisions about their work methods and pace, and it also encouraged more frequent team interactions—shifting from monthly to weekly meetings. These practices piqued the interest of the software industry, and in 1991, the term "Scrum" was coined. By 2001, a systematic agile development model was formally proposed, rapidly becoming the dominant model in the industry and extending its influence across different sectors.

Software products distinguish themselves from hardware products in several key ways:

1) They are easier to modify,
2) They have lower capital and equipment costs apart from labor,
3) For the same function, there are multiple possible software solutions. The motivation, initiative, and creativity of programmers significantly affect the quality of programming.

Figure 3.13 illustrates how Agile Development operates.

The left side of Figure 3.13 assumes that a complete software should deliver several (N) functions.

Figure 3.14 illustrates how Agile Development operates. It shows how functions are developed sequentially, and each function goes through iterative cycles of testing, debugging, and improvement until a satisfactory performance is reached. This process continues until all functional modules are integrated, resulting in the first version of a complete, functioning software. Then, the developer shares this software with the customer for feedback and further improvements, all while being transparent on a shared development platform.

The right side of Figure 3.14 describes the work rhythm. There are often interactive meetings (Scrum), and there are also short cycles (Sprint), usually two weeks long, and the duration of each short cycle is timed so a portion of the work can be completed and shown to other team members. Agile development has become the mainstream approach in software development, and many products are now a combination of hardware and software. Despite this, there are still notable differences between hardware and software:

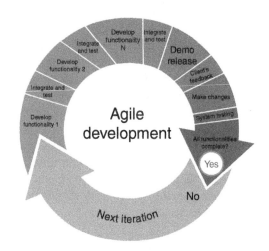

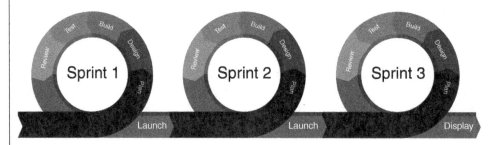

Figure 3.14   Agile Development Model

1) In hardware, elements such as physical shape, spatial structure, and interface with other components need to be finalized earlier, making significant changes in the later design stages challenging.
2) Hardware may have numerous properties and features that require testing and verification, which demands substantial cost and time.
3) Compared with software, the number of engineering changes and tests for hardware cannot be excessive, especially in terms of physical testing.

However, with the rapid advancement of computer simulation technology, many hardware parts and subsystems can be simulated with a high degree of accuracy. This development allows for the adoption of short cycle, interactive, and iterative development methods even in hardware design. Figure 3.15 depicts an agile development process for a product that incorporates both hardware and software elements, often referred to as an embedded system. This type of product is quite common and includes items like mobile phones and cars. The corresponding hardware and software are designed in parallel, with the hardware's short cycle (Sprint) length being approximately twice as long as that of the software.

Agile QA refers to the incorporation of quality assurance principles and practices within an Agile software development context. Agile methodologies, such as Scrum, emphasize iterative development, frequent feedback, and close collaboration among team members. In an Agile QA

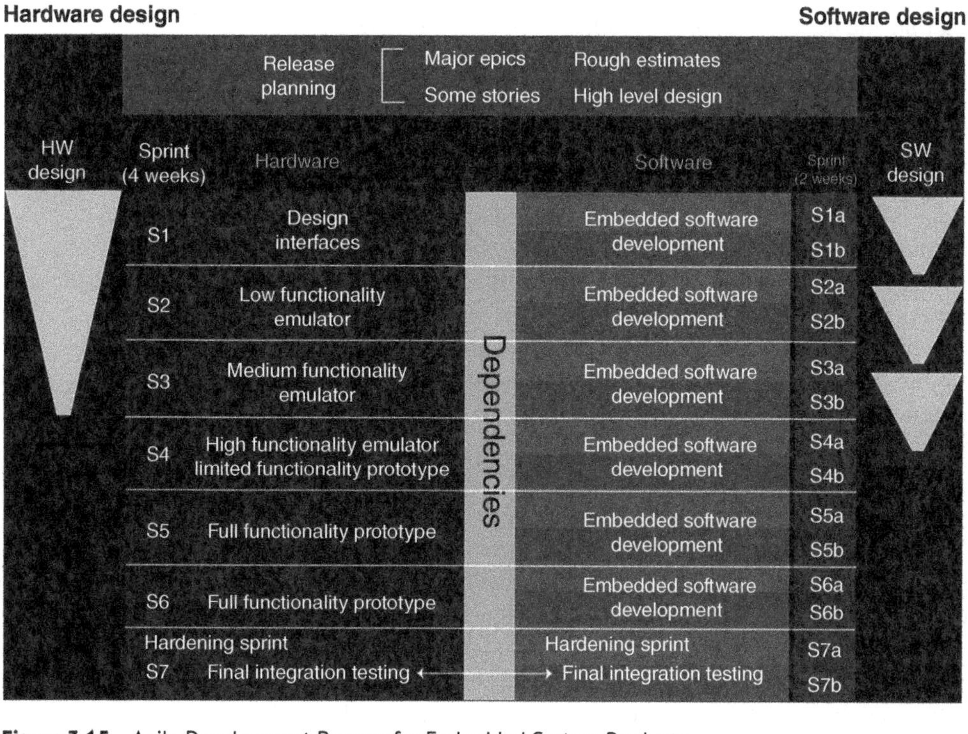

Figure 3.15 Agile Development Process for Embedded System Products

framework, the conventional role of QA is adapted to align with the Agile paradigm. Here are some salient characteristics of Agile QA:

1) **Early Involvement**: QA experts are part of the project from its inception, collaborating with developers, product owners, and other stakeholders to comprehend requirements, define acceptance criteria, and contribute to user story development.
2) **Continuous Testing**: Testing is an ongoing process carried out throughout the project, as opposed to being a singular event at the end of the development cycle. It encompasses automated and manual testing activities to verify the software's functionality, performance, and user-friendliness.
3) **Test Automation**: In Agile QA, test automation is paramount. Automated tests ensure quick and reliable regression testing, facilitating frequent code alterations without compromising quality.
4) **Cross-Functional Teams**: QA professionals closely collaborate with developers and other team members, promoting shared responsibility for quality. This collaboration aids in the early identification and resolution of issues, thereby reducing the likelihood of undetected defects.
5) **Agile Metrics**: Agile QA teams commonly utilize specific metrics to assess and track the quality of the software development process. These metrics can include parameters such as defect rates, test coverage, customer satisfaction, and velocity.
6) **Continuous Improvement**: Emphasis is placed on continuous learning and improvement in Agile QA teams. At the end of each iteration, retrospectives are conducted to reflect on the successes and identify potential areas for improvement in the testing and QA processes.

This integration of agile development and Agile QA exemplifies how teams focused on value creation and quality control can work in seamless harmony.

Quality professionals can contribute to the product development stage in several ways. For example, they can:

- Help establish quality objectives and standards for the product.
- Conduct design reviews to identify and address quality issues in the design stage.
- Develop quality control plans to ensure that quality standards are met throughout the development process.
- Conduct risk assessments to identify potential quality issues and develop plans to mitigate them.
- Ensure that design changes are properly reviewed and tested before being implemented to avoid unintended quality issues.

3.6.4 Overview of Risk Management for New Product Development

In the early stage of product development, extensive customer usage of the product is still absent, hence our understanding of the product's actual quality is limited. Nonetheless, it is critical for quality professionals to be involved at this stage. If the product concept is flawed or the design process is improperly executed, a multitude of quality issues may arise throughout the product life cycle. The earlier we can identify and mitigate potential problems, the more straightforward the solution and the better the outcomes. Our aim is to preemptively prevent quality issues from occurring, which leads us to focus on risk management in the early phase of the product life cycle.

Risk management is a significant and expanding field, especially in the context of Industry 4.0, and warrants a comprehensive chapter dedicated to it. For quality professionals, this topic is becoming increasingly important. Despite the scarcity of books and articles specifically addressing risk management for new product development, this subsection will provide an overview and outline a few effective strategies.

3.6.4.1 Framework for Risk Management in New Product Development

A risk management framework provides a structured approach for identifying, assessing, mitigating, and monitoring risks across an organization. While specific frameworks may vary, they typically involve the following key steps:

1) **Risk Identification**: The first step is to identify and document potential risks that may impact the product.
2) **Risk Assessment**: Once risks are identified, they need to be assessed to understand their potential impact and likelihood. Risk assessment involves analyzing the severity of the risk and the likelihood of it occurring. This step helps prioritize risks based on their significance and determine where to allocate resources for risk mitigation.
3) **Risk Mitigation**: Risk mitigation involves developing and implementing strategies and controls to reduce the probability or impact of identified risks.
4) **Risk Monitoring**: After implementing risk mitigation measures, it is important to continuously monitor the effectiveness of these measures and the overall risk landscape.

For a new product development process, the following risks are most relevant for quality professionals.

- **Market Risk**: This involves uncertainty about whether the product will meet customer needs and achieve sufficient demand in the market. Conducting thorough market research and validation can help mitigate this risk.
- **Technical Risk**: Refers to the uncertainty associated with the product's design, development, and technology. Ensuring that the technical aspects are feasible, scalable, and meet quality standards is crucial. Prototyping, testing, and involving experts can help manage this risk.
- **Regulatory and Compliance Risk**: Pertains to the potential noncompliance with legal and regulatory requirements. Understanding and adhering to relevant laws, regulations, safety standards, and industry-specific guidelines is crucial to managing this risk effectively.

Next, I will outline the basic frameworks for managing these three kinds of risks.

Market Risk Management   Market risks in new product development refer to the uncertainties and challenges associated with the product's reception and success in the target market. Addressing market risks is crucial to ensure that the product meets customer needs, achieves sufficient demand, and generates a positive return on investment. Here are key considerations for managing market risks:

- **Market Research**: Conduct appropriate market research to understand the target market, customer preferences, trends, and competitive landscape. Gather data on customer needs, pain points, and preferences. This research provides insights to guide product development and positioning.
- **Customer Validation**: Validate the product concept and value proposition with potential customers before proceeding with full-scale development. This can be done through prototype testing, pilot programs, or prelaunch marketing campaigns. Gather feedback and iterate based on customer insights to ensure market fit.

- **Competitive Analysis**: Analyze the competitive landscape to understand existing products, pricing, distribution channels, and marketing strategies. Identify key competitors and assess their strengths and weaknesses. Differentiate your product by offering unique features, superior value, or innovative solutions.
- **Target Market Segmentation**: Clearly define the target market segments and tailor the product's features, pricing, and marketing messages accordingly. Segmenting the market helps focus resources, refine product positioning, and better address specific customer needs.
- **Pricing Strategy**: Determine the optimal pricing strategy based on factors such as production costs, value provided, competitive pricing, and customer WTP. Consider pricing models, discounts, bundling options, and any pricing regulations or constraints in the target market.

Regulatory and Compliance Risk Management   Regulatory and compliance risks in new product development refer to the potential noncompliance with legal and regulatory requirements governing the industry or market in which the product will be introduced. Failure to manage these risks can lead to legal issues, penalties, product recalls, reputational damage, and even the inability to bring the product to market. Here are some key aspects to consider when addressing regulatory and compliance risks:

- **Regulatory Framework**: Understand the regulatory landscape and requirements that apply to your product. This includes industry-specific regulations, safety standards, labeling requirements, certifications, permits, and any other applicable laws or regulations.
- **Compliance Planning**: Develop a compliance strategy and plan early in the product development process. Identify the key regulatory requirements, assess their impact on the product, and define the necessary steps to achieve compliance. Integrate compliance activities into the project timeline to ensure timely implementation.
- **Product Safety**: Safety regulations are particularly important for products that can potentially harm consumers or the environment. Conduct thorough risk assessments, hazard analyses, and safety testing to ensure that the product meets safety standards and minimizes risks to users.
- **Documentation and Record Keeping**: Maintain accurate and comprehensive documentation throughout the product development lifecycle.
- **Quality Management Systems**: Implement robust quality management systems to ensure compliance with quality standards and regulations. This includes establishing standard operating procedures, conducting internal audits, implementing corrective and preventive actions, and maintaining a strong focus on product quality and consistency.

Technology Risk Management   Technical risks in new product development refer to the uncertainties associated with the product's design, development, and technology. These risks can arise from various factors, including:

- **Feasibility**: There may be uncertainties regarding the technical feasibility of implementing certain features, functionalities, or performance requirements of the product. It is crucial to assess the feasibility early on through research, analysis, and prototype development to identify and address any potential challenges.
- **Technology Readiness**: If the product relies on new or emerging technologies, there may be risks associated with their readiness, stability, or compatibility with existing systems.
- **Complexity**: Products with complex designs, intricate functionalities, or advanced technologies may have a higher risk of technical challenges. Managing complexity requires system assessment, robust engineering processes, and continuous testing and validation throughout the development cycle.

- **Scalability**: When developing a new product, it is essential to consider its scalability to meet future demand. Technical risks may arise if the product's design or infrastructure cannot handle increased usage, larger customer bases, or expansion into new markets. Considering scalability from the beginning and building flexibility and robustness into the design can help mitigate these risks.
- **Performance and Reliability**: Technical risks can arise if the product does not meet performance requirements or fails to function reliably in different environments or usage scenarios. Thorough testing, simulation, and validation procedures are critical to ensure that the product performs as expected and meets customer expectations.
- **Integration and Interoperability**: If the new product needs to integrate with existing systems, devices, or platforms, there may be risks associated with compatibility, data exchange, or interoperability. Rigorous testing, collaboration with partners or vendors, and adherence to industry standards can help mitigate integration risks.
- **Manufacturing and Supply Chain**: Technical risks can extend beyond the design and development phase to manufacturing and the broader supply chain. Challenges related to production processes, quality control, component sourcing, or supplier reliability can impact the product's success. Effective supply chain management, supplier evaluation, and QA processes are essential to manage these risks.

Besides these risk management frameworks, I will also describe a few good or emerging methodologies for new product development risk management.

3.6.4.2 New Content Risk Analysis and Management

New Content Risk Analysis and Management (NCRA) [86, 87] is a systematic methodology to quantify the reliability risk, of the new product, resulting from product changes. Assessment of the NCRA helps New Product Development (NPD) teams to focus on product functional groups with major risks, and to develop plans to mitigate such risks. New content risk analysis, in the context of NPD, refers to the evaluation and management of risks associated with introducing new or novel content in a product. It focuses on assessing the potential challenges, uncertainties, and impacts that arise when incorporating innovative or untested content elements into a product.

The following are the key aspects involved in NCRA:

1) **Identification of New Content**: The first step is to identify the specific new content elements that will be incorporated into the product. This could include new technologies, features, materials, design elements, or any other content that deviates from existing products or industry norms. Some people used categories of new, unique, different, and difficult or new, unique, and different, to describe the components, materials, subsystems, software modules, mechanisms, etc., in the new design that are new, unique, different, and difficult in nature.
2) **Risk Assessment**: Assess the potential risks associated with the new content elements. This involves evaluating factors such as technical feasibility, performance reliability, compatibility with existing systems or components, potential legal or regulatory constraints, and any known limitations or challenges related to the new content.
3) **Technical Feasibility**: Evaluate the technical feasibility of integrating the new content into the product. Assess the readiness of the content, its compatibility with existing components or systems, and the availability of necessary expertise or resources for implementation. Identify any technical risks or dependencies that could impact the successful integration of the new content.
4) **Market Impact**: Analyze the potential impact of the new content on the target market and customer acceptance. Assess customer needs and preferences, market trends, and the competitive landscape to understand how the new content may influence market adoption, demand, and competitive positioning. Identify any market risks or uncertainties related to the new content.

5) **Regulatory and Compliance Considerations**: Evaluate any regulatory or compliance requirements associated with the new content. Determine if the introduction of new content elements may trigger additional compliance obligations, certifications, or safety considerations. Assess the potential impact on intellectual property rights, legal constraints, and product liability associated with the new content.

6) **Supply Chain and Manufacturing Risks**: Consider any risks or challenges related to the supply chain and manufacturing processes associated with the new content. Assess the availability of necessary resources, components, or expertise required for manufacturing. Evaluate the impact on production costs, quality control, or production timelines. Identify any potential risks or bottlenecks in the supply chain due to the introduction of new content.

7) **Mitigation Strategies**: Develop strategies to mitigate identified risks and challenges associated with the new content. This may involve conducting further research and development, prototyping, testing, collaboration with external partners or experts, ensuring regulatory compliance, or adjusting the design or implementation approach. Implement risk mitigation measures to increase the likelihood of successful integration and market acceptance.

It is worth noting that Toyota typically aims for evolutionary rather than revolutionary changes in their vehicle designs [4, 88], thus tends to limit the new contents in each new model design to a certain extent, to maintain a certain level of familiarity and reliability among their models. In author's view, this approach also has some downsides, such as lack of innovation, lack of excitement, and slow in catching up new technology, on the other hand, Tesla is doing a great job on adopting new contents in its vehicle design.

3.6.4.3 Robust Technology Development

Robust technology development [89] aims to build robustness into newly developed generic technology, or new technology at its infant stage. Examples of such new generic technology include new memory chips, new electronic bonding technology, new materials, and so on. New technologies are usually developed at research laboratories with ideal conditions, small batches, and small scale. After a generic new technology is developed, product developers will try to integrate it into new products. But usually, there are a lot of hiccups in this integration process; the new technology that works well in the lab may not work well after integration, and its performance may not be stable and up to people's expectations. It usually takes many trials and errors to "make it work right."

Robust technology development is a strategy that tries to reduce the hiccups and make new technology integrate with product and production faster. Robust technology development proposes to conduct robust parameter design on the new technology when the new technology is still in the research lab. This robust design must be a dynamic, robust parameter design that focuses on improving its main, generic function of the new technology.

In a regular lab environment, there are very little noise factors. Robust technology development study will artificially bring many noise factors into the environment. Design and redesign will take place for the new technology until robustness in functional performance against noise factors is achieved. If a new technology is "tested out" in the lab for many potential noise factors, its success in future products and future production will be greatly improved.

3.6.4.4 Risk Management by Complexity Theory

Complexity Theory, a branch of mathematics that focuses on the study of complex systems and behaviors that emerge from simple interactions, has found its applications across various disciplines since its inception.

Historically, Complexity Theory evolved from the field of computational theory in the mid-20th century. In essence, it attempted to answer the fundamental question: "What can be efficiently computed?" Over time, researchers started noticing its potential beyond just computational systems, leading to its expansion into areas like physics, biology, and social sciences.

The 1980s marked significant contributions from researchers like Mitchell Feigenbaum, who developed mathematical tools to explain complex phenomena in chaotic systems [90]. At the same time, the Santa Fe Institute in New Mexico became a key hub for Complexity Theory [91], bringing together scientists from different fields to study complex systems.

In the context of management and organizational theory, scholars started applying complexity principles in the late 20th century. Scholars such as Stacey [92] and Anderson [93] started to examine organizations as complex adaptive systems, leading to new insights into organizational dynamics and change. There are also publications related to the applications of Complexity Theory in engineering and risk management [94, 95].

Now let us delve into how Complexity Theory can be leveraged to manage design vulnerabilities:

1) **Understanding Interconnectedness**: Complexity Theory underscores the interdependence and interconnectedness of components within a system. Recognizing and understanding the relationships between various design elements can illuminate potential vulnerabilities stemming from complex dependencies or coupling. This analysis can aid in simplifying or decoupling the design to diminish vulnerabilities.

2) **Managing Emergent Behaviors**: Complexity Theory acknowledges the emergent behaviors often exhibited by complex systems—unanticipated properties or patterns stemming from the interaction of their constituents. These emergent behaviors could lead to vulnerabilities or unforeseen consequences. Hence, the theory advocates for the monitoring and managing of emergent behaviors via modeling, simulation, and real-time feedback.

3) **Adaptability and Resilience**: Emphasizing adaptability and resilience in the face of change and uncertainty, Complexity Theory can aid in identifying design vulnerabilities that may surface from an absence of adaptability or robustness to changing conditions or requirements. Its principles foster the creation of systems capable of adapting, self-organizing, or reconfiguring to maintain performance in dynamic environments.

4) **Systemic Risk Management**: Complexity Theory illuminates how design vulnerabilities can propagate throughout a system, influencing various interconnected components or subsystems. This highlights the need for a systemic approach to managing design vulnerabilities, considering the interactions, and feedback loops within the system.

5) **Iterative and Adaptive Design Processes**: Promoting iterative and adaptive design processes that embrace uncertainty, Complexity Theory encourages learning through experimentation and feedback. Such processes enable the identification and resolution of vulnerabilities as they arise, thus reducing the likelihood of significant issues in the later stages.

References

1 Fredriksson, B. (1994). *Holistic Systems Engineering in Product Development*. Linkoping, Sweden: The Saab-Scania Griffin, 1994, Saab-Scania, AB.

2 Ulrich, K.T., Eppinger, S.D., and Yang, M.C. (2008). *Product Design and Development*, vol. 4. Boston: McGraw-Hill higher education.

3 Fleischer, M. and Liker, J.K. (1992). The hidden professionals: product designers and their impact on design quality. *IEEE Transactions on Engineering Management* 39 (3): 254–264.

4 Morgan, J. and Liker, J.K. (2020). *The Toyota Product Development System: Integrating People, Process, and Technology.* CRC Press.

5 Taguchi, G. (1995). Quality engineering (Taguchi methods) for the development of electronic circuit technology. *IEEE Transactions on Reliability* 44 (2): 225–229.

6 Kackar, R.N. (1985). Off-line quality control, parameter design, and the Taguchi method. *Journal of Quality Technology* 17 (4): 176–188.

7 Yang, K. and El-Haik, B.S. (2009). *Design for Six Sigma: A Roadmap for Product Development.* McGraw-Hill Education.

8 Jenab, K., Wu, C., and Moslehpour, S. (2018). Design for six sigma: a review. *Management Science Letters* 8 (1): 1–18.

9 Watson, G.H. and DeYong, C.F. (2010). Design for Six Sigma: caveat emptor. *International Journal of Lean Six Sigma* 1 (1): 66–84.

10 Chowdhury, S. (2002). *Design for Six Sigma.* Kaplan Business.

11 Chang, S.-J. (2011). *Sony Vs Samsung: The Inside Story of the Electronics giants' Battle for Global Supremacy.* Wiley.

12 Song, J. and Lee, K. (2014). *The Samsung Way: Transformational Management Strategies from the World Leader in Innovation and Design.* McGraw Hill Professional.

13 Lee, K.-C. and Choi, B. (2006). Six Sigma management activities and their influence on corporate competitiveness. *Total Quality Management & Business Excellence* 17 (7): 893–911.

14 Shahin, A. (2008). Design for Six Sigma (DFSS): lessons learned from world-class companies. *International Journal of Six Sigma and Competitive Advantage* 4 (1): 48–59.

15 Yang, K. and Cai, X. (2009). The integration of DFSS, lean product development and lean knowledge management. *International Journal of Six Sigma and Competitive Advantage* 5 (1): 75–99.

16 Al'tshuller, G.S. (1999). *The Innovation Algorithm: TRIZ, Systematic Innovation and Technical Creativity.* Technical Innovation Center, Inc.

17 Willyard, C.H. and McClees, C.W. (1987). Motorola's technology roadmap process. *Research Management* 30 (5): 13–19.

18 Heiss, M. and Jankowsky, J. (2001). The technology tree concept-an evolutionary approach to technology management in a rapidly changing market. In: *IEMC'01 Proceedings. Change Management and the New Industrial Revolution.* IEMC-2001 (Cat. No. 01CH37286), 37–43. IEEE.

19 Ilevbare, I.M., Probert, D., and Phaal, R. (2013). A review of TRIZ, and its benefits and challenges in practice. *Technovation* 33 (2–3): 30–37.

20 Cheong, S., Lenyashin, V.A., Kynin, A.T. et al. (2008). TRIZ and innovation culture at Samsung Electro-Mechanics Company. In: *Proceedings of the Fourth TRIZ Symposium in Japan*, Osaka, Japan (10–12 September 2008). http://www.ogjc.osaka-gu.ac.jp/php/nakagawa/TRIZ/eTRIZ/ epapers/e2009Papers/eCheongTRIZSymp2008/09eS-Cheong-TRIZSymp2008-090709.pdf (accessed 23 August 2023).

21 Kim, H.J. (2003). *How to Apply TRIZ at SAMSUNG.* Samsung Advanced Institute of Technology (SAIT).

22 Park, S. and Gil, Y. (2006). How Samsung transformed its corporate R&D center. *Research-Technology Management* 49 (4): 24–29.

23 Krasnoslobodtsev, V. and Langevin, R. (2006). Applied TRIZ in high-tech industry. In: *Proceedings of the TRIZCON*, Milwaukee, WI, USA. https://www.aitriz.org/articles/TRIZFeatures/6B7261736E 6F736C6F626F64747365762D31323033.pdf (accessed 23 August 2023).

24 Kim, W.C. and Mauborgne, R. (2014). *Blue Ocean Strategy, Expanded Edition: How to Create Uncontested Market Space and Make the Competition Irrelevant.* Harvard Business Review Press.

25 Johansson, F. (2017). *The Medici Effect, with a New Preface and Discussion Guide: What Elephants and Epidemics Can Teach Us About Innovation*. Harvard Business Review Press.

26 Rowe, P.G. (1991). *Design Thinking*. MIT Press.

27 Lockwood, T. (2010). *Design Thinking: Integrating Innovation, Customer Experience, and Brand Value*. Simon and Schuster.

28 Lashinsky, A. (2012). *Inside Apple: How America's Most Admired – and Secretive – Company Really Works*. Hachette UK.

29 Gallo, C. (2011). *Innovation Secrets of Steve Jobs: Insanely Different Principles for Breakthrough Success*. McGraw-Hill Education.

30 Elliot, Jay, and William Simon. *The Steve Jobs Way: iLeadership for a New Generation*. Vol. 33, no. 9. Vanguard, 2011.

31 Goldenberg, J. and Mazursky, D. (1999). The voice of the product: templates of new product emergence. *Creativity and Innovation Management* 8 (3): 157–164.

32 Han, W. and Park, Y. (2010). Mapping the relations between technology, product, and service: Case of Apple Inc. In: *2010 IEEE International Conference on Industrial Engineering and Engineering Management*, 127–131. IEEE.

33 Ulrich, K.T. and Pearson, S. (1998). Assessing the importance of design through product archaeology. *Management Science* 44 (3): 352–369.

34 Walsh, V. (1992). *Winning by Design: Technology, Product Design, and International Competitiveness*. Wiley.

35 Marshall, A. (2018). Tesla's Quick Fix for Its Braking System Came From the Ether, Wired, Transportation, May 30, 2018 7:46 PM. https://www.wired.com/story/tesla-model3-braking-software-update-consumer-reports (accessed 27 June 2023).

36 Dans, E. (2019). What Is It That Really Sets Tesla Apart From The Competition? Forbes, December 2, 2019. https://www.forbes.com/sites/enriquedans/2019/12/02/what-is-it-that-really-sets-tesla-apart-from-the-competition/?sh=cfa9fdf741bb (accessed 27 June 2023).

37 Gale, B. and Wood, R.C. (1994). *Managing Customer Value: Creating Quality and Service that Customers Can See*. Simon and Schuster.

38 Bowers, K. and Pickerel, T.V. (2019). Vox Populi 4.0. *Quality Progress* 52 (3): 32–39.

39 Barravecchia, F., Mastrogiacomo, L., and Franceschini, F. (2022). Digital voice-of-customer processing by topic modelling algorithms: insights to validate empirical results. *International Journal of Quality & Reliability Management* 39 (6): 1453–1470.

40 Holmlund, M., Van Vaerenbergh, Y., Ciuchita, R. et al. (2020). Customer experience management in the age of big data analytics: a strategic framework. *Journal of Business Research* 116: 356–365.

41 McColl-Kennedy, J.R., Zaki, M., Lemon, K.N. et al. (2019). Gaining customer experience insights that matter. *Journal of Service Research* 22 (1): 8–26.

42 Hildebrand, C., Efthymiou, F., Busquet, F. et al. (2020). Voice analytics in business research: conceptual foundations, acoustic feature extraction, and applications. *Journal of Business Research* 121: 364–374.

43 Rosenbaum, M.S., Otalora, M.L., and Ramírez, G.C. (2017). How to create a realistic customer journey map. *Business Horizons* 60 (1): 143–150.

44 Lee, S.M. and Lee, D.H. (2020). "Untact": a new customer service strategy in the digital age. *Service Business* 14 (1): 1–22.

45 Moghaddam, M.T., Muccini, H., Dugdale, J., and Kjægaard, M.B. (2022). Designing internet of behaviors systems. In: *2022 IEEE 19th International Conference on Software Architecture (ICSA)*, 124–134. IEEE.

46 Rainchwar, P., Mate, R., Wattamwar, S. et al. (2023). Analytical study on consumer social and behavioral psychology and its influence on online purchasing. In: *Internet of Behaviors (IoB)* (ed. R. Dhaya and R. Kanthavel), 41–60. CRC Press.

47 Stewart, M.C. and Arnold, C.L. (2018). Defining social listening: recognizing an emerging dimension of listening. *International Journal of Listening* 32 (2): 85–100.

48 Stewart, M.C., Atilano, M., and Arnold, C.L. (2017). Improving customer relations with social listening: a case study of an American academic library. *International Journal of Customer Relationship Marketing and Management (IJCRMM)* 8 (1): 49–63.

49 Breidert, C., Hahsler, M., and Reutterer, T. (2006). A review of methods for measuring willingness-to-pay. *Innovative Marketing* 2 (4): 8–32.

50 Homburg, C., Koschate, N., and Hoyer, W.D. (2005). Do satisfied customers really pay more? A study of the relationship between customer satisfaction and willingness to pay. *Journal of Marketing* 69 (2): 84–96.

51 Baehre, S., O'Dwyer, M., O'Malley, L., and Lee, N. (2022). The use of Net Promoter Score (NPS) to predict sales growth: insights from an empirical investigation. *Journal of the Academy of Marketing Science* 1–18.

52 Mandal, P.C. (2014). Net promoter score: a conceptual analysis. *International Journal of Management Concepts and Philosophy* 8 (4): 209–219.

53 Jain, D. and Singh, S.S. (2002). Customer lifetime value research in marketing: a review and future directions. *Journal of Interactive Marketing* 16 (2): 34–46.

54 Dwyer, F.R. (1997). Customer lifetime valuation to support marketing decision making. *Journal of Direct Marketing* 11 (4): 6–13.

55 Ulrich, D. and Brockbank, W. (2005). *The HR Value Proposition*. Harvard Business Press.

56 Osterwalder, A., Pigneur, Y., Bernarda, G., and Smith, A. (2015). *Value Proposition Design: How to Create Products and Services Customers Want*. Wiley.

57 Turner, J.S. and Soar, R.C. (2008). Beyond biomimicry: what termites can tell us about realizing the living building. In: *First International Conference on Industrialized, Intelligent Construction at Loughborough University*, 1–18. I3CON in collaboration with BSRIA (www.bsria.co.uk). ISBN 978-0-86022-698-7.

58 Gallagher, L. (2017). *The Airbnb Story: How Three Guys Disrupted an Industry, Made Billions of Dollars. . . and Plenty of Enemies*. Random House.

59 Kelley, T. (2001). *The Art of Innovation: Lessons in Creativity from IDEO, America's Leading Design Firm*. Currency.

60 Brown, T. (2008). Design thinking. *Harvard Business Review* 86 (6): 8.

61 Rowe, P. (1998). *Design Thinking*. Cambridge, MA: MIT Press.

62 Kelley, T. (2001). Prototyping is the shorthand of innovation. *Design Management Journal (Former Series)* 12 (3): 35–42.

63 West, H. (2014). A chain of innovation: the creation of Swiffer. *Research-Technology Management* 57 (3): 20–23.

64 Prahalad, C.K. and Ramaswamy, V. (2004). Co-creation experiences: the next practice in value creation. *Journal of Interactive Marketing* 18 (3): 5–14.

65 Sawhney, M., Verona, G., and Prandelli, E. (2005). Collaborating to create: the internet as a platform for customer engagement in product innovation. *Journal of Interactive Marketing* 19 (4): 4–17.

66 Frow, P. and Payne, A. (2011). A stakeholder perspective of the value proposition concept. *European Journal of Marketing* 45 (1/2): 223–240.

67 Yang, K. (2020). Hyper linked: "how an ecosystem including the internet of things will change product innovation and quality". *Quality Progress* 53 (8): 14–21.

68 Yang, K., Mejabi, B., and Rogers, J. (2022). Paradigm shift: the new model for quality in industry 4.0 is predicated on individualized experiential and relationship-based quality. *Quality Progress* 55 (10): 28–35.

69 Yang, K., Krishnan, S., and Siegel, J. (2021). Individualized customer value: where hyper-targeting and hyper-tailoring meet. https://papers.ssrn.com/sol3/papers.cfm?abstract_id=3661753 (accessed 23 August 2023).

70 Kuhl, J. and Krause, D. (2019). Strategies for customer satisfaction and customer requirement fulfillment within the trend of individualization. *Procedia CIRP* 84: 130–135.

71 Mani, M., Kavanagh, D.J., Hides, L., and Stoyanov, S.R. (2015). Review and evaluation of mindfulness-based iPhone apps. *JMIR mHealth and uHealth* 3 (3): e4328.

72 "Li Ziqi breaks YouTube subscribers record for Chinese language channel". Guinness World Records. Guinness World Records Limited. 3 February 2021. Retrieved 10 February 2021. Chinese vlogger Li Ziqi set a new record for "Most subscribers for a Chinese language channel on YouTube" with 14.5 million subscribers, Guinness World Records announced on Tuesday.

73 Brunnström, K., Beker, S.A., De Moor, K., et al. (2013). Qualinet White Paper on Definitions of Quality of Experience. https://hal.science/file/index/docid/977812/filename/QoE_whitepaper_v1.2.pdf (accessed 23 August 2023).

74 Möller, S. and Raake, A. (ed.) (2014). *Quality of Experience: Advanced Concepts, Applications and Methods*. Springer.

75 Hassenzahl, M. (2010). Experience design: technology for all the right reasons. *Synthesis Lectures on Human-centered Informatics* 3 (1): 1–95.

76 Allanwood, G. and Beare, P. (2014). *Basics Interactive Design: User Experience Design: Creating Designs Users Really Love*. A&C Black.

77 Desmet, P.M.A. and Pohlmeyer, A.E. (2013). Positive design: an introduction to design for subjective well-being. *International Journal of Design* 7 (3): 5–19.

78 Turner, A. (2015). Generation Z: technology and social interest. *The Journal of Individual Psychology* 71 (2): 103–113.

79 Seemiller, C. and Grace, M. (2018). *Generation Z: A Century in the Making*. Abingdon: Routledge.

80 Hu, J. (2020). An auto quality audit method base on simulated user scenario. *Auto Engineer* 7: 35–38.

81 Pontes, J. (2022). Top 20 electric cars in the world — April 2022 (charts). https://cleantechnica.com/2022/06/04/top-20-electric-cars-in-the-world-april-2022-charts (accessed 28 February 2023).

82 Watson, G.H. (2002). Peter F. Drucker: Delivering value to customers. *Quality Progress* 35 (5): 55–61.

83 Deming, W.E. (2018). *Out of the Crisis, Reissue*. MIT press.

84 Stach, J. (2021). Siemens Xcelerator Adds Quality Management to Its Portfolio, March 8, 2021. http://engineering.com, https://www.engineering.com/story/siemens-xcelerator-adds-quality-management-to-its-portfolio).

85 Takeuchi, H. and Nonaka, I. (1986). The new product development game. *Harvard Business Review* 64 (1): 137–146.

86 Bartos, J., Thomas, M., and Roter, M. (2017). Risk assessment of new programs as a basis for reliability planning. In: *2017 Annual Reliability and Maintainability Symposium (RAMS)*, Orlando, FL, USA, 1–8. IEEE https://doi.org/10.1109/RAM.2017.7889716.

87 Powell, J. and Nemu, T. (2016). The implementation of the NUD process to design reliable products such as Lennox iComfort S30 Smart Thermostat. In: *Applied Reliability Symposium*, North America. IEEE.

88 May, M.E. (2007). *The Elegant Solution: Toyota's Formula for Mastering Innovation*. Simon and Schuster.

89 Taguchi, G. (1992). *Taguchi on Robust Technology Development: Bringing Quality Engineering Upstream*. ASME Press (American Society of Mechanical Engineers).

90 Feigenbaum, M.J. (1980). *Universal Behavior in Nonlinear Systems*, 49–84. Universality in Chaos.

91 Anderson, P.W. (2018). *The Economy as an Evolving Complex System*. CRC Press.

92 Stacey, R.D. (1995). The science of complexity: an alternative perspective for strategic change processes. *Strategic Management Journal* 16 (6): 477–495.

93 Anderson, P., Meyer, A., Eisenhardt, K. et al. (1999). Introduction to the special issue: applications of complexity theory to organization science. *Organization Science* 10 (3): 233–236.

94 Sheard, S.A. and Mostashari, A. (2009). Principles of complex systems for systems engineering. *Systems Engineering* 12 (4): 295–311.

95 Groš, S. (2011). Complex systems and risk management. In: *2011 Proceedings of the 34th International Convention MIPRO*, 1522–1527. IEEE.

4

Quality Management in the Era of Industry 4.0

4.1 Introduction

The evolution of quality management systems (QMS) toward Industry 4.0 has been significantly propelled by technological advancements and an increasing need for efficient, automated, and customer-focused approaches. Traditional QMS, while effective in maintaining quality standards, have often been reactive, dealing with quality issues after their occurrence. However, the advent of Industry 4.0 and its myriad technologies have allowed for a shift toward a more proactive approach, integrating quality management into the fabric of operations and spanning the entire product lifecycle.

This paradigm shift in quality management has been well-documented in various studies. For instance, Zawadzki and Żywicki discussed how Industry 4.0 could transform traditional QMS into a data-driven, highly integrated system [1]. Similarly, Moeuf et al. emphasized the role of artificial intelligence (AI) and the Internet of Things (IoT) in enhancing QMS within the Industry 4.0 framework [2]. There is plenty of literature discussing about QMS in the era of Industry 4.0 [3, 4]. Now, the QMS in Industry 4.0 utilizes the latest digital technologies, proactively predicting and preventing defects, as well as addressing quality issues that do occur.

Key capabilities and benefits of a QMS in Industry 4.0 include:

1) **Data Integration**: The QMS amalgamates data from a variety of sources, such as IoT devices, sensors, and software systems, enabling real-time tracking and analysis of quality metrics for informed decision-making.
2) **Automated Inspections**: Leveraging AI technologies like computer vision to automate the inspection process, increasing precision and efficiency.
3) **Predictive Quality Control**: With the use of advanced analytics, AI, and machine learning, the QMS can predict potential defects before they occur, saving costs and ensuring superior product quality.
4) **Real-Time Monitoring and Control**: QMS can provide instant detection and rectification of deviations from set standards using Industry 4.0 technologies, ensuring uniformity in quality.
5) **Process Optimization**: Machine learning can identify production inefficiencies, optimizing parameters in real-time for improved quality outcomes.
6) **Customer Value Creation**: The QMS gathers and analyzes customer feedback and usage data to inform product design and improvement efforts, creating more value for customers.
7) **Supply Chain Quality Management**: The QMS monitors supplier quality, helping to prevent upstream quality issues and maintain supply chain consistency.

Quality in the Era of Industry 4.0: Integrating Tradition and Innovation in the Age of Data and AI,
First Edition. Kai Yang.
© 2024 John Wiley & Sons, Inc. Published 2024 by John Wiley & Sons, Inc.

8) **Automated Documentation, Reporting, and Compliance**: AI-driven QMS generates comprehensive quality reports and dashboards and automates quality-related documentation, aiding compliance demonstration.
9) **Continuous Improvement**: The QMS promotes a culture of continuous improvement by identifying process enhancement opportunities.
10) **Interoperability**: Industry 4.0 fosters system integration and interoperability for overall operational efficiency.
11) **Risk Management**: The QMS identifies potential quality risks and their root causes for effective risk management.

This QMS in Industry 4.0 framework covers the entire product lifecycle, from concept to end-of-life, and integrates internal and external stakeholders. This system incorporates the following stages:

1) **Concept and Design Phase**: The QMS integrates product innovation concepts, customer feedback, market research, and industry trends to meet customer expectations and ensure compliance and risk mitigation.
2) **Supply Chain and Procurement**: The QMS ensures supplier materials meet necessary standards, tracking performance to reduce supply chain disruptions.
3) **Manufacturing Process**: The QMS provides real-time monitoring and control during manufacturing, reducing waste and ensuring product quality.
4) **Inspection and Testing**: The QMS streamlines inspection and testing processes, reducing defect risks.
5) **Distribution and After-Sales Service**: The QMS monitors product quality postsale, using customer feedback for future improvements.
6) **End-of-Life Management**: The QMS manages product disposal or recycling to meet regulatory requirements and minimize environmental impact.

In the subsequent sections, I will detail the emerging quality management approaches in various business sectors, including smart factories, supplier quality management, customer service quality management, and the impact of Industry 4.0 techniques on quality management in the service sector. The final section will focus on the digital QMS.

4.2 Smart Factory

4.2.1 What Is a Smart Factory?

A smart factory [5, 6] is a highly digitalized and connected production facility that relies on smart manufacturing. It is driven by technology such as AI, the IoT, data analytics, advanced robotics, and cloud computing, among others. These technologies enable the smart factory to operate and evolve autonomously by learning from data and using that data to drive improvements in efficiency, quality, and agility.

The term "smart factory" was first introduced around the 2010s with the German government's high-tech strategy project, Industry 4.0. It represents a leap forward from more traditional automation to fully connected and flexible systems that can use constant stream of data from connected operations and production systems to learn and adapt to new demands.

Several researchers have delved into the evolution and importance of smart factories. For instance, Lee et al. [7] discussed cyber-physical systems, the Internet of Things (IoT), and cloud computing as key drivers of the smart factory concept. They highlighted how these elements contribute to improved efficiency and flexibility in manufacturing processes.

Moreover, Lasi et al. [8] underscored the role of Industry 4.0 technologies in transforming factories into intelligent environments. They explained how real-time data and interconnectivity could lead to more autonomous and efficient manufacturing processes.

Meanwhile, Zhong et al. [9] examined the integration of cyber-physical systems in smart factories, emphasizing their role in facilitating real-time monitoring, decentralized decision-making, and the creation of a virtual copy of the physical world for simulation and analysis.

Here are some of the main features of a smart factory:

1) **Interconnectivity**: Smart factories leverage the IoT to connect machines, devices, and systems. This connectivity allows for seamless communication and data sharing, enabling real-time monitoring and control of manufacturing processes.

2) **Automation**: In a smart factory, many traditional manual processes are automated. This can increase efficiency, reduce errors, and free up human workers to focus on more complex tasks. Automation in a smart factory goes beyond physical tasks and includes automation of decision-making processes using AI.

3) **Artificial Intelligence and Machine Learning**: These technologies are used to analyze the vast amounts of data generated in a smart factory. They provide predictive insights, optimize operations, and can even make autonomous decisions to improve efficiency and productivity.

4) **Real-Time Data Analysis**: Smart factories collect data from various sources, including machines, products, and workers. This data is analyzed in real time to monitor performance, detect issues, and guide decision-making.

5) **Predictive Maintenance**: Instead of following a set maintenance schedule or reacting to equipment failures, smart factories use data to predict when maintenance will be needed. This can reduce downtime and extend the lifespan of equipment.

6) **Cyber-Physical Systems**: These systems, which are a combination of software, sensors, and physical components, are a key feature of smart factories. They can monitor physical processes, create a virtual copy of the physical world, and make decentralized decisions.

7) **Digital Twin Technology**: This involves creating a digital replica of a physical asset or system. It allows for simulations and what-if analyses, helping to optimize performance and predict potential issues.

8) **Flexible and Agile Production**: Smart factories can quickly adapt to changes in demand or product design. They can switch between different products without significant downtime, allowing for a more responsive and customer-focused approach to manufacturing.

9) **Resource Efficiency**: Smart factories optimize the use of resources, including materials, energy, and human resources. This can reduce waste and environmental impact, as well as save costs.

10) **Improved Safety**: By using technologies like robotics and automation, smart factories can reduce the need for human involvement in potentially hazardous environments, improving worker safety.

The World Economic Forum (WEF), in collaboration with McKinsey & Company, has played a significant role in identifying and promoting the concept of "smart factories" as part of its "Global Lighthouse Network."

The Global Lighthouse Network is a project by the WEF that showcases factories, often referred to as "Lighthouses," that are leaders in the adoption of the Fourth Industrial Revolution (Industry 4.0) technologies. These facilities are called Lighthouses because they light the way for other companies in the journey toward Industry 4.0.

The Lighthouse Certification is awarded by WEF [10] to manufacturers that use Industry 4.0 technologies to improve their factories, value chains, and business models for increased levels of sustainability, improved productivity and to empower workforces. These Lighthouse factories demonstrate the significant benefits of adopting digital technologies and smart manufacturing techniques. The improvements are not just in productivity and efficiency but also in sustainability, supply chain resilience, design customization, workforce development, and customer experience. The Global Lighthouse Network aims to share knowledge, best practices, and lessons learned from these Lighthouse factories with the broader manufacturing community. This is to help accelerate the adoption of Industry 4.0 technologies and drive the digital transformation of manufacturing at a global scale. As of January 2023, the global lighthouse network is comprised of 132 lighthouses all over the world, see Figure 4.1.

Smart factories have significant advantages over the traditional factory in terms of productivity and flexibility. With advanced technologies and integrated systems, they offer substantial improvements in various areas of manufacturing including flexibility, connectivity, speed of changeover, and customization. Here is a detailed breakdown:

1) **Flexibility**: Smart factories are highly flexible due to their use of digital technologies and automated systems. They can easily adapt to changes in demand, product design, or production processes. For instance, with digital twin technology, changes can be simulated and optimized in a virtual environment before being implemented in the physical factory. This allows for rapid, cost-effective adaptation. Automation and AI further enhance flexibility by enabling machines to handle a wide variety of tasks and adjust to new processes quickly.

2) **Connectivity**: The IoT and cyber-physical systems provide the foundation for connectivity in smart factories. Machines, devices, and systems are interconnected, allowing for seamless communication and data sharing. Real-time data from all parts of the factory can be collected, analyzed, and used to monitor and control manufacturing processes. This connectivity also extends beyond the factory to include suppliers, customers, and other external stakeholders, enabling an integrated and responsive supply chain.

3) **Extremely Fast Changeover**: Smart factories can achieve extremely fast changeovers due to their high levels of automation and digitization. Automated systems can be quickly reconfigured for different tasks, and digital technologies can guide these changeovers to ensure they are done correctly and efficiently. AI can also predict and manage the impact of changeovers on production schedules, minimizing downtime.

4) **Ability for Customization**: Smart factories can cater to the increasing demand for customized products. Advanced technologies like AI and data analytics can understand customer preferences and inform product design. Automated and flexible production systems can then produce these customized products at scale. Additive manufacturing or 3D printing is another technology that enables cost-effective customization in smart factories.

With these capabilities, smart factories are more responsive to market changes and customer needs, increase operational efficiency, and create more value. They also enable smart factories to be more resilient and adaptable in the face of disruptions, which is increasingly important in today's volatile and uncertain business environment. These capabilities together ensure that smart factories can deliver high-quality products at the right time and at the right cost, while also being sustainable and worker-friendly.

Besides the above capabilities in productivity and flexibility, which are important in creating customer values for a diverse consumer population, the smart factory also adopted some new quality control methods to insure extremely low defects and high dependability for manufactured goods.

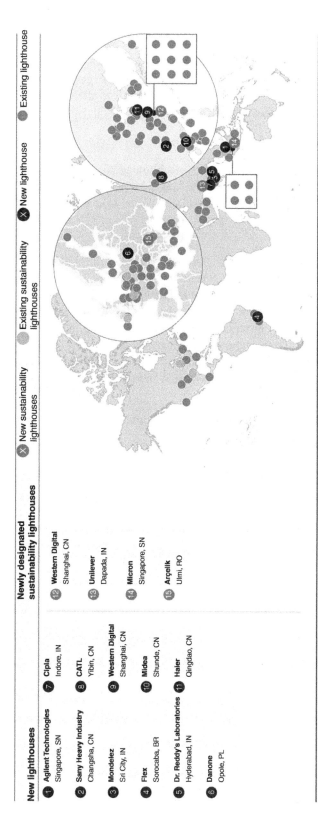

New lighthouses

1. **Agilent Technologies**
   Singapore, SN
2. **Sany Heavy Industry**
   Changsha, CN
3. **Mondelez**
   Sri City, IN
4. **Flex**
   Sorocaba, BR
5. **Dr. Reddy's Laboratories**
   Hyderabad, IN
6. **Danone**
   Opole, PL

7. **Cipla**
   Indore, IN
8. **CATL**
   Yibin, CN
9. **Western Digital**
   Shanghai, CN
10. **Midea**
    Shunde, CN
11. **Haier**
    Qingdao, CN

**Newly designated
sustainability lighthouses**

12. **Western Digital**
    Shanghai, CN
13. **Unilever**
    Dapada, IN
14. **Micron**
    Singapore, SN
15. **Arçelik**
    Ulmi, RO

(X) New sustainability lighthouses

(●) Existing sustainability lighthouses

(X) New lighthouse

(●) Existing lighthouse

Figure 4.1 Certified Global Lighthouse Smart Factories in 2023

4.2.2 Several New Quality Control Methods in Smart Factory

4.2.2.1 Real-Time Monitoring and Control

IoT devices and sensors are commonly used in smart factories to continuously monitor manufacturing processes. This real-time data can be analyzed instantly to identify deviations from expected results or to detect potential issues before they become a problem. Here I give a few relevant technologies which are very effective:

Real-Time 3D Scanning It is a revolutionary technology for quality detection and control, particularly in manufacturing industries [11, 12]. This technology uses various methods to capture the geometry of physical objects, creating a digital representation of the object in three dimensions.

Here is how it works:

1) **Scanning:** The scanner emits a light (often a laser) toward the object. The light bounces back to the scanner after hitting the object. Depending on the technology used, this can be done from one or multiple angles to capture the entire geometry of the object.
2) **Data Processing:** The scanner measures the time it takes for the light to return and uses this information to calculate the distance between the scanner and the object at each point. This creates a "point cloud" of data, representing the object's shape in three dimensions.
3) **3D Modeling:** The point cloud data is then processed using computer software to create a 3D model of the object. This model can be analyzed and compared to the design specifications to check for any deviations.

Real-time 3D scanning offers several advantages for quality control:

- **Speed and Efficiency:** 3D scanning is a noncontact and nondestructive process, making it much faster and more efficient than traditional inspection methods. It can capture millions of data points in a matter of seconds, providing a highly detailed representation of the object.
- **Accuracy:** 3D scanners can capture the geometry of objects with a high degree of accuracy, allowing for precise quality control. They can detect even minor deviations from the design specifications, which might be missed by human inspection.
- **Versatility:** 3D scanners can be used with a wide range of materials and object sizes. They can also capture the geometry of complex shapes that would be difficult to measure with traditional methods.
- **Digital Archiving:** The 3D models created from the scans can be stored digitally for future reference. This can be useful for tracking changes in the object over time, or for comparing different batches of the same product.
- **Real-Time Feedback:** The real-time nature of 3D scanning means that quality issues can be detected immediately, allowing for quick corrective action. This can reduce the cost and time associated with rework and waste.

Real-time 3D scanning has become a key technology in Industry 4.0, with applications in various sectors including automotive, aerospace, electronics, and more. As technology continues to advance, it is likely to play an even larger role in quality control. Figure 4.2 is an example image of a 3-D scanning of an automotive part, and we can see it indicated precise deviations of the part dimensions with design and their precise locations.

Besides Real-Time 3D scanning, there are several other technologies that offer similar capabilities to real-time 3D scanning for quality control and inspection. Here are a few:

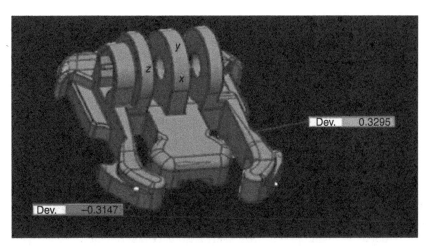

Figure 4.2 A 3D Scanning Image of an Automotive Part with Deviation Spotted

- **Computed Tomography (CT) Scanning:** CT scanning, also known as industrial X-ray scanning, is a nondestructive testing (NDT) technique that uses X-rays to capture the internal structure of an object in 3D. It is particularly useful for inspecting components with complex geometries or internal features that cannot be accessed with other inspection methods [13].
- **Structured Light 3D Scanning:** This technology uses a series of linear patterns of light projected onto an object. The way these patterns deform when projected onto the object allows the scanner to calculate the object's depth and surface information [14].
- **Laser Scanning Microscopy (LSM):** LSM is a noncontact technology that can capture highly detailed 3D images of surfaces. It is often used for quality control of micro-scale components or features [15].
- **Optical Metrology:** Optical metrology uses various light-based technologies to measure objects. This includes methods such as interferometry, photogrammetry, and fringe projection, which can provide highly accurate measurements for quality control [16].
- **Laser Tracker Systems:** Laser trackers are highly accurate measuring systems that use laser light to accurately measure and inspect large-scale components and assemblies [17].
- **Terrestrial Laser Scanning (TLS):** TLS is often used in construction and civil engineering to capture detailed 3D data of large structures or landscapes [18].
- **Ultrasonic Testing (UT):** UT is a nondestructive testing method that uses high-frequency sound waves to detect defects or measure material thickness. It is commonly used in industries such as aerospace and automotive, where it is important to inspect materials without causing damage [19].

These technologies, like real-time 3D scanning, provide noncontact, highly accurate, and detailed inspection capabilities, making them valuable tools for quality control in a variety of industries.

Machine Vision Systems Machine vision uses cameras and image processing software to inspect and analyze physical objects and processes [20, 21]. It has become a game-changer in quality control, enabling nondestructive, high-speed, and highly accurate inspections. Machine vision systems can detect defects that might be missed by human inspection, and they can operate continuously without fatigue. Machine vision is a broader field involving using cameras and computers to capture and analyze images than real-time 3-D scanning. Machine vision systems can inspect, identify, and measure objects and guide robots and other machinery. Machine vision can be used with 2D images or 3D data, and it can capture a variety of object attributes, including shape, color, and texture.

In an industrial setting, these two technologies might be used together. For example, a 3D scanner could be used to capture the shape of a part, and then a machine vision system could use that data to guide a robot in assembling the part. Alternatively, a machine vision system could use a 2D camera to inspect parts for defects, using color and texture data that a 3D scanner would not capture.

4.2.2.2 Predictive Quality Assurance (PQA)

It is a proactive approach to quality management that leverages AI, machine learning (ML), data analytics, and IoT technologies to predict potential quality issues before they occur. It represents a significant shift from traditional reactive or preventive Quality Assurance (QA) methods, which typically involve identifying and addressing quality issues after they have occurred or are about to occur.

Predictive QA involves the continuous collection, monitoring, and analysis of data from various sources throughout the production process. This can include data from machines, sensors, and other IoT devices, as well as data from external sources like supplier information, environmental conditions, and even market trends.

Here is how it works:

1) **Data Collection**: IoT devices and sensors installed in machines and equipment continuously collect data during the production process. This can include data on temperature, pressure, vibration, speed, and other variables that can affect product quality.
2) **Data Analysis**: The collected data is analyzed using ML algorithms and advanced analytics tools. These algorithms can identify patterns and trends in the data that may not be apparent to human observers.
3) **Predictive Modeling**: Based on the analysis, predictive models are built to forecast potential quality issues. For example, the system may predict that a certain machine is likely to produce defective parts if it continues to operate under the current conditions.
4) **Preemptive Action**: Once a potential quality issue is predicted, preemptive action can be taken to prevent the issue from occurring. This might involve adjusting machine parameters, scheduling maintenance, changing a process, or alerting human operators.
5) **Continuous Improvement**: Predictive QA is a continuous process. The system constantly learns from new data and adjusts its predictive models accordingly. This enables it to become more accurate and effective over time.

The primary advantage of predictive QA lies in its potential to markedly diminish defects, waste, and subsequent rework, leading to enhanced product quality and significant cost reductions. This approach also fosters operational efficiency by curtailing downtime and fine-tuning production processes. Moreover, predictive QA is critical in boosting customer satisfaction, ensuring consistent product quality, and minimizing the likelihood of recalls or service disruptions. Given the significance of this concept in the context of QMSs in the Industry 4.0 era, I will dedicate the entire Chapter 5 to a comprehensive exploration of this topic.

4.2.2.3 Electronic/Digital Poka Yoke Methods

Poka Yoke is a Japanese term that means "mistake-proofing." It is a mechanism in a lean manufacturing process that helps an equipment operator avoid (yokeru) mistakes (poka). Its purpose is to eliminate product defects by preventing, correcting, or drawing attention to human errors as they occur.

In modern manufacturing environments, many Poka Yoke measures are digital or electronic [22, 23]. Here are some examples:

1) **Vision Systems:** Advanced camera systems paired with image recognition software can automatically inspect parts for defects. These systems can check for a variety of issues, such as missing or misaligned components, and alert operators if a problem is detected.
2) **RFID Systems:** Radio Frequency Identification (RFID) can be used to ensure the correct components are used in a process. For example, an RFID reader could verify that the correct part is being used before allowing a machine to proceed with its operation.
3) **Barcode Scanners:** Similar to RFID, barcode scanners can be used to verify the correct parts or tools are being used. They can also be used to ensure steps are performed in the correct order, as the system can be programmed to expect a certain sequence of barcodes.
4) **Sensors and Limit Switches:** Various types of sensors (like proximity sensors, photoelectric sensors, etc.) and switches can be used to ensure that parts are correctly positioned before a machine operation is performed. If a part is not where it is supposed to be, the process can be stopped, or an alert can be triggered.
5) **Programmable Logic Controllers (PLCs):** These are industrial digital computers that can be programmed to control manufacturing processes. By programming certain rules or conditions into the PLC, you can prevent operations from proceeding if conditions are not right, thereby preventing mistakes.
6) **Software Controls:** In many cases, software controls can be built into computer-controlled equipment to prevent mistakes. For example, a Computerized Numerical Control (CNC) machine might be programmed to verify that the correct program is loaded before starting a job.
7) **Electronic Checklists:** In complex processes, operators may use electronic checklists that guide them through the process step by step, only allowing them to proceed to the next step once they have confirmed that the current step is completed correctly.

4.2.2.4 Tesla's "Giga Press"

Tesla's innovative approach to automobile body assembly offers a remarkable shift from the traditional manufacturing process. Conventionally, car body frames are assembled from a multitude of smaller parts, each of which requires individual casting or pressing, sorting, and subsequent welding together. This traditional methodology is not only time-consuming and labor-intensive but also necessitates impeccable precision to ensure the correct fit of all parts, and Figure 4.3 illustrates this process.

Moreover, with the multitude of steps involved, assembly errors tend to compound at each station, thereby complicating the control over the dimensional quality of the car body. For a significant period, the US automotive industry has struggled to limit the deviation between the actual body dimension and the design to within a range of 2 mm.

Elon Musk attributed this problem to the error accumulation from the vast number of parts involved and, in response, championed the philosophy that "the best number of parts is no part." Consequently, he introduced the Giga Press, an enormous die-casting machine that enables the manufacturing of large frame segments in a single piece, see Figure 4.4.

Produced by the Italian company IDRA, these machines are among the largest high-pressure die-casting machines globally. By casting larger components as single pieces, the Giga Press drastically simplifies the assembly process, minimizing potential errors and enhancing the overall dimensional accuracy of the vehicle's body [24].

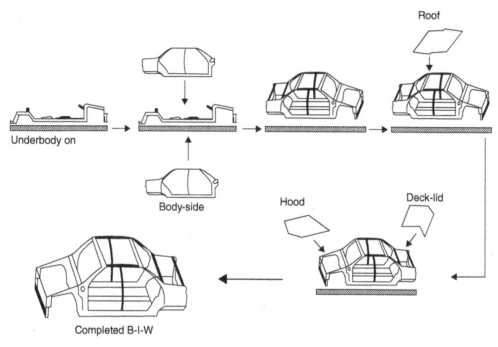

Figure 4.3 Traditional Automotive Body Assembly Process

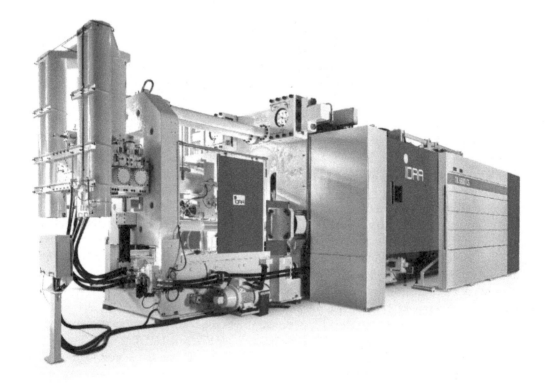

Figure 4.4 Tesla's Giga Press. *Source:* IDRA

The Giga Press technology aims to simplify this process by die-casting a very small number of large pieces of the car body, reducing the number of parts and steps required in the assembly process, from hundreds of pieces to three pieces. Here are some of the key advantages of this technology:

1) **Simplicity and Speed:** The Giga Press allows for the production of large, complex vehicle parts in a single step, reducing the number of parts and assembly operations, which can lead to faster production times.
2) **Quality and Consistency:** By reducing the number of parts and welds, there is less chance of assembly errors or weaknesses in the structure of the vehicle. This can lead to improvements in the overall quality and consistency of the vehicles produced.
3) **Cost-Efficiency:** Fewer parts and steps in the manufacturing process can lead to cost savings, both in terms of the materials used and the labor required for assembly.
4) **Weight Reduction:** Die-casting can produce lighter parts compared to traditional manufacturing techniques, which can contribute to greater energy efficiency in the final vehicle.
5) **Structural Integrity:** Producing larger single-piece components can also enhance the structural rigidity and crashworthiness of the vehicle, potentially improving safety.
6) **Sustainability:** The Giga Press, by reducing the number of processes and components required, can lower the environmental impact of the manufacturing process.

However, it is worth noting that the Giga Press approach also presents some challenges. The machines are very large and require significant upfront investment. There can also be complexities in managing the die-casting process at this scale, particularly in ensuring consistent quality. Nevertheless, Tesla's adoption of this technology illustrates its commitment to innovation and efficiency in its manufacturing processes.

4.2.3 Collaboration of Manufacturing, Engineering, and Quality in Smart Factory

Collaboration is a critical aspect of a smart factory in Industry 4.0, where various departments, like manufacturing, engineering, and QA need to work together seamlessly. Here is a general framework for their collaboration:

1) **Data Integration and Standardization**: The first step in effective collaboration is ensuring that all departments are working from the same data. This requires integrating data from various sources and standardizing it so it can be easily understood and used by everyone. This could involve using a common data platform or data lake and establishing data governance policies and procedures.
2) **Digital Twinning**: Digital twins are virtual replicas of physical assets or processes. They allow different departments to simulate, visualize, and understand the manufacturing process and make changes or improvements without disrupting the actual production. This enhances collaboration by enabling all departments to work from a common, real-time understanding of the factory's operations.
3) **Collaborative Platforms and Tools**: Various software platforms and tools can facilitate collaboration by providing a shared space for communication, project management, and data analysis. These could include collaborative engineering software, QMSs, project management tools, and data analytics platforms.
4) **Real-Time Communication**: IoT, cloud computing, and other Industry 4.0 technologies enable real-time communication and data sharing across departments. This can help identify and resolve issues more quickly, allowing for more proactive and coordinated decision-making.

5) **Cross-Functional Teams and Agile Methods**: Establishing cross-functional teams that include representatives from manufacturing, engineering, and QA can enhance collaboration and ensure that all perspectives are considered in decision-making. Using agile methodologies can further facilitate collaboration by promoting iterative development, regular communication, and responsiveness to change.

6) **Continuous Learning and Improvement**: In a smart factory, learning and improvement are ongoing processes. This involves regularly reviewing performance data, identifying opportunities for improvement, and adjusting processes and strategies based on what is learned. This requires close collaboration between all departments.

4.2.4 Predictive Maintenance in Smart Factory

Predictive Maintenance (PdM) is a crucial component of Industry 4.0 and the smart factory concept [25, 26]. It leverages technologies such as the IoT, AI, ML, and big data analytics to predict potential equipment failures and schedule maintenance before these failures occur. Here is a general framework for predictive maintenance in a smart factory:

1) **Data Collection**: IoT sensors embedded in machines and equipment collect real-time data such as temperature, vibration, pressure, speed, and more. This data is crucial for monitoring the state of equipment and detecting early signs of potential issues.

2) **Data Aggregation and Preprocessing**: The collected data is aggregated and preprocessed for further analysis. This may involve data cleaning, normalization, and feature extraction to ensure it is in a suitable format for the predictive models.

3) **Predictive Modeling**: ML algorithms are trained on the preprocessed data to create predictive models. These models can identify patterns and trends that may indicate an impending equipment failure. Different types of algorithms can be used, depending on the nature of the data and the specific problem to be solved.

4) **Maintenance Prediction**: The predictive models are used to forecast future equipment failures. The system can provide an estimated timeframe for when a failure might occur, giving maintenance teams enough time to intervene.

5) **Maintenance Scheduling**: Upon receiving a prediction of a potential failure, maintenance can be scheduled proactively to prevent the failure from occurring. This helps to avoid unexpected downtime and extends the lifespan of the equipment.

6) **Continuous Improvement**: Predictive maintenance is a continuous process. New data is constantly being fed into the predictive models, improving their accuracy and effectiveness over time. This continuous learning and adjustment is a key aspect of AI and ML technologies.

7) **Integration with Other Systems**: Predictive maintenance should be integrated with other systems within the smart factory, such as production planning, QA, and supply chain management. This can help to minimize the impact of maintenance activities on overall factory operations.

4.3 Quality Management for Smart Supply Chain

4.3.1 Understanding the Smart Supply Chain

The concept of the smart supply chain is a recent evolution, deeply interwoven with the onset of the Fourth Industrial Revolution, also known as Industry 4.0. Rooted in the fundamental principles of traditional supply chain management, smart supply chains herald an era of enhanced operational efficiency, facilitated by advanced technologies.

Tracing back the evolution of supply chain management, we see the initial stages marked by the introduction of simple tracking systems and computerized transactions in the 1960s and 1970s. These systems gradually developed into more sophisticated Enterprise Resource Planning (ERP) systems in the 1980s. The late 1990s saw the advent of Internet-based supply chain management, providing a precursor to today's smart supply chains.

Smart supply chains truly began to take shape with the popularization of IoT technologies in the late 2000s and early 2010s. The application of AI and ML, followed by the integration of blockchain technologies and the employment of robotics, has given further momentum to this revolution [27,28].

The Smart supply chain, deeply embedded in the fabric of Industry 4.0, is a highly integrated, digitized network that employs a multitude of advanced technologies like AI, ML, IoT, big data analytics, blockchain, and robotics. The intent is to drive operational efficiency, augment decision-making processes, curtail expenses, and elevate customer satisfaction levels.

Below are the critical components that constitute a smart supply chain:

1) **Connectivity**: Arguably the core of a smart supply chain, connectivity enables an uninterrupted flow of communication by interlinking devices, machinery, and systems. Primarily facilitated by IoT technologies, it allows real-time monitoring and instant reaction to alterations or complications in the supply chain.

2) **Automation**: A significant number of tasks and processes, including warehouse operations like picking, packing, inventory management, and demand forecasting, are automated, primarily leveraging robotics and AI.

3) **Intelligence**: AI and ML algorithms are employed to sift through and analyze the substantial data generated by the supply chain. This analysis provides valuable insights, predictive analytics, and strategic decision-making support, allowing for demand anticipation, inventory optimization, and preemptive problem identification.

4) **Transparency**: Technologies like blockchain bolster transparency, providing end-to-end visibility across the supply chain. It assures traceability of products and transactions, mitigating fraud, enhancing accountability, and boosting customer trust.

5) **Flexibility**: Smart supply chains are inherently flexible and adaptable. They can swiftly respond to demand variations, supply disruptions, or other unexpected events. This is accomplished through technologies such as digital twins, which simulate various scenarios and strategies in a virtual environment devoid of real-world risks.

6) **Customer Centricity**: Advanced analytics and real-time data enable smart supply chains to offer a personalized customer experience, improved service levels, and superior satisfaction. This can span from customized products to precise delivery time predictions.

Example 4.1 Supply Chain of Amazon

Amazon supply chain [29, 30] essentially refers to Amazon's entire process from product warehousing to inventory management, pricing, delivery, and more. Amazon has optimized each of these elements to ensure that everything is working smoothly and efficiently. When it comes to Amazon's supply chain, there are two main elements critical to its success—Amazon Fulfillment Centers and the Amazon Delivery Fleet. Amazon has more than 175 fulfillment operating centers globally in more than 150 million square feet of space. Amazon has an airplane fleet that helps with delivery. These airplanes fly from more than 20 airports around the United States and can carry about 30 or more containers. On the ground, Amazon uses trucks, vans, bikes. Drones will be a big part of the future of Amazon. Amazon is a great example of utilizing a smart supply chain in the following aspects:

Connectivity: Amazon's supply chain is fully integrated, with all devices and systems interconnected. This facilitates real-time monitoring and immediate response to changes or issues in the supply chain. For instance, Amazon uses IoT technology to track their delivery trucks' locations and conditions, ensuring that goods are delivered in the most efficient manner.

Automation: Amazon uses robotics extensively in their warehouses. The robots are responsible for picking up items and moving them around the warehouse, making the order fulfillment process faster and more efficient. This automation reduces human error and increases productivity.

Intelligence: Amazon uses advanced ML algorithms and AI to analyze vast amounts of data for demand forecasting, order fulfillment, and transportation logistics. Their AI system also optimizes routes for delivery based on factors like traffic, weather, and package type.

Transparency: Amazon provides end-to-end visibility of the order process to its customers. When a customer places an order, they can track it in real-time from the warehouse to their doorstep. This transparency enhances customer trust and satisfaction.

Flexibility: Amazon's supply chain is designed to be flexible and adaptable. They are well-known for their ability to handle massive demand spikes during peak shopping periods like Black Friday or Christmas. They can quickly scale up their operations to meet increased demand, thanks to their advanced predictive analytics and robust supply chain infrastructure.

Customer Centricity: Everything in Amazon's supply chain is designed with the customer in mind. For instance, their predictive analytics not only help ensure that products are in stock when customers want them, but they also help anticipate what customers might want in the future. They also offer personalized recommendations based on customers' past purchases and browsing history.

In the manufacturing industry, Siemens has an Industry 4.0-empowered supply chain.

Example 4.2   Siemens Smart Supply Chain

Siemens has been a pioneer in implementing Industry 4.0 and smart supply chain principles [31].

Siemens' smart supply chain has the following features:

Connectivity: Siemens relies heavily on IoT technologies to interconnect all of its machines, devices, and systems across its global operations. Their MindSphere, a cloud-based, open IoT operating system, enables them to connect their devices and analyze vast amounts of data across their operations. (Figure 4.5)

Automation: Siemens uses automation throughout its supply chain. In their Amberg Electronics Plant in Germany, most of the production process is automated, with machines and computers handling 75% of the value chain without human intervention. Machines in the factory are self-coordinated for tasks like material restocking and predictive maintenance.

Intelligence: Siemens uses AI and advanced analytics for predictive maintenance, process optimization, and resource allocation. They use ML algorithms to predict potential malfunctions in their machinery before they occur, which reduces downtime and increases efficiency.

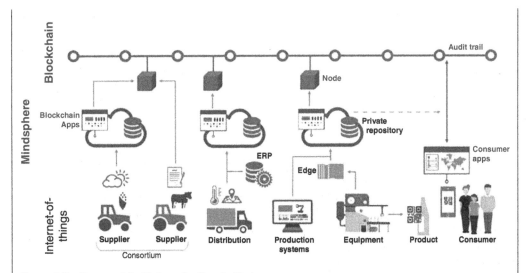

Figure 4.5 Siemens MindSphere for Supply Chain

Transparency: Siemens uses digital twin technology to create a virtual replica of their entire production process. This allows them to monitor the production process in real-time and make adjustments as needed. It also enhances transparency by providing a detailed, real-time view of the production process.

Flexibility: Siemens' supply chain is highly flexible, thanks to its use of digital twins and simulation tools. They can simulate different scenarios, test new strategies, and optimize their production processes without disrupting their actual operations.

Customer Centricity: Siemens uses data from its interconnected devices to provide better services to its customers. For instance, they use data-driven insights to predict when a customer's equipment might need maintenance and provide timely service, which reduces downtime and saves costs for their customers.

4.3.2 Overview of Supplier Quality Management and Capabilities Brought by Industry 4.0

Supplier quality is a supplier's ability to deliver goods or services that will satisfy customers' needs. Supplier quality management (SQM) is defined as the system in which supplier quality is managed by using a proactive and collaborative approach. SQM is very important because:

1) **Product Quality**: The quality of your suppliers directly impacts the quality of your products or services. If a supplier provides substandard materials or components, the end product's quality will likely suffer, which can lead to customer dissatisfaction, returns, and a damaged reputation.

2) **Regulatory Compliance**: Many industries have stringent regulations regarding the quality of products. These regulations often extend to the supply chain, meaning companies are responsible for ensuring their suppliers also comply. An effective SQM system can help manage and document this compliance.

3) **Cost-Efficiency**: Poor quality goods from suppliers can lead to inefficiencies and increased costs, such as rework, returns, and wasted materials. By managing and improving supplier quality, companies can avoid these costs and enhance their overall efficiency.

4) **Risk Mitigation**: SQM can help to identify and mitigate risks in the supply chain. This might include risks associated with supply chain disruption, supplier financial stability, or quality issues. By identifying these risks early, companies can take proactive measures to mitigate them.

SQM comprises several key steps that allow organizations to assess, monitor, and improve the quality of their suppliers' products or services. It is worth noting that while the specific processes may vary depending on the industry and the company, the following steps generally form the backbone of an effective SQM framework:

1) **Supplier Selection**: The first step involves identifying potential suppliers and evaluating them based on a set of predetermined criteria, such as cost, quality, delivery time, and capacity. This process may involve a formal Request for Proposal (RFP) or Request for Quotation (RFQ).
2) **Supplier Evaluation and Certification**: Once potential suppliers have been identified, they must be evaluated more deeply. This could involve on-site audits, quality assessments, and reviews of financial and operational stability. Suppliers that meet the company's requirements can then be certified as approved suppliers.
3) **Contract Negotiation**: The next step is to negotiate contracts with the approved suppliers. These contracts should specify the terms and conditions of the relationship, including pricing, delivery schedules, and quality standards.
4) **Quality Assurance**: Once the supplier is onboard, there needs to be a continuous QA process in place. This includes regular monitoring of supplier performance against quality standards and contract requirements. Tools like scorecards and key performance indicators (KPIs) can be used to track and measure performance.
5) **Continuous Improvement**: The SQM process does not end once a supplier has been approved and is delivering as per the contract. Continuous improvement is a key component of any SQM framework. This involves regular reviews of supplier performance, feedback sessions, and joint efforts to identify and implement improvements in cost, quality, delivery, and service.
6) **Risk Management**: Part of ongoing supplier management involves identifying and managing risks. This could be supply chain risks (such as disruptions due to natural disasters or political instability), quality risks, or financial risks. Companies need to have contingency plans in place to manage these risks.
7) **Communication and Collaboration**: Throughout all of these steps, open and effective communication with suppliers is crucial. This can be facilitated by technology, such as supplier portals or other collaborative platforms.

SQM is a crucial component of modern supply chains, and under Industry 4.0, it has undergone a significant transformation. The advent of new technologies such as big data, IoT, AI, ML, and blockchain has revolutionized how companies manage and interact with their suppliers, leading to improved quality, efficiency, and collaboration.

Here are some key advancements in SQM in the context of Industry 4.0:

1) **Real-Time Monitoring**: IoT and other connected technologies allow for real-time monitoring of supplier operations. Sensors can track everything from the temperature conditions during transport to the operational status of production machines. This means that potential issues can be identified and corrected more quickly, leading to less waste and higher overall quality.
2) **Predictive Analytics**: AI and ML can analyze vast amounts of data to predict potential supplier issues before they become serious problems. This might include predicting equipment failures, supply delays, or quality issues based on patterns in the data. Early warning of such problems allows for proactive measures to prevent or mitigate the impact.

3) **Blockchain for Traceability**: Blockchain technology offers the potential to create a secure, transparent record of all transactions and interactions with suppliers. This can greatly enhance traceability, allowing companies to quickly identify and resolve any issues that arise. It also offers a higher level of security and fraud prevention.
4) **Advanced Supplier Evaluation**: AI and ML algorithms can be used to evaluate suppliers more effectively. This could involve analyzing performance data to assess a supplier's reliability, quality, and efficiency, or using predictive models to assess the risk of future problems. This allows companies to make more informed decisions about which suppliers to work with.
5) **Collaborative Platforms**: Industry 4.0 also involves greater collaboration between companies and their suppliers. This might involve shared digital platforms where data is exchanged in real time, enabling more effective communication and coordination. This not only improves efficiency but also helps to build stronger, more cooperative relationships with suppliers.
6) **Digital Twins for Quality Control**: Digital twin technology can be used to simulate and optimize supplier processes. This can help to identify potential quality issues and test solutions without having to disrupt real-world operations.

4.3.3 Contemporary Collaboration Models Between Producers and Suppliers in Quality Management

For manufacturing entities, the bond between producers and suppliers is profound and enduring. The collaboration begins as early as the product development stage and sustains throughout the entirety of the product life cycle. Industry professionals frequently adopt two primary collaboration models.

These models are Advanced Product Quality Planning (APQP) and Integrated Product Development (IPD).

4.3.3.1 APQP and PPAP

APQP and Production Part Approval Process (PPAP) [32, 33] are two fundamental methodologies used in the automotive industry to ensure quality and reliability in the development of new products and processes, and they give a detailed operational framework for collaborative quality management with all tiers of suppliers in product development and production, here are the outlines:

Advanced Product Quality Planning (APQP) This is a structured process aimed at ensuring customer satisfaction with new products or processes. It was developed by the Automotive Industry Action Group (AIAG) and is utilized by the automotive industry but can be applied to many other industries as well.

APQP consists of five key phases, including:

1) **Plan and Define Program**: Identify the customers' needs, expectations, and requirements. Develop a plan to meet those requirements, including defining the overall project plan and quality metrics.
2) **Product Design and Development Verification**: Design the product to meet the customers' requirements. Validate the design through reviews and testing to ensure it meets specifications.
3) **Process Design and Development Verification**: Design and verify the production process. This includes creating process flow diagrams, conducting failure mode and effects analysis (FMEA), and developing control plans.

4) **Product and Process Validation**: Validate the production process and the product itself through testing and evaluation. This includes production trial runs, capability studies, and product validation testing.

5) **Feedback, Assessment, and Corrective Action**: Collect and analyze data from production and field trials to assess the product and process. Make any necessary adjustments and implement corrective actions as needed.

Production Part Approval Process (PPAP) PPAP is a standardized process in the automotive and aerospace industries that helps manufacturers and suppliers communicate and approve production designs and processes before, during, and after manufacture. It is an output of APQP and formally verifies that all design and specification requirements are understood by the supplier and that the process has the potential to produce a product consistently meeting these requirements during an actual production run at the quoted production rate.

PPAP requires the submission of several documents, including:

1) Design Records and Engineering Change Documents
2) Customer Engineering Approval
3) Design Failure Modes and Effects Analysis (DFMEA)
4) Process Flow Diagram
5) Process Failure Modes and Effects Analysis (PFMEA)
6) Control Plan
7) Measurement System Analysis Studies (MSA)
8) Dimensional Results
9) Records of Material/Performance Tests
10) Initial Process Studies
11) Qualified Laboratory Documentation
12) Appearance Approval Report (AAR)
13) Sample Production Parts
14) Master Sample
15) Checking Aids
16) Customer-Specific Requirements
17) Part Submission Warrant (PSW)

Together, APQP and PPAP provide a framework for product quality planning and validation, ensuring that products meet customer requirements and production processes are capable of producing the product consistently.

4.3.3.2 Integrated Product Development (IPD)

IPD is a specific approach to new product development that emphasizes the integration of various functions and disciplines throughout the product development process. In traditional product development, different functions (like design, manufacturing, and marketing) may work in silos, which can lead to inefficiencies and misalignments. I introduced this model in Chapter 3; in this section, I will focus more on its collaborative approach to supply quality management.

IPD aims to overcome these challenges by fostering collaboration and coordination across functions. This can involve concurrent engineering, cross-functional teams, early supplier and customer involvement, and other collaborative strategies.

Here are some key elements of IPD:

1) **Cross-Functional Teams**: Teams are composed of members from various functional areas such as design, engineering, manufacturing, marketing, and sales. This allows for a holistic approach to product development, ensuring all aspects are considered from the outset.
2) **Concurrent Engineering**: Rather than working in a linear fashion, with one task being completed before the next begins, tasks overlap in concurrent engineering. This means that while the design team is still working, the manufacturing team can begin their work. This can significantly speed up the product development process.
3) **Customer Focus**: The customer's needs and feedback are integrated throughout the product development process. This can involve techniques such as Quality Function Deployment (QFD) and Voice of the Customer (VOC).
4) **Early Supplier Involvement**: Suppliers are involved early in the product development process, which can help to ensure that materials and components are available when needed and that they meet the necessary specifications.
5) **Risk Management**: Risks are identified and managed throughout the product development process. This can involve techniques such as FMEA.

IPD approach includes a substantial focus on supply quality management. This focus helps ensure that all components and materials used in a product meet the necessary standards and specifications. Here is a closer look at how supply quality management fits into the IPD process:

Early Supplier Involvement (ESI): In an IPD approach, suppliers are involved early on in the development process. This is a key part of supply quality management. ESI allows potential quality issues to be identified and addressed before they become significant problems. It also ensures that suppliers understand the product requirements and can meet them.

Cross-Functional Teams: Supply quality management is a cross-functional responsibility in an IPD approach. This means that individuals from various departments, including procurement, design, manufacturing, and QA, work together to manage supply quality. This collaboration helps to ensure that all aspects of supply quality are considered and managed effectively.

Supplier Quality Audits and Assessments: Suppliers are regularly audited and assessed to ensure they can meet the company's quality standards. These audits can be a part of the supplier selection process and can also be conducted on an ongoing basis for existing suppliers.

Risk Management: Managing risk is a key part of supply quality management in an IPD approach. This involves identifying potential risks that could impact supply quality, such as supplier reliability or the availability of key materials, and developing strategies to mitigate these risks.

Continuous Improvement: Finally, the IPD approach encourages continuous improvement. This means that supply quality management processes are regularly reviewed and improved based on feedback, performance data, and changing business needs.

4.3.4 Leveraging Industry 4.0 for Supply Quality Management Enhancement

In recent times, the rise and maturation of numerous enabling technologies, notably digital platforms and assorted Industry 4.0 technologies, have opened up a plethora of avenues to enhance contemporary supply quality management. Here, we will discuss a few promising strategies.

4.3.4.1 Early Supplier Involvement During the Product Development Stage

Traditional product development processes tend to incorporate suppliers late into the design stage, which could result in inefficiencies and discordance. Industry 4.0 promotes real-time collaboration and data sharing, allowing suppliers to participate from the earliest stages of design. The initial stages of product development often encompass innovation, market research, and engineering design, where the team may encounter risks associated with new concepts, producibility issues with existing technology, and economic viability of the product concept, among others. These challenges must be addressed promptly and frequently require the involvement of relevant suppliers. Industry 4.0 technologies can augment collaboration between manufacturers and suppliers. Cloud-based platforms and other digital tools facilitate real-time communication and data sharing, simplifying the management and coordination of product development processes.

Once the concept is approved, the product design phase ensues, necessitating the participation of relevant suppliers to ensure the product can be economically produced, deliver functional performance, and maintain high quality. Industry 4.0 technologies can foster co-innovation between suppliers and manufacturers. Suppliers, for example, could employ virtual reality (VR) or augmented reality (AR) technologies for collaboration on product designs or utilize additive manufacturing (3D printing) to quickly develop prototypes for testing and validation. Suppliers can also make use of digital twin technology to construct a virtual replica of their parts or components. These digital twins can be integrated into the overall product design, allowing for simulation and testing before physical prototypes are created. This approach can expedite the design process, minimize the necessity for physical testing, and permit early identification and resolution of potential issues.

4.3.4.2 Upgrading the Supplier Quality Validation Process Via Industry 4.0 Technology

The aim of APQP and the PPAP is to implement a well-structured supplier quality validation process, albeit one part at a time, one supplier at a time. Traditional processes can be slow and time-consuming.

Industry 4.0 technologies, when used in conjunction with connected platforms, can significantly boost APQP and PPAP in several ways:

- **Collaboration and Communication**: Cloud-based platforms can streamline information exchange and enhance collaboration among various teams involved in APQP and PPAP processes. For instance, design teams, manufacturing teams, suppliers, and QA teams can all access and contribute to a shared data platform, enabling quicker decision-making and problem resolution.
- **Automated Documentation**: Connected platforms can automatically record every step of APQP and PPAP processes, generating comprehensive and precise records that can be used for future reference, compliance, and continuous improvement initiatives.
- **Data Integration and Real-Time Monitoring**: IoT devices can gather data from every stage of the production process in real time, providing a more complete and immediate snapshot of product quality and process efficiency. This can bolster APQP by allowing for timelier and more accurate modifications to the product design and production process. In the context of PPAP, real-time data can provide immediate confirmation of part conformity and process stability.
- **Integration with Other Systems**: Connected platforms can be integrated with other systems, such as ERP and Customer Relationship Management (CRM), offering a more comprehensive view of the entire product lifecycle. This can yield valuable insights for optimizing APQP and PPAP processes.
- **Traceability**: Connected platforms enable tracking and tracing of every part, batch, or process in a product's lifecycle. This level of traceability is especially beneficial in PPAP when validating the conformance of a part or assembly.

4.4 Quality Management in After-Sale Customer Service

4.4.1 Introduction

Many producers have an after-sale customer service department, which provides regular after-sale customer service, such as installation, repair, maintenance, etc. It can also collect and process vast amount of information about product usage and customer experiences, so this information can be sent back to producers to improve relevant stakeholders in the product life cycle, such as design engineers, manufacturing engineers, etc. People in the customer service department have extensive and sometimes deep interactions with customers; they can also contribute to product innovation practices. I will describe the roles of after-sales customer service as follows.

4.4.1.1 Regular After-Sale Customer Service

It refers to the range of services that a business provides to its customers after they have purchased a product or service. The scope of the regular after-sale service varies from business to business, but it generally covers the following areas:

- **Product Installation/Setup Assistance:** This service helps customers install or set up the product they purchased. This is especially common for tech products, appliances, or furniture.
- **Troubleshooting and Technical Support:** This involves assisting customers when they encounter problems or difficulties in using a product or service. Support can be provided through various channels, such as phone calls, emails, live chat, or social media.
- **Product Maintenance and Repair:** Some businesses offer services for maintaining and repairing the product. This can include scheduled maintenance, repairs, parts replacement, and in some cases, product recalls.
- **Warranty and Guarantee Services:** This covers the repair or replacement of products within the warranty period or under the terms of a guarantee.
- **Product Returns and Exchanges:** This involves managing the return or exchange of products that are faulty, not as described, or unsuitable for the customer's needs.
- **Customer Training:** Some companies offer training sessions or tutorials to help customers better understand and use their products or services.
- **Customer Feedback and Complaint Handling:** This includes collecting and addressing customer feedback and complaints about a product or service.

After-sale customer service generally works as follows:

1) **Initiation:** The process usually starts when a customer contacts the company with a question, concern, or problem related to a product or service they have purchased.
2) **Identification:** The customer service representative identifies the customer's issue. This could involve asking further questions, looking up the customer's purchase history, or consulting with other team members.
3) **Resolution:** The representative then works to resolve the issue. This could involve giving advice, arranging for a product to be repaired or replaced, scheduling a maintenance appointment, or processing a return or refund.
4) **Follow-Up:** After the issue has been addressed, the company often follows up with the customer to ensure they are satisfied with the solution and to ask for feedback.
5) **Analysis and Improvement:** The company analyzes the data from all customer interactions to identify trends, common issues, or areas for improvement. They then use this information to improve their products, services, and customer service processes.

Good after-sale service can greatly enhance customer satisfaction, build loyalty, and encourage repeat business. It also provides valuable insights that can help the business improve its products and services.

4.4.1.2 Users Feedback Management

After-sale service offers an invaluable source of real-world, user-generated feedback that can be used to improve product design and service. Here is how this can work:

1) **Collecting Feedback**: After-sale customer service involves interacting with customers who have used the product and may have run into issues, discovered flaws, or identified potential areas for improvement. Feedback can be collected via various channels such as phone calls, emails, live chat, social media, surveys, or product review sections on the website.

2) **Organizing and Analyzing Feedback**: Once the feedback is collected, it needs to be organized and analyzed. This can be done manually or through CRM software or other data analytics tools. The goal is to identify common themes, trends, or recurring issues that indicate a problem or opportunity for improvement.

3) **Sharing Feedback with Relevant Teams**: The insights gained from this analysis should then be shared with the relevant teams in the organization. For example, if customers are consistently having trouble with a certain feature, that feedback should be passed along to the product design team. If customers are unhappy with the delivery times, that information should go to the logistics or operations team.

4) **Making Improvements**: Based on the feedback, the relevant teams can then make improvements. The product design team might refine the design or add new features, while the customer service team might create new resources to help customers with common issues.

5) **Testing and Implementing Changes**: Any changes should be thoroughly tested before they are implemented. Depending on the change, this might involve prototyping and user testing (for a product design change), pilot testing (for a new service procedure), or A/B testing (for a new website feature).

6) **Communicating Changes to Customers**: Once improvements are made, it is crucial to communicate these changes to customers. This could be done through email updates, blog posts, social media announcements, or notifications within the product itself. This not only ensures customers are aware of the improvements but also shows that the company values their feedback.

7) **Continued Monitoring**: Even after improvements are made, customer service teams should continue to collect and analyze feedback to monitor the effectiveness of the changes and identify any new issues or opportunities for improvement.

By integrating after-sale service feedback into the product design and service improvement process, companies can create a cycle of continuous improvement that is directly driven by customer needs and experiences. This can lead to higher-quality products, more efficient service, and ultimately, higher customer satisfaction and loyalty.

4.4.1.3 Product Innovation

After-sale service teams, due to their direct interaction with customers, are a valuable source of insights that can contribute to product innovation. Here is a process that a company could follow:

1) **Collecting Customer Feedback:** This is the first step in the process, where the after-sales service teams engage with customers and collect feedback about their experiences, likes and

dislikes, pain points, and features they find exciting. This feedback can come from various channels such as customer support calls, emails, social media interactions, surveys, and user forums.

2) **Categorizing Feedback:** The collected feedback needs to be categorized based on different themes. These themes could be related to usability, functionality, design, pricing, etc. Categorization aids in the systematic analysis of the feedback.

3) **Analysis of Feedback:** Once the feedback is categorized, it needs to be analyzed to extract meaningful insights. Advanced data analysis techniques and tools can be used to spot trends, patterns, and correlations in the feedback data. The goal here is to understand the underlying needs, desires, and frustrations of the customers.

4) **Generating Ideas:** Based on the insights derived from the analysis, the team can brainstorm new ideas for product innovation. This could involve enhancing existing features, creating new functionalities, improving user interfaces, or any other changes that could enhance the customer experience.

5) **Prototype and Testing:** The most promising ideas are then turned into prototypes, which are tested internally and/or with a selected group of customers. This allows the company to gather initial feedback and assess the feasibility of the ideas.

6) **Refining Ideas:** Based on the feedback from testing, the ideas are refined and adjusted. This could involve fine-tuning certain features or completely rethinking certain aspects of the product.

7) **Implementation:** Once the ideas are fully developed and tested, they are implemented in the product. This could involve updating the existing product or launching a completely new product.

8) **Feedback Loop:** After the new or updated product is launched, the after-sale service team once again plays a critical role in gathering customer feedback, thus closing the loop and starting the process over again.

By incorporating the insights gained from after-sale service into the product innovation process, companies can ensure that their new products or features are directly informed by the needs, preferences, and experiences of their customers. This can lead to more successful and customer-centric product innovations.

4.4.2 Upgrading After-Sale Customer Services with Industry 4.0

Industry 4.0, also known as the Fourth Industrial Revolution, introduces advanced digital technologies such as IoT, AI, ML, big data, and cloud computing into the industrial sector. It promises to revolutionize many aspects of business operations, including after-sale services. Here is how:

1) **Product Installation:**
   - **Augmented Reality (AR) and Virtual Reality (VR):** These technologies can be used to guide customers through product installation processes. For example, AR can overlay digital instructions onto the physical world, showing customers exactly what to do.
   - **IoT and Connectivity:** Smart, connected devices can communicate with each other and with the manufacturer, making installation easier. For example, a device could automatically download and install the latest software as soon as it is connected to the internet.
2) **Maintenance and Repair:**
   - **Predictive Maintenance:** IoT devices can continuously monitor product performance and send data back to the manufacturer. Using ML algorithms, the manufacturer can analyze this

data to predict when a device is likely to fail or need maintenance, and proactively schedule repairs.

- **Remote Diagnostics and Repair:** In some cases, issues can be diagnosed and even repaired remotely. For example, software issues could be fixed through remote updates.
- **AR/VR Assistance:** AR can assist technicians in performing complex maintenance tasks by providing them with real-time information and guidance. In some cases, it might even be possible for an expert to guide a customer or less experienced technician through a repair process using VR.

3) **Customer Support:**
- **Chatbots and AI:** AI-powered chatbots can handle routine customer queries, freeing up human customer service agents to handle more complex issues. These bots can be available 24/7, providing instant responses to customers.
- **Personalized Service:** With the help of big data and AI, businesses can offer personalized customer service. For example, when a customer contacts support, the representative could instantly have access to the customer's product usage data, purchase history, and previous interactions with the company, allowing them to provide more targeted and effective assistance.

4) **Training:**
- **AR/VR Training:** These technologies can be used to provide immersive, interactive training experiences for both customers and technicians. This can be particularly useful for complex products or procedures.
- **Online Learning Platforms:** Cloud-based learning platforms can provide customers and technicians with access to a wealth of training resources, which they can access at their own pace from anywhere.

5) **Over-The-Air Update:** Over-The-Air (OTA) updates are an important part of after-sales service in the era of Industry 4.0, especially for connected devices. OTA updates involve delivering software updates wirelessly, rather than requiring a physical connection (like a USB) or manual installation. With the advent of the IoT, many devices now have built-in internet connectivity, which allows them to download and install updates directly from the manufacturer.

6) **Benefits of Over-The-Air Updates:**
- **Ease of Use:** One of the biggest advantages of OTA updates is that they can be delivered directly to the device, without requiring any action from the user. This makes it easy to keep devices up to date with the latest software.
- **Faster Updates:** Because OTA updates can be delivered as soon as they are ready, they can be rolled out much more quickly than manual updates. This is especially important for security updates, which need to be implemented as quickly as possible to protect against threats.
- **Reduced Support Costs:** OTA updates can reduce the need for physical service visits, which can be costly and inconvenient. If a problem can be fixed with a software update, it can be resolved quickly and easily through an OTA update.
- **Improved Product Functionality and Lifespan:** OTA updates allow manufacturers to continually improve their products even after they have been sold. They can add new features, improve performance, or fix bugs. This can enhance the user experience and extend the useful life of the product.
- **Enhanced Customer Satisfaction:** By ensuring that devices are always running the latest and best software and resolving issues quickly and conveniently, OTA updates can greatly enhance customer satisfaction.

There are numerous examples of using OTA for after-sale services, such as Apple, Samsung for their smartphones, and other connected services. John Deere and other agricultural equipment manufacturers also use OTA updates to push software updates to their large farm machinery [34, 35]. This allows them to add new features and improve the performance of their machines without requiring a service visit.

Overall, Industry 4.0 technologies can make after-sale services more proactive, efficient, and customer-centric. They can reduce downtime, improve first-time fix rates, and enhance customer satisfaction. However, they also require significant investments in technology and skills and raise new challenges around data security and privacy.

4.4.3 Upgrading User Feedback Management with Industry 4.0

Industry 4.0, characterized by integrating of digital technologies into industrial practices, offers many opportunities to improve after-sales information collection and analysis. Here is how:

1) **IoT and Connectivity:** The IoT allows for smart, connected devices that can send real-time usage data back to the manufacturer. This can include how often the product is used, which features are used most and least, and when and where the product is used. If the product malfunctions or fails, the device can send diagnostic data to help identify the cause of the problem. If a product is being used in a way that it was not designed for (i.e., irregular use), the data can show this. Similarly, if there are consistent issues or malfunctions, this could point to a weakness in the product.

2) **Big Data Analytics:** The data collected from IoT devices can be vast and complex. Big Data analytics tools can help to analyze this data and extract meaningful insights. For example, ML algorithms could identify patterns or trends in the data that indicate common user behaviors or recurring problems. On the other hand, if a certain component consistently fails after a specific amount of usage, this could indicate a product weakness. If usage patterns deviate from the expected norm, this could indicate irregular use.

3) **Predictive Analytics:** Based on the patterns identified in the data, predictive analytics can be used to anticipate future customer needs or potential product issues. This can enable proactive customer service and preventive maintenance, improving the customer experience and reducing the risk of product failures. If a product consistently fails sooner than predicted, this could suggest a product weakness.

4) **Customer Feedback Channels:** Industry 4.0 also offers new ways to collect and analyze customer feedback. For example, AI-powered chatbots can ask customers for feedback during interactions, and natural language processing (NLP) can be used to analyze this feedback to identify common themes or sentiments.

5) **Augmented Reality (AR) and Virtual Reality (VR):** AR and VR can be used to provide immersive product demonstrations or training and can also be used to collect feedback. For example, eye-tracking data could be used to identify which parts of a product demonstration a customer found most interesting or confusing. They can also collect feedback on user interactions, which can reveal if users are struggling with certain aspects of the product or using it in unintended ways.

6) **Digital Twin Technology:** This involves creating a digital replica of a physical product, which can be used to simulate and analyze product performance under different conditions. This can help to identify design flaws or opportunities for improvement.

By harnessing these Industry 4.0 technologies, manufacturers can gain a deeper and more nuanced understanding of how their products are actually used in the real world. This can inform product design and service improvements, leading to higher-quality products, more efficient service, and increased customer satisfaction.

4.4.4   Upgrading User Feedback Management with Social Listening

Social listening, also known as social media monitoring, involves tracking mentions and conversations about a brand, its products, or related keywords across social media platforms and other online forums. It is an important tool for understanding customer feedback in a real-time, unfiltered manner. Here is how social listening can be used in managing user feedback and identifying product weaknesses or irregular use:

1) **Identifying Issues and Trends:** Social listening tools can alert companies to negative feedback or complaints about their products. If multiple customers are reporting the same issue, it could indicate a product weakness. Similarly, if customers discuss unconventional ways they use a product, it could signal irregular use.

2) **Immediate Feedback:** Unlike traditional surveys or customer feedback forms, social listening provides immediate feedback. As soon as a customer posts a comment or review online, companies can see it. This allows them to respond more quickly to issues and potentially prevent small problems from becoming major ones.

3) **Understanding Customer Sentiment:** Social listening tools often include sentiment analysis, which can help companies understand how customers feel about their products. If sentiment is generally negative, it could indicate a product weakness; if it is mixed, it could suggest that customers are using the product in different ways, some of which might be irregular.

4) **Engaging with Customers:** By monitoring social media conversations, companies can identify opportunities to engage with customers. They can respond to complaints, answer questions, and provide information on proper product use. This can improve customer relations and also provide valuable insights into product strengths and weaknesses.

5) **Influencer Identification:** Social listening can help companies identify influencers in their industry or product category. By monitoring these influencers' conversations and feedback about the product, companies can gain valuable insights into potential product weaknesses or irregular use.

6) **Competitor Analysis:** Companies can use social listening to monitor conversations about competitors' products. This can provide insights into potential weaknesses in their own products, or reveal opportunities to differentiate their products.

In summary, social listening is a powerful tool for managing user feedback in the era of Industry 4.0. By monitoring online conversations, companies can gain valuable insights into product weaknesses and irregular use and respond quickly to address these issues.

4.4.5   Upgrading User Feedback Management with Quality of Experience Mining and Analysis

In Chapter 3 of this book, I described the importance of Quality of Experience (QoE) for new generations of customers in making buying decisions and forming their opinions on the product. So it is also useful to mine and analyze the QoE of customers on current products and send this feedback to product redesign and improvements.

QoE is a measure of a customer's perceptions and satisfaction with a product or service. It goes beyond technical performance metrics to consider the entire customer experience, including factors like ease of use, reliability, and overall satisfaction.

QoE indeed covers a holistic view of a customer's interaction with a product or service, capturing not just the negative aspects but also the positive ones. By evaluating these experiences, companies can understand the strengths and weaknesses of their products and identify areas for improvement. Here is how:

1) **Understanding User Satisfaction:** QoE analysis can reveal how satisfied users are with various aspects of the product. This includes product features, reliability, performance, and even customer support. High satisfaction levels in certain areas can signify product strengths.

2) **Identifying Key Features:** Through QoE analysis, companies can identify which features are most appreciated by users. These could be features that users frequently mention positively in reviews or surveys. Understanding these key features can help companies recognize their product's strengths and focus on these areas in future development.

3) **Analyzing Usage Patterns:** QoE can provide insights into how users are interacting with the product. This can help identify which parts of the product are most used and appreciated and which parts are ignored or used less frequently, indicating potential areas for improvement.

4) **Gauging Performance and Reliability:** QoE analysis can reveal whether a product consistently meets users' expectations in terms of performance and reliability. High performance and reliability are often seen as key strengths of a product.

5) **Feedback on Customer Support:** QoE can also provide feedback on the quality of customer support, which is a crucial aspect of the overall user experience. If users report positive experiences with customer support, this is a strength. If not, it is an area that needs improvement.

6) **Net Promoter Score (NPS):** A high NPS can be seen as an overall strength, suggesting that users are likely to recommend the product to others. If the NPS is low, it is a sign that improvements are needed to enhance the customer experience.

7) **Sentiment Analysis:** This can help gauge the overall sentiment toward different aspects of the product, highlighting both strengths and weaknesses.

By using QoE to evaluate both the strengths and weaknesses of a product, businesses can get a more complete picture of how their product is performing and where improvements can be made. It is a powerful tool for enhancing the user experience and driving product innovation.

4.4.6 Improving After-Sale Customer Service Team's Contribution in Product Innovation by Industry 4.0

Industry 4.0 technologies can significantly enhance the process of gathering insights from after-sale service teams and contribute to product innovation. Here is how:

1) **Digital Communication Platforms:** Using digital communication tools like chatbots, messaging apps, and social media channels, after-sale service teams can quickly and efficiently gather customer feedback, complaints, and suggestions. This information can then be shared with product development teams in real time, accelerating the process of identifying areas for improvement.

2) **Customer Relationship Management (CRM) Systems:** Integrating after-sales service data with CRM systems allows for better tracking and management of customer interactions. This way, product development teams can access customer feedback and service history, enabling them to make data-driven decisions on product improvements.

3) **AI-Powered Analytics:** AI and ML algorithms can analyze the data collected by after-sale service teams to identify patterns and trends in customer feedback. This can help pinpoint specific product features or issues that need improvement and provide insights into potential innovations.

4) **IoT and Remote Monitoring:** By leveraging IoT devices and remote monitoring, after-sale service teams can gather real-time data on product usage, performance, and potential issues. This data can be shared with product development teams to inform design improvements and predict maintenance needs.

5) **Augmented Reality (AR) and Virtual Reality (VR):** AR and VR technologies can be used to enhance after-sale services, such as remote assistance or training. These tools can also help collect valuable data on customer interactions, which can be shared with product development teams to inform product innovation.

6) **Predictive Maintenance:** Industry 4.0 technologies enable predictive maintenance, which anticipates potential product failures based on usage data. This information can be used to improve product design and maintenance processes, reducing downtime and improving customer satisfaction.

7) **Knowledge Management Systems:** Implementing a knowledge management system can help centralize and organize the information collected by after-sale service teams, making it easily accessible to product development teams. This enables efficient sharing of insights and best practices, fostering a culture of continuous improvement and innovation.

By integrating Industry 4.0 technologies into the after-sales service process, companies can optimize the collection, analysis, and sharing of customer insights, leading to more informed product innovation and an enhanced customer experience.

4.5 Quality Management for Service Industry

4.5.1 What Are the Differences in Quality Management Between Service and Manufacturing Industry

Quality management in the service industry focuses on ensuring that services meet or exceed customer expectations. While the principles of quality management are similar across industries, there are some key differences between quality management in the service industry and in the manufacturing industry due to the nature of services. Here is an overview:

Nature of Output: Services are intangible, heterogenous, and produced and consumed simultaneously. This makes quality management for services more complex and subjective than for manufacturing, where the output is a tangible product that can be easily inspected and measured for defects.

Customer Involvement: In many services, the customer is actively involved in the production process (for example, a haircut or a consultation), which means that the customer's input and interaction can significantly affect the quality of the service. This is in contrast to manufacturing, where the product is produced independently of the customer.

Measuring Quality: Because services are intangible and often customized to the individual customer, measuring quality can be more challenging. Service quality is often evaluated based on the customer's perception of the service delivery process and the outcome. This involves criteria such as responsiveness, empathy, assurance, reliability, and tangibles (physical evidence of the service).

Quality Improvement: In the service industry, quality improvement often involves enhancing the customer service experience, improving the delivery process, training staff, or adjusting the service offering to better meet customer needs. In manufacturing, quality improvement often involves reducing defects, improving the production process, or enhancing the physical characteristics of the product.

Quality Management Techniques in the Service Industry: Techniques such as Service Quality (SERVQUAL) model [36], which measures the gap between customer expectations and perceptions, or Lean and Six Sigma methodologies adapted for service industries are commonly used in the service sector. These techniques focus on improving processes, reducing waste, and increasing customer satisfaction.

In conclusion, while the goal of quality management in both the service and manufacturing industries is to meet or exceed customer expectations, the methods and challenges involved can be quite different due to the unique characteristics of services.

4.5.2 What Industry 4.0 Can Help in Service Quality Management

Industry 4.0, with its advanced technologies such as IoT, AI, big data, ML, and more, can greatly enhance service quality management in several ways:

1) **Customer Personalization:** Industry 4.0 technologies enable businesses to collect and analyze customer data in real time, providing insights that can be used to customize and personalize services to individual customers' needs and preferences, improving overall service quality.
2) **Predictive Analytics:** ML algorithms can analyze large datasets to identify patterns and trends, predict customer behavior and preferences, and anticipate future service needs. This allows for proactive service management and enhances service quality by meeting customer needs more effectively.
3) **Service Automation:** AI and automation technologies can streamline and automate routine service tasks, improving efficiency and reducing the likelihood of human error. This allows service employees to focus on more complex tasks and enhances overall service quality.
4) **Real-Time Monitoring and Feedback:** IoT technologies enable real-time monitoring of service delivery and instant customer feedback. This allows for quick adjustments to improve service quality and address any issues promptly.
5) **Advanced Training Tools:** Virtual Reality (VR) and Augmented Reality (AR) can provide immersive training experiences for service employees, enhancing their skills and knowledge and improving service quality.
6) **Enhanced Communication:** Digital platforms and tools can enhance communication between service providers and customers, improving service coordination and customer satisfaction.
7) **Resource Optimization:** Big data and analytics can help optimize resource allocation and scheduling, ensuring that services are delivered efficiently and effectively.
8) **Improved Decision Making:** The insights gained from big data analytics, AI, and machine learning can support better decision making at all levels of the service organization, enhancing service strategies and outcomes.

By leveraging these Industry 4.0 technologies, service organizations can enhance their quality management efforts, improve service delivery, and exceed customer expectations.

4.5.3 Industry 4.0 and Individualized Services

Industry 4.0 and its associated technologies can enable individualized services in numerous ways, transforming the service industry landscape and elevating customer experiences. Here's how:

1) **Data Analytics:** Industry 4.0 brings with it the power of data analytics. Service providers can collect and analyze vast amounts of customer data to understand individual preferences,

behaviors, and needs. This understanding enables businesses to tailor their services to each customer, providing a truly individualized experience.

2) **Artificial Intelligence (AI):** AI and machine learning algorithms can process and analyze large volumes of data to identify patterns and trends, make predictions, and provide personalized recommendations. For example, AI can be used in online shopping platforms to recommend products based on a customer's browsing and purchasing history.

3) **Internet of Things (IoT):** IoT devices can collect real-time data on user behavior and preferences, allowing services to be adjusted and personalized in real time. For instance, smart home devices can learn a user's routine and preferences to control lighting, temperature, and other settings for each individual in the household.

4) **Digital Twin Technology:** Digital twins are virtual replicas of physical devices that data scientists and IT pros can use to run simulations before actual devices are built and deployed. They can model how a service will perform for an individual customer, allowing for fine-tuning and personalization.

5) **Automation and Robotics:** Automation technology can be used to customize services. For example, automated manufacturing systems can produce custom-made products based on individual customer specifications.

6) **Blockchain:** Blockchain can create secure, decentralized records of customer preferences and transactions, enabling highly personalized service delivery.

7) **Augmented Reality (AR) and Virtual Reality (VR):** AR and VR can offer personalized experiences for customers in areas like gaming, shopping, and learning. For instance, VR can offer individualized virtual tours, while AR can provide personalized shopping experiences.

Through these technologies, Industry 4.0 paves the way for individualized services, where services are no longer one-size-fits-all but uniquely tailored to each customer's needs and preferences, enhancing customer satisfaction and loyalty.

Example 4.3 Personalized Digital Dentistry

Digital dentistry [37, 38] refers to the use of dental technologies or devices that incorporate digital or computer-controlled components to carry out dental procedures rather than using mechanical or electrical tools. The use of 3D scanning and printing technologies has revolutionized the dental industry, particularly in creating dental prosthetics like dentures, crowns, and implants.

Here is a general overview of the process steps (Figure 4.6):

1) **3D Dental Scanning:** The first step involves creating a 3D digital model of the patient's mouth. This is typically done using an intraoral scanner, which captures thousands of images of the mouth and uses these to construct a highly accurate 3D model. This process is usually quicker, contact-free, and more comfortable for the patient than traditional methods of taking dental impressions.

2) **Designing the Prosthetic:** Using specialized computer-aided design (CAD) software, the dentist or a dental technician uses the 3D model to design the dental prosthetic. This allows for a high degree of customization, as the prosthetic can be tailored precisely to the patient's unique anatomy.

3) **3D Printing the Prosthetic:** Once the design is finalized, it is sent to a 3D printer. The printer uses a special dental resin or other material to build the prosthetic layer by layer. The result is a highly accurate, custom-made dental prosthetic.

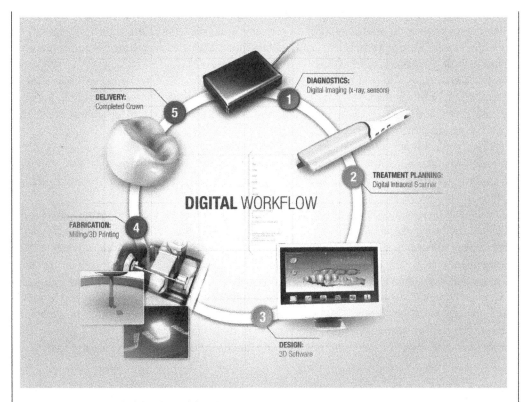

Figure 4.6 Workflow of 3-D Digital Dentistry

4) **Fitting the Prosthetic:** The printed prosthetic is then checked for fit and comfort in the patient's mouth. Because the prosthetic is based on a highly accurate 3D model of the patient's mouth, it typically requires fewer adjustments than a prosthetic made using traditional methods. Fittings are customized for each client, and the dental office follows up one-on-one with each patient through its CRM system and provides after-sales service that is highly customized based on data from customer reviews on social media.

4.6 Digital Quality Management System Under Industry 4.0

4.6.1 Introduction

The concept of QMS originated in the manufacturing sector during the 20th century, driven by the need to ensure product quality and safety. With the advent of standards such as ISO 9001, the application of QMS expanded across various industries, including healthcare, food, and pharmaceuticals.

The journey from a QMS to an eQMS began with the digital revolution in the late 20th and early 21st century. Initially, quality management relied heavily on paper-based systems for documentation and record-keeping. However, these manual systems were prone to human error and inefficiencies, leading to the rise of electronic record management and, eventually, fully fledged eQMS.

The first eQMS were simple software systems designed to digitize paper-based processes. These early systems primarily focused on document control and training management. However, as technology advanced, so did the capabilities of eQMS. The rise of the internet facilitated global collaboration, leading to enhancements in the features of eQMS. Around the same time, regulations like FDA 21 CFR Part 11 were introduced, which allowed for electronic records and signatures, further boosting the adoption of eQMS.

In the 2000s, eQMS evolved to include modules for auditing, nonconformance management, corrective and preventive actions (CAPAs), and more. The emergence of cloud-based platforms in the late 2000s and early 2010s opened up new possibilities for eQMS, providing scalability, flexibility, and real-time data access.

There are continuous advancements in eQMS, driven by new technologies such as AI, ML, and the IoT. These technologies promise to further enhance the capabilities of eQMS, particularly in areas such as predictive analytics, risk management, and real-time quality monitoring [39, 40].

The digital QMS, also known as an eQMS, represents a software-based solution designed for the efficient and effective management of quality processes and documentation. This system streamlines, automates, and connects all quality processes, providing a more efficient quality management system. Common modules included in an eQMS encompass document control, audit management, nonconformances, corrective actions, change control, training, and more.

An effective eQMS can significantly improve an organization's capacity to uphold high-quality standards across its operations. This system is designed to automate and streamline processes, enhance traceability, minimize errors, and foster continuous improvement.

Here is an overview of the structure, features, and functionalities that characterize a robust eQMS:

4.6.1.1 Structure

A well-constructed eQMS typically exhibits a modular design that can be tailored to an organization's specific needs. This design generally incorporates core modules for key quality processes such as document control, audits, nonconformance tracking, CAPAs, and change management. Additional modules may be available for other processes, such as SQM, risk management, and training management. These modules should be interconnected to ensure seamless data sharing and workflow coordination across various processes.

4.6.1.2 Functionalities and Features

1) **Document Control**: An eQMS should provide robust document control features, including version control, approval workflows, and automatic archiving of obsolete documents, ensuring that everyone is utilizing the most recent, approved document versions.
2) **Audit Management**: The system should facilitate the planning, execution, and follow-up of audits, including features such as audit schedules, checklists, and tracking of audit findings and corrective actions.
3) **Nonconformance and CAPA**: An effective eQMS should enable the tracking and management of nonconformances and corrective and preventive actions, including the documentation of nonconformances, investigation of their causes, planning and implementation of corrective actions, and verification of their effectiveness.
4) **Change Management**: The eQMS should support the management of changes to processes, products, or the QMS itself, including the assessment of change impacts, acquisition of necessary approvals, and assurance of effective change implementation.
5) **Supplier Quality Management**: For organizations reliant on suppliers, the eQMS might include features for managing supplier quality, such as supplier evaluations, tracking of supplier nonconformances, and supplier scorecards.

6) **Risk Management**: Many eQMSs include features for identifying and assessing risks, planning and implementing risk mitigation measures, and monitoring the effectiveness of these measures.

7) **Training Management**: The system may also include features for managing employee training, such as tracking training requirements and completion, and maintaining training records.

8) **Analytics and Reporting**: An efficient eQMS should provide robust analytics and reporting capabilities, allowing for monitoring of key quality metrics and identification of trends and opportunities for improvement.

9) **Integration**: The eQMS should be capable of integration with other systems utilized by the organization, such as ERP or CRM systems, to ensure seamless data sharing and workflow coordination.

10) **User-Friendly Interface**: An intuitive, user-friendly interface can enhance the usability of the eQMS, facilitating easier adoption and use of the system by employees.

11) **Compliance Features**: The eQMS should support compliance with relevant quality standards and regulations, such as ISO 9001, FDA 21 CFR Part 11, or EU Annex 11. This may include features such as electronic signatures, audit trails, and data security and integrity controls.

While an eQMS is a powerful tool for managing and improving quality, it is crucial to remember that its effectiveness relies not only on the system itself but also on its implementation and use within the organization. Factors such as training, user support, and a culture of quality are all vital to the success of an eQMS.

There were several eQMS software options that are highly rated and widely used in various industries. Below are a few examples:

- **MasterControl:** MasterControl offers a comprehensive suite of quality management tools that can help automate key processes such as document control, change management, training management, and audits. It is used in a variety of highly regulated industries, including pharmaceuticals, medical devices, and food and beverage.

- **Qualio:** Qualio is an eQMS designed specifically for the life sciences industry. It offers robust document management, training management, and CAPA capabilities, among others.

- **Greenlight Guru:** Greenlight Guru is another eQMS tailored for the medical device industry. It offers a variety of modules, including design control, risk management, and postmarket surveillance.

- **ETQ Reliance:** Excellence Through Quality (ETQ) offers a flexible, modular eQMS solution that can be tailored to the needs of a wide range of industries. Its modules include document control, SQM, and risk management.

- **Intellect eQMS:** Intellect offers a highly customizable eQMS solution that includes modules for document control, change management, training management, and more.

Many of the modern eQMS solutions have started to leverage elements of Industry 4.0, AI, and big data to enhance their offerings. However, the level of integration and the specific features may vary between systems. Here are some ways these technologies are being utilized:

1) **Industry 4.0:** This primarily involves the integration of IoT devices and sensors with the eQMS. This can allow for real-time data collection and monitoring of production processes, which can enhance the ability to detect and respond to quality issues. Some eQMS vendors offer integrations with manufacturing execution systems (MES) and other Industry 4.0 technologies.

2) **AI and Machine Learning:** AI can be used in an eQMS for a range of purposes, such as predictive analytics, NLP, and anomaly detection. For example, AI can analyze historical data to predict potential quality issues, allowing for proactive quality management. ML can also help in automating and enhancing various QMS processes. Some eQMS providers may offer AI-powered analytics or automated process capabilities.

3) **Big Data:** In the context of an eQMS, big data primarily relates to the ability to handle and analyze large volumes of quality data. This can provide deeper insights into quality performance and help identify patterns and trends that might not be visible with a smaller data set. eQMS solutions often provide analytics and reporting tools to help make sense of this data.

Specifically, vendors like MasterControl, Veeva, and others have been incorporating these advanced technologies into their eQMS solutions, enhancing their capability to manage quality processes more efficiently and effectively.

In this book, I have elucidated that quality exhibits two facets: customer value creation and QA. While customer value creation is primarily achieved through innovation and design, QA is necessary at all stages of the product lifecycle.

Superior quality is a blend of exceptional customer value and minimized quality loss, safeguarded by QA. This process begins at the inception of product concept definition and continues throughout all stages of the product lifecycle. Such achievements are accomplished through the collaborative work of various professionals. To facilitate this collaboration, a robust infrastructure is essential.

4.6.2 Cloud-Based Master Platforms that Integrate eQMS with Other Business Applications

There are several prominent cloud-based platforms that host a variety of business applications, including electronic Quality Management Systems (eQMS) and other enterprise systems. Here are a few examples:

1) **Amazon Web Services (AWS):** AWS is a widely used cloud platform that provides a broad set of services and features, including computing power, storage, and database solutions. Many software vendors host their applications on AWS, and it supports a wide range of integration options.

2) **Microsoft Azure:** Azure is Microsoft's cloud platform, offering a comprehensive suite of cloud services including those for computing, analytics, storage, and networking. Like AWS, it supports a wide range of applications and integration options.

3) **Google Cloud Platform (GCP):** GCP is another major cloud platform that hosts a variety of applications. Its services include computing, data storage, data analytics, and ML.

4) **IBM Cloud:** IBM Cloud offers a range of cloud services, including infrastructure as a service (IaaS), platform as a service (PaaS), and software as a service (SaaS). IBM's cloud is particularly strong in areas like AI and blockchain.

5) **Oracle Cloud:** Oracle Cloud offers a full range of cloud services, including those for AI, blockchain, IoT, integration, and data management. Its services are often used to host Oracle's own applications, as well as those from other vendors.

6) **Salesforce:** Salesforce is best known for its CRM software, but it also offers a broader platform (Salesforce Cloud) for developing and hosting a wide range of business applications. Salesforce is particularly well-known for its integration capabilities, which are supported by a large ecosystem of partners, integrations, and add-ons.

These platforms all support a variety of business applications, including eQMS and other enterprise systems. When selecting a cloud platform for your applications, it is important to consider factors like its compatibility with your existing systems, the availability of needed services and features, the platform's reliability and performance, data security and compliance features, cost, and the quality of customer support.

Most modern eQMS software solutions are designed to work well with cloud-based platforms and to integrate with other enterprise software systems. This is a key requirement for many organizations, as integration can improve data consistency, streamline workflows, and provide a more comprehensive view of quality across different processes and functions. Here is how eQMS software can typically integrate with other types of systems:

1) **Product Lifecycle Management (PLM) Systems:** eQMS and PLM systems often need to share information about product specifications, design changes, and nonconformances. For example, a change in a product's design (managed in the PLM system) might trigger a change in a quality process (managed in the eQMS). Similarly, a nonconformance identified in the eQMS might require a change to the product's design.

2) **Manufacturing Execution Systems (MES):** Integration between eQMS and MES systems can help to ensure that quality requirements are effectively communicated and implemented in the manufacturing process. For example, an MES might use quality specifications from the eQMS to guide production processes and report back production data that can be used for quality monitoring and control.

3) **Enterprise Resource Planning (ERP) Systems:** ERP systems manage various business processes and data, including purchasing, inventory, sales, and finance. Integration with an eQMS can help to ensure that quality considerations are integrated into these processes. For example, purchasing decisions can consider a supplier's quality performance, and sales processes can include checks for product quality.

When considering different eQMS solutions, it is important to assess their integration capabilities. This includes not only the technical capability to integrate with your specific systems but also the quality of the integration—how well it supports your workflows, how reliably it operates, and how easy it is to set up and manage. Different eQMS solutions may offer different types of integrations (e.g., API-based, prebuilt connectors, etc.), and they may support different levels of integration (e.g., one-way data syncing versus two-way data sharing, real-time versus batch updates, etc.).

4.6.3 Collaborative Work on Quality Through Product Life Cycle

Now I am going to talk about how to leverage these technologies, platforms, and necessary organizational restructuring to link and manage quality-related work throughout all stages of the product life cycle, such as quality in innovation, design quality, and quality control in production seamlessly.

Linking and managing quality in innovation, design quality, and quality control in production requires a systemic and strategic approach. Here is how emerging technologies, cloud-based platforms, and organizational restructuring can be used to achieve this:

1) **Cloud-Based Platforms:** Cloud-based platforms provide a central place for all quality-related information. By migrating all data to a single, accessible platform, organizations can ensure that everyone has access to the same information, reducing inconsistencies and confusion. They enable seamless integration of quality management in design, innovation, and production.

- **Design:** CAD and PLM software hosted on the cloud can be used to design products and manage their lifecycle, ensuring all changes are updated in real-time across the organization.
- **Innovation:** Cloud-based collaboration tools can be used to generate, share, and refine innovative ideas, while cloud-based project management tools can be used to manage the innovation process.
- **Production:** Cloud-based ERP and MES can be used to manage production processes and ensure consistent quality.

2) **Emerging Technologies:** AI, big data, IoT, and other emerging technologies can be used to gather, analyze, and act on quality-related data across all stages of the product lifecycle.
  - **Design:** AI can be used to predict and optimize product performance, reducing the number of design iterations required.
  - **Innovation:** Data analytics can be used to identify market trends and customer needs, guiding the innovation process.
  - **Production:** IoT devices can be used to monitor production processes in real-time and identify potential quality issues before they occur.

3) **Organizational Restructuring:** Organizational restructuring can help break down silos and foster collaboration, ensuring quality is managed effectively across all stages of the product lifecycle.
  - **Cross-Functional Teams:** Create cross-functional teams comprising members from the design, innovation, and production departments. This can promote knowledge sharing and collaboration, ensuring quality considerations are incorporated at all stages.
  - **Quality Champions:** Appoint quality champions in each department to drive quality initiatives and ensure they are aligned with the overall business strategy.
  - **Quality Management System (QMS):** Implement a company-wide QMS to provide a structured framework for managing quality across all departments.

In conclusion, effective management of quality in design, innovation, and production requires a combination of modern technology and organizational changes. The goal is to create an integrated and collaborative environment where quality is everyone's responsibility.

4.6.4 Enhance Digital Quality Management System by Industry 4.0 Technologies

Industry 4.0 and emerging technologies like AI and big data are transforming QMS in a number of ways, enhancing efficiency, reliability, and overall productivity.

1) **Data-Driven Decision-Making:** Big data can provide vast amounts of information that can be used to monitor, control, and improve quality. Traditional quality management often relies on a limited set of data, while big data can provide a more comprehensive picture. By analyzing big data, companies can identify hidden patterns, correlations, and insights to make more informed decisions about quality management.
2) **Predictive Analytics:** One of the most significant ways that AI and big data are transforming quality management is through predictive analytics. AI can analyze past and present data to predict future outcomes. This can help companies anticipate quality issues before they occur and take preventive measures to mitigate them. This predictive capability can lead to reduced waste, increased efficiency, and better product quality.
3) **Automation of Quality Inspections:** AI and ML technologies can be used to automate routine quality inspections, reducing the risk of human error and increasing efficiency. This

can involve automated visual inspections using computer vision technology, or the use of ML algorithms to analyze product data and identify anomalies that could indicate a quality issue.

4) **Real-Time Monitoring and Feedback:** IoT devices and sensors can provide real-time data about product quality and production processes. This can allow for immediate adjustments and corrections, reducing the risk of producing substandard products. This real-time data can be combined with AI and ML to provide predictive insights and proactive quality management.

5) **Integration and Interoperability:** Industry 4.0 emphasizes the integration of different systems and technologies. This can lead to improved interoperability between different QMSs, making it easier to monitor and manage quality across different stages of the production process.

6) **Improving Traceability:** Blockchain, another technology associated with Industry 4.0, can greatly enhance traceability, a crucial aspect of quality management. It provides a transparent, immutable record of every transaction and interaction, making it much easier to track and verify the quality of raw materials, components, and final products.

In summary, these technologies bring about an evolutionary change in the QMS by enabling predictive quality management, real-time monitoring, automation, and advanced data analytics. They hold the promise of not just identifying and solving quality issues when they occur but also predicting and preventing them from happening in the first place. In summary, these technologies bring about an evolutionary change in the QMS by enabling predictive quality management, real-time monitoring, automation, and advanced data analytics. They hold the promise of not just identifying and solving quality issues.

4.6.5 Unified Quality Management System

A Unified Quality Management System (UQMS) is a centralized framework for managing quality across an organization. It integrates all quality processes, procedures, responsibilities, and performance metrics into one cohesive system. The goal of a UQMS is to create an environment where every aspect of quality—from product design and development, to production, to customer service—is monitored and managed consistently.

Organizations can structure their UQMS in a variety of ways, depending on their specific needs and goals. However, some common features might include:

1) **Centralized Platform:** An UQMS typically uses a centralized software platform that allows all quality-related data to be stored, accessed, and analyzed in one place. This can help to ensure that everyone is working from the same information and that quality issues are quickly identified and addressed.

2) **Cross-Functional Teams:** To effectively manage quality across an organization, it can be helpful to form cross-functional teams that include representatives from different departments, such as product development, production, and customer service. These teams can work together to identify and address quality issues and to ensure that quality standards are consistently met.

3) **Defined Roles and Responsibilities:** Within a UQMS, it is important to clearly define the roles and responsibilities of each person or team. This can help to ensure that everyone knows what they are supposed to be doing and that nothing falls through the cracks.

4) **Performance Metrics:** An UQMS should also include performance metrics that allow the organization to track its progress toward its quality goals. These metrics can help identify areas where improvements are needed and provide a benchmark for evaluating the success of quality initiatives.

5) **Continuous Improvement:** One of the core principles of a UQMS is continuous improvement. This means regularly evaluating and updating processes, procedures, and standards to ensure the organization always strives to improve its quality.
6) **Compliance Management:** An UQMS should also have mechanisms for managing compliance with industry standards and regulations. This could include processes for conducting audits, tracking nonconformances, and implementing CAPAs.

In terms of organizational structure, it can vary widely based on the size and complexity of the company. However, it is common to have a Quality Management or QA department headed by a Quality Manager or Director who oversees the UQMS. Under this leader, teams or individuals can be responsible for different aspects of quality, like Product Assurance, Production Quality Control, and Customer Service Quality. Despite the structure, it is crucial to ensure cross-departmental communication and collaboration for the UQMS to function effectively.

4.6.6 Collaborations of Professionals in Unified Quality Management System

4.6.6.1 Collaboration Among Quality Professionals in Different Sectors

In a manufacturing or producer company, it is important that there is a close and cooperative relationship among product assurance engineers, production quality control personnel, and customer service quality professionals. Here is how their relationships might ideally be structured:

1) **Product Assurance Engineers and Production Quality Control Personnel:** Product assurance engineers are usually responsible for ensuring that a product is designed and developed in a way that ensures high quality and reliability. They should work closely with production quality control personnel, who are responsible for ensuring that the actual manufacturing processes align with the quality standards. Both groups should collaborate on identifying potential quality issues, formulating corrective actions, and implementing preventive measures.
2) **Product Assurance Engineers and Customer Service Quality Professionals:** Product assurance engineers should communicate with customer service quality professionals. The latter group often has direct contact with customers and is privy to valuable feedback about the product's real-world performance, which can help product assurance engineers understand how the product is used and identify potential areas for improvement.
3) **Production Quality Control Personnel and Customer Service Quality Professionals:** Customer service quality professionals can relay customer feedback to production quality control personnel, who can then adjust production processes to rectify any issues identified. In turn, production quality control personnel can inform customer service quality professionals about any production changes or issues that may affect customers.

In all these interactions, the shared goal should be ensuring the highest quality products and experiences for customers. Regular meetings, shared documentation, and collaborative problem-solving can all help to build strong relationships among these teams.

Furthermore, implementing a UQMS can be extremely beneficial in maintaining these relationships. A shared platform allows for better communication and visibility and ensures everyone is working from the same information. The insights and data generated from a centralized QMS can help all parties involved to effectively do their jobs and contribute toward a culture of continuous improvement.

Creating an environment where open communication and collaboration is encouraged is critical to establishing the right relationships among these groups, leading to better QA and customer satisfaction.

4.6.6.2 Collaboration Between Quality Professionals and Others

Quality professionals in successful companies typically work very closely with other professionals across various roles and departments. The exact nature of these relationships can vary depending on the organization's structure and culture, but some common themes include:

1) **Cross-Functional Teams:** Quality professionals often work as part of cross-functional teams that might include representatives from design, engineering, production, supply chain, and customer service, among others. These teams work together on projects and issues impacting multiple business areas. This collaboration ensures that all perspectives are considered and that decisions made support quality goals.
2) **Shared Responsibilities:** In many successful organizations, quality is seen as everyone's responsibility, not just the quality department's. This means that quality professionals often work with others to help them understand and implement quality standards. They may provide training, resources, and support to help others contribute to the organization's quality goals.
3) **Spread Over All Sectors:** Quality professionals might be embedded in different sectors or departments of the organization. For example, a large company might have quality professionals within its production, design, and customer service departments. This allows quality standards and practices to be tailored to the specific needs of each area, while still maintaining a consistent overall approach to quality.
4) **Communication and Collaboration:** Quality professionals often act as liaisons or facilitators, helping to ensure clear communication and collaboration between different areas of the organization. This might involve arranging and leading meetings, managing shared documentation, and coordinating joint efforts.
5) **Project Management:** Quality professionals often work with other teams on specific projects, such as implementing a new QMS, carrying out a process improvement project, or addressing a specific quality issue. This can involve a range of project management tasks, such as planning, coordinating activities, tracking progress, and reporting on results.
6) **Continuous Improvement:** Quality professionals typically play a key role in driving continuous improvement efforts. This might involve working with other professionals to identify opportunities for improvement, plan and implement changes, and evaluate the results.

In summary, quality professionals in successful companies tend to work very closely with others across the organization, fostering a collaborative culture where quality is seen as everyone's responsibility. They use their expertise to support others in meeting quality standards, and they work to foster open communication and continuous improvement.

References

1 Zawadzki, P. and Żywicki, K. (2016). Smart product design and production control for effective mass customization in the Industry 4.0 concept. *Management and Production Engineering Review*.
2 Moeuf, A., Pellerin, R., Lamouri, S. et al. (2018). The industrial management of SMEs in the era of Industry 4.0. *International Journal of Production Research* 56 (3): 1118–1136.
3 Zaidin, N.H.M., Diah, M.N.M., and Sorooshian, S. (2018). Quality management in Industry 4.0 era. *Journal of Management and Science* 8 (2): 182–191.
4 Souza, F.F.D., Corsi, A., Pagani, R.N. et al. (2022). Total quality management 4.0: adapting quality management to Industry 4.0. *The TQM Journal* 34 (4): 749–769.

5 Wang, S., Wan, J., Li, D., and Zhang, C. (2016). Implementing smart factory of Industrie 4.0: an outlook. *International Journal of Distributed Sensor Networks* 12 (1): 3159805.

6 Chen, B., Wan, J., Shu, L. et al. (2017). Smart factory of Industry 4.0: key technologies, application case, and challenges. *IEEE Access* 6: 6505–6519.

7 Lee, J., Ardakani, H.D., Yang, S., and Bagheri, B. (2015). Industrial big data analytics and cyber-physical systems for future maintenance & service innovation. *Procedia CIRP* 38: 3–7.

8 Lasi, H., Fettke, P., Kemper, H.G. et al. (2014). Industry 4.0. *Business & Information Systems Engineering* 6: 239–242.

9 Zhong, R.Y., Xu, X., Klotz, E., and Newman, S.T. (2017). Intelligent manufacturing in the context of industry 4.0: a review. *Engineering* 3 (5): 616–630.

10 https://initiatives.weforum.org/global-lighthouse-network/home (accessed 15 May 2023).

11 Kim, M.-K., Wang, Q., Park, J.-W. et al. (2016). Automated dimensional quality assurance of full-scale precast concrete elements using laser scanning and BIM. *Automation in Construction* 72: 102–114.

12 Haleem, A., Javaid, M., Singh, R.P. et al. (2022). Exploring the potential of 3D scanning in Industry 4.0: an overview. *International Journal of Cognitive Computing in Engineering*.

13 Kruth, J.P., Bartscher, M., Carmignato, S. et al. (2011). Computed tomography for dimensional metrology. *CIRP Annals* 60 (2): 821–842.

14 Wang, R., Law, A.C., Garcia, D. et al. (2021). Development of structured light 3D-scanner with high spatial resolution and its applications for additive manufacturing quality assurance. *The International Journal of Advanced Manufacturing Technology* 117: 845–862.

15 Leopold, J., Günther, H., and Leopold, R. (2003). New developments in fast 3D-surface quality control. *Measurement* 33 (2): 179–187.

16 Catalucci, S., Thompson, A., Piano, S. et al. (2022). Optical metrology for digital manufacturing: a review. *The International Journal of Advanced Manufacturing Technology* 120 (7-8): 4271–4290.

17 Luo, S., Li, W., Cai, D., and Li, J. (2021). Application of laser tracker in the industrial measurement field. *Journal of Physics: Conference Series* 1820 (1): 012119.

18 Wu, C., Yuan, Y., Tang, Y., and Tian, B. (2021). Application of terrestrial laser scanning (TLS) in the architecture, engineering and construction (AEC) industry. *Sensors* 22 (1): 265.

19 Washer, G., Fuchs, P., Rezai, A., and Ghasemi, H. (2006). Ultrasonic testing for quality control of ultra-high performance concrete. In: *Structures Congress 2006: Structural Engineering and Public Safety*, 1–8.

20 Golnabi, H. and Asadpour, A. (2007). Design and application of industrial machine vision systems. *Robotics and Computer-Integrated Manufacturing* 23 (6): 630–637.

21 Anand, S. and Priya, L. (2019). *A Guide for Machine Vision in Quality Control*. CRC Press.

22 Widjajanto, S., Purba, H.H., and Jaqin, S.C. (2020). Novel POKA-YOKE approaching toward Industry-4.0: a literature review. *Operational Research in Engineering Sciences: Theory and Applications* 3 (3): 65–83.

23 Zhang, A. (2014). Quality improvement through Poka–Yoke: from engineering design to information system design. *International Journal of Six Sigma and Competitive Advantage* 8 (2): 147–159.

24 Morris, J. (2023). Tesla next generation platform: everything we know so far. *Forbes* (7 January). https://www.forbes.com/sites/jamesmorris/2023/01/07/tesla-next-generation-platform-everything-we-know-so-far/?sh=76224e927aaa (accessed 4 June 2023).

25 Pech, M., Vrchota, J., and Bednář, J. (2021). Predictive maintenance and intelligent sensors in smart factory. *Sensors* 21 (4): 1470.

26 Wang, K. (2016). Intelligent predictive maintenance (IPdM) system–Industry 4.0 scenario. *WIT Transactions on Engineering Sciences* 113: 259–268.

27 Koh, L., Orzes, G., and Jia, F.J. (2019). The fourth industrial revolution (Industry 4.0): technologies disruption on operations and supply chain management. *International Journal of Operations & Production Management* 39 (6/7/8): 817–828.

28 Tirkolaee, E.B., Sadeghi, S., Mooseloo, F.M. et al. (2021). Application of machine learning in supply chain management: a comprehensive overview of the main areas. *Mathematical Problems in Engineering* 2021: 1–14.

29 Bauer, M. (2021). How Amazon supply chain works. *Teikametrics* (23 February). https://www.teikametrics.com/blog/how-amazon-supply-chain-works.

30 Banker, S. (2021). Amazon supply chain innovation continues. *Forbes* (1 April). https://www.forbes.com/sites/stevebanker/2021/04/01/amazon-supply-chain-innovation-continues/?sh=949a2ac77e6f, (accessed 16 May 2023).

31 Petrisor, I. and Cozmiuc, D. (2020). Global supply chain management organization at siemens in the advent of industry 4.0. In: *Supply Chain and Logistics Management: Concepts, Methodologies, Tools, and Applications*, 1095–1114. IGI Global.

32 Chrysler, Ford, GM, and AI (1995). *Advanced Product Quality Planning and Control Plan (APQP), Reference Manual*. Automotive Industry Action Group (AIAG).

33 Stamatis, D.H. (2018). *Advanced Product Quality Planning: The Road to Success*. CRC Press.

34 Firouzi, F., Farahani, B., Weinberger, M. et al. (2020). IoT fundamentals: Definitions, architectures, challenges, and promises. In: *Intelligent Internet of Things: From Device to Fog and Cloud*, 3–50.

35 Vogt, W. (2022). Platform takes equipment updates to air. *Farm Progress* (18 April). https://www.farmprogress.com/farming-equipment/platform-takes-equipment-updates-to-air (accessed 17 May 2023).

36 Daniel, C.N. and Berinyuy, L.P. (2010). Using the SERVQUAL model to assess service quality and customer satisfaction. An empirical study of grocery Stores in Umea. *Umeå Universitet* 15 (1): 1–66.

37 Rekow, E.D. (2020). Digital dentistry: the new state of the art—is it disruptive or destructive? *Dental Materials* 36 (1): 9–24.

38 Joda, T. and Zitzmann, N.U. (2022). Personalized workflows in reconstructive dentistry—current possibilities and future opportunities. *Clinical Oral Investigations* 26 (6): 4283–4290.

39 Lee, S.M., Lee, D., and Kim, Y.S. (2019). The quality management ecosystem for predictive maintenance in the Industry 4.0 era. *International Journal of Quality Innovation* 5: 1–11.

40 Salimova, T., Vatolkina, N., Makolov, V., and Anikina, N. (2020). The perspective of quality management system development in the era of industry 4.0. *Humanities & Social Sciences Reviews* 8 (4): 483–495.

5

Predictive Quality

5.1 Introduction

Predictive Quality is a relatively new paradigm that has emerged with the advent and convergence of data analytics, machine learning, and quality assurance methodologies. Its roots trace back to the early applications of statistics in quality control during the early 20th century. However, it was not until the late 20th century, with the rise of computing power and the digitization of manufacturing processes, that the concept began to take shape in a more concrete form.

The initial applications of Predictive Quality were rudimentary, relying primarily on simple statistical models to predict future outcomes based on historical data. However, with the advent of more advanced data analytics techniques and the exponential growth of available data in the 21st century, these models have become increasingly sophisticated.

Over the past few decades, predictive analytics [1, 2] has become a key component of quality management in a variety of sectors, from manufacturing to healthcare. With the advent of Industry 4.0, the increased digitization and connectivity of industrial processes have enabled real-time data collection and analysis on an unprecedented scale. This, in turn, has facilitated the rise of Predictive Quality [3, 4], making it possible to predict and prevent quality issues before they arise.

Research on Predictive Quality has focused on developing more advanced and accurate prediction models, exploring the use of machine learning and artificial intelligence (AI) techniques, and understanding how these models can be effectively integrated into quality management systems. Studies have also highlighted the importance of data quality and data management practices in ensuring the accuracy and reliability of Predictive Quality models.

Predictive Quality represents a groundbreaking development at the intersection of quality assurance, data analytics, and machine learning, fundamentally reshaping industrial operations. By harnessing potent algorithms and immense data volumes, predictive quality analytics equip businesses with invaluable insights into the future performance and quality of their products, processes, and services. This forward-thinking approach enables organizations to anticipate potential issues, refine quality control strategies, and ultimately elevate their overall performance and customer satisfaction. The driving force behind this strategy is not merely to identify and respond to problems but to foresee and mitigate them before they transpire. The ensuing discourse will explore the intricacies of predictive quality, examining its importance, components, application, and implications for the future.

5.1.1 Definition and Importance

5.1.1.1 Definition

Predictive Quality, as the term suggests, refers to the utilization of predictive analytics techniques to forecast the quality of products, services, or processes. This is typically accomplished through the application of statistical and machine learning algorithms to historical and real-time data to generate accurate future predictions.

Predictive Quality extends beyond traditional quality control methods that are often reactive, only identifying and addressing issues after they have occurred. Instead, it provides a proactive approach, enabling potential problems to be anticipated and addressed ahead of time, thereby averting detrimental impacts.

5.1.1.2 Importance

The importance of Predictive Quality in today's dynamic and competitive business environment cannot be overstated. Here is why:

1) **Preemptive Quality Control:** Predictive Quality allows companies to detect potential flaws or deviations in their products or processes before they manifest. This proactive approach reduces waste, saves costs, and prevents any negative impact on customer satisfaction.
2) **Optimized Processes:** By predicting where and when quality issues are likely to occur, businesses can optimize their operations and supply chain processes. This efficiency can lead to significant cost savings and increased profitability.
3) **Improved Decision-Making:** Predictive Quality provides valuable insights that can guide decision-making processes. When businesses know what to expect, they can make more informed, data-driven decisions that bolster their competitive advantage.
4) **Enhanced Customer Satisfaction:** Delivering consistent, high-quality products is key to maintaining customer satisfaction and loyalty. Predictive Quality ensures that companies meet and exceed customer expectations, enhancing brand reputation and driving business growth.
5) **Risk Mitigation:** Predictive Quality analytics can identify potential risks and vulnerabilities in the system, allowing businesses to take preventive measures and reduce the likelihood of severe, unexpected disruptions.

In conclusion, Predictive Quality is a potent tool that has the potential to revolutionize business operations across diverse industries. By enabling companies to anticipate quality issues and take preemptive measures, Predictive Quality transforms the traditional reactive approach into a proactive strategy, enhancing overall business performance and resilience.

5.1.2 Historical Perspective

The concept of Predictive Quality, in its modern understanding, has its roots in the advent of big data, machine learning, and AI. While the concept of using data and statistical methods to improve quality has been around for decades, it was not until the 21st century that technology enabled us to move from simple statistical analysis to complex predictive modeling and use predictive modeling to estimate quality metrics and proactively manage quality.

The earliest instances of predictive quality methodologies can be traced back to the mid-to-late 2000s when businesses started leveraging advanced predictive analytics techniques to anticipate and prevent quality issues. These techniques were initially applied in industries such as manufacturing, where small improvements in product quality could yield significant cost savings [5–7].

With the rapid advancements in data storage, processing capabilities, and AI algorithms in the last couple of decades, Predictive modeling has seen wider application across diverse sectors. Industries such as healthcare, finance, IT, and retail have begun to see the potential benefits of predictive modeling and predictive quality. In healthcare [8], for instance, predictive modeling techniques are used to forecast patient outcomes and prevent medical errors [9]. In the finance sector, these methodologies can predict the likelihood of loan defaults or fraudulent transactions [10, 11].

As of now, the quality community sees Predictive Quality as a powerful tool for improving processes, products, and services. While traditional quality methodologies focus on detecting and correcting defects, Predictive Quality focuses on preventing defects from occurring in the first place. This shift in paradigm has been welcomed by quality professionals as it aligns with the overarching goal of continuous improvement.

However, it is also recognized that Predictive Quality comes with its own set of challenges, such as the need for high-quality data, model interpretability, and data privacy concerns. Despite these challenges, the consensus in the quality community is that the potential benefits of Predictive Quality far outweigh the hurdles. The future of quality management is seen as increasingly data-driven, with Predictive Quality playing a central role in shaping this future.

5.1.3 Current Trends

In the context of Predictive Quality, several trends are emerging as technology and data analytics continue to evolve:

1) **Integration of IoT and Predictive Quality:** The Internet of Things (IoT) is increasingly significant in Predictive Quality. Sensors embedded in machines, products, and processes collect real-time data, which can be analyzed to predict potential quality issues. This real-time, continuous data collection allows for more timely and accurate predictions.
2) **Machine Learning and AI-Driven Predictive Models:** Machine learning and AI algorithms are becoming more sophisticated and capable of analyzing complex datasets. These algorithms can identify patterns and relationships that would be virtually impossible for humans to detect, leading to more accurate predictions and insights.
3) **Emphasis on Data Quality:** As the saying goes, "garbage in, garbage out." Organizations are realizing that the accuracy of Predictive Quality models is heavily dependent on the quality of data being used. This has led to an increased focus on data cleaning, preprocessing, and governance to ensure reliable predictions.
4) **Predictive Quality in the Cloud:** With the growth of cloud computing, more and more companies are utilizing cloud-based platforms for their predictive quality analytics. These platforms provide scalable computing resources, allowing companies to handle large datasets and complex algorithms without the need for extensive on-premises infrastructure.
5) **Integration of Predictive Quality in Quality Management Systems (QMS):** Predictive Quality is increasingly integrated into QMS, making predictive analytics a part of the overall quality strategy. This enables a more proactive approach to quality management and aligns with the broader organizational goal of continuous improvement.
6) **Focus on Model Interpretability:** As Predictive Quality models become more complex, there is an increasing emphasis on model interpretability. Understanding why a model makes a certain prediction can provide valuable insights and improve trust in the model's predictions.

These trends are driving the evolution of Predictive Quality, enabling more proactive and data-driven approaches to quality management. As technology advances, we can expect these trends to evolve and new ones to emerge.

5.2 Elements of Predictive Quality

Predictive Quality is a potent blend of several interrelated components, each playing a critical role in ensuring the efficacy of the outcomes. As we delve into the world of Predictive Quality, we must pay heed to the key elements that form the backbone of this sophisticated approach. When effectively integrated, these components can provide insightful foresight that pave the way for proactive quality management. The four crucial elements of Predictive Quality that we will be discussing in this section are Data Collection, Data Quality, Data Analysis, and Predictive Models. Each of these elements presents its own set of challenges and opportunities, and their successful orchestration is pivotal to leveraging the true potential of Predictive Quality.

5.2.1 Data Collection

Data collection is the first and arguably the most crucial step in the Predictive Quality process. Without relevant, accurate, and sufficient data, even the most sophisticated predictive models can fail to deliver accurate predictions.

In the context of Predictive Quality, data collection involves identifying and gathering relevant data that can help predict the quality of products, services, or processes. This data can come from various sources, including production machines, quality control systems, customer feedback systems, supply chain management systems, and more. The advent of IoT technology has made it possible to collect real-time data from machines and processes, enhancing the scope and accuracy of Predictive Quality.

When collecting data, several factors need to be considered:

1) **Relevance:** The collected data should be relevant to the quality attribute or parameter you are trying to predict. Irrelevant data can lead to noise and inaccuracies in predictions.
2) **Volume:** The amount of data collected should be sufficient to train predictive models effectively. Insufficient data can lead to underfitting, where the model fails to capture the underlying relationships in the data.
3) **Variety:** Data from diverse sources and of different types (structured, unstructured) can provide a more comprehensive view of the factors affecting quality.
4) **Velocity:** In certain cases, the speed at which data is generated and processed can be crucial, especially in real-time Predictive Quality applications.
5) **Timeliness:** Data should be collected and used in a timely manner. Outdated data may no longer be relevant and can lead to inaccurate predictions.

5.2.2 Data Quality

Data Quality [12, 13] is a critical component of Predictive Quality. The success of any predictive model hinges on the quality of the data being used to train and test the model. If the data is riddled with errors, inconsistencies, or missing values, the predictive model's accuracy can be significantly compromised, leading to suboptimal or even erroneous predictions.

Data Quality refers to the condition of a set of values of qualitative or quantitative variables. There are several dimensions to data quality, including:

1) **Accuracy:** The data should be free from errors and accurately represent the real-world scenario they are supposed to capture. Inaccurate data can lead to false predictions and insights.
2) **Completeness:** The dataset should not have missing or incomplete values. Missing data can create gaps in the predictive model's understanding and affect its performance.
3) **Consistency:** The data should be consistent, meaning the format, units, scale, etc., should be the same across the entire dataset. Inconsistent data can confuse the predictive model and affect its learning process.
4) **Timeliness:** The data should be up-to-date and relevant. Old data might not accurately represent the current situation and can lead to outdated predictions.
5) **Validity:** The data should conform to the specified format and value ranges. Invalid data can mislead the predictive model.
6) **Uniqueness:** There should not be unnecessary duplication in the dataset as it can skew the model toward certain values.

Ensuring data quality often involves data cleaning and preprocessing steps, such as handling missing values, removing duplicates, resolving inconsistencies, and validating the accuracy of the data. By ensuring high data quality, organizations can enhance the accuracy and reliability of their Predictive Quality models.

Calibration and measurement error analysis are essential to ensuring Data Quality, especially in Predictive Quality applications where data is often collected through various measuring devices or sensors. Let us delve into the following topics:

7) **Calibration:**

Calibration is the process of adjusting a device to ensure its readings are accurate and reliable. It involves comparing the measurements of a device under test with those of a standard or reference device. Calibration is important in Predictive Quality to ensure that the data collected from different devices is accurate and consistent.

Without proper calibration, measuring devices might produce erroneous data, which can significantly impact the performance of predictive models. Regular calibration checks should be performed to maintain the integrity of the data and ensure reliable predictions.

8) **Measurement Error Analysis:**

Measurement error analysis refers to the process of identifying, quantifying, and minimizing the errors associated with measurements. Measurement errors can be systematic (bias), random (noise), or due to gross errors (such as incorrect data entry).

Systematic errors can be identified and corrected through calibration or other correction methods. Random errors can be reduced by averaging multiple measurements. Gross errors need to be identified and removed from the dataset.

In the context of Predictive Quality, measurement error analysis is important to ensure that the variations in the data are due to the actual changes in the quality parameters and not due to measurement errors. By minimizing measurement errors, we can enhance the accuracy of the data and, in turn, the reliability of the predictive models.

In conclusion, both calibration and measurement error analysis are essential for maintaining high data quality in Predictive Quality applications. They ensure that the data used for building and testing predictive models accurately represents the real-world scenario and is free from avoidable errors and inconsistencies.

Gauge R&R is a statistical tool used to measure the amount of variation in a measurement system caused by the measurement device itself (repeatability) and the individuals taking the measurement (reproducibility).

When considering Predictive Quality, the data used to feed predictive models often comes from various measurement systems. The accuracy and reliability of these measurement systems can directly impact the quality of data and, subsequently, the performance of the predictive models.

In Predictive Quality, Gauge R&R can be an excellent tool to ensure the consistency and reliability of data collected from various measurement systems. Here is how:

- **Repeatability:** By assessing repeatability, Gauge R&R helps confirm that a specific measurement system provides consistent results for the repeated measurement of the same characteristic under the same conditions. If a measurement system shows poor repeatability, it could introduce unnecessary variance into the data, making it more challenging for predictive models to accurately detect underlying patterns.
- **Reproducibility:** Gauge R&R's reproducibility component helps ensure that different operators using the same measurement system on the same characteristic produce consistent results. If a measurement system is highly sensitive to who is performing the measurement, it could again introduce variance that can distort the predictive model's learning process.

5.2.3 Data Analysis

Data analysis is the process of inspecting, cleaning, transforming, and modeling data to discover useful information, draw conclusions, and support decision-making. In the context of Predictive Quality, data analysis serves as the bridge between data collection and the creation of predictive models.

Here are the key stages of data analysis in Predictive Quality:

1) **Data Cleaning and Preprocessing:** This step involves handling missing values, removing duplicates, resolving inconsistencies, and ensuring the data is in a suitable format for analysis. This stage is crucial for enhancing data quality and ensuring reliable predictions.
2) **Exploratory Data Analysis (EDA):** EDA is used to understand the structure and patterns in the data. It involves visualizing the data using plots and charts, calculating descriptive statistics, and identifying correlations between variables. EDA can provide valuable insights that guide the selection of predictive models.
3) **Feature Selection and Extraction:** Not all data collected will be relevant for predicting quality. Feature selection involves identifying the most relevant variables (features) that significantly impact quality. Feature extraction involves creating new features that may better represent the underlying patterns in the data.
4) **Outlier Detection:** Outliers are data points that significantly deviate from other observations. Outliers can be due to measurement errors or data entry errors, or they can represent rare events. Outlier detection is important as outliers can affect the performance of predictive models.
5) **Data Transformation:** Some predictive models require the data to be in a specific format or distribution. Data transformation involves changing the scale, distribution, or shape of the data to meet these requirements.

Once the data analysis stage is complete, the data will be ready for creating predictive models. By thoroughly analyzing the data, organizations can ensure that their predictive models are based on a sound understanding of the data, leading to more accurate and reliable predictions.

5.2.4 Predictive Models

Predictive models serve as the linchpin in predictive quality analysis. These models harness the data procured and processed from various sources to predict potential outcomes. In its simplest form, a predictive model can be represented as follows:

$$Y = f(\mathbf{X}) = F(X_1, X_2, ..., X_n) \tag{5.1}$$

In this equation, Y may represent a quality metric or a collection of such metrics. Depending on the context, Y might be a continuous variable like strength, yield, or speed, or it could be a categorical variable. For instance, a Y value of 1 may signify good quality, whereas a value of 0 could indicate poor quality.

$\mathbf{X} = (X_1, X_2, ..., X_n)$ are input variables or features that significantly influence the value of Y. $\mathbf{X}$ is often referred to as the predictor variable. These $\mathbf{X}$ values can be structured variables like numerical values or unstructured variables such as images, data streams, social media messages, or tweets.

These models are instrumental in leveraging diverse statistical and machine learning methods. These techniques help discern patterns and relationships within the data, which can then be utilized to make elaborate predictions.

Here are some key points to consider regarding predictive models in Predictive Quality:

1) **Selection of Appropriate Models:** Depending upon the nature and complexity of the data, different types of predictive models may be used. These can range from relatively simple linear regression models to more complex machine learning models like support vector machines, neural networks, decision trees, and ensemble methods. The selection of an appropriate model depends on the characteristics of the data, the problem at hand, and the required level of interpretability.

2) **Model Training and Validation:** Once a model is chosen, it is trained using a portion of the data (training set). The model learns to map the inputs (features) to the output (quality parameter) during this training process. The model's performance is then validated on a different portion of the data (validation set) to ensure that it can generalize well to unseen data.

3) **Model Evaluation:** Various metrics are used to evaluate the performance of the model, such as accuracy, precision, recall, F1 score, or mean squared error, depending on whether it is a classification or regression task. The goal is to have a model that performs well according to these metrics on both the training and validation sets.

4) **Model Tuning:** Almost all predictive models have hyperparameters that control the model's complexity and learning process. Tuning these hyperparameters can help improve the model's performance. Techniques such as grid search or random search are often used for hyperparameter tuning.

5) **Model Deployment and Monitoring:** Once a model is trained, validated, and tuned, it can be deployed to predict real-world data. It is important to continuously monitor the model's performance as new data comes in and update or retrain the model as necessary.

In the context of Predictive Quality, predictive models enable organizations to anticipate quality issues before they occur, allowing them to take proactive measures to prevent these issues, reduce waste, improve efficiency, and enhance customer satisfaction.

5.3 Exploration of Predictive Quality Models

The domain of Predictive Quality has expanded considerably over the years, birthing an extensive array of models that are skillfully designed to forecast quality-oriented outcomes. The sophistication and diversity of these models encapsulate the multifaceted scenarios typically encountered in the realm of quality management. These models range from traditional statistical frameworks to cutting-edge AI approaches, each specifically tailored to address various data configurations and quality prediction endeavors.

This section delves deeper into the intricacies of the four predominant categories of Predictive Quality models: Regression Models, Time Series Models, Machine Learning Models, and Deep Learning Models. These categories each hold a unique set of benefits and underlying principles. By understanding these models adequately, we can make a more informed decision when choosing the best-fitting model for any given Predictive Quality task.

Further, each model type's intricacies, capabilities, and practical applications will be explored, providing readers with an adequate understanding of their unique characteristics and usage in Predictive Quality tasks and practical insight needed to apply these models effectively in real-world scenarios.

5.3.1 Regression Models

Regression models form the foundation of many predictive modeling techniques, including those used in Predictive Quality. They estimate the relationship between a dependent variable (the variable we want to predict, often a quality-related parameter) and one or more independent variables (the variables we use to make the prediction, often process parameters or environmental conditions).

1) **Simple Linear Regression:** This is the simplest form of regression with one independent variable and one dependent variable. The relationship is expressed with the equation:

$$Y = \beta_0 + \beta_1 X + \varepsilon \tag{5.2}$$

Here,
- Y is the dependent variable (quality parameter)
- X is the independent variable
- β_0 is the intercept (value of Y when $X = 0$)
- β_1 is the slope (change in Y for a one-unit change in X)
- ε is the error term (difference between the observed and predicted value of Y) and is usually assumed to be a normally distributed random variable.

2) **Multiple Linear Regression:** This extends simple linear regression to include two or more independent variables. The equation is as follows:

$$Y = \beta_0 + \beta_1 X_1 + \beta_2 X_2 + \ldots + \beta_n X_n + \varepsilon \tag{5.3}$$

Here,
- $X_1, X_2, \ldots, X_n$ are the independent variables,
- $\beta_1, \beta_2, \ldots, \beta_n$ are the coefficients of the independent variables, where β_i represents the change in Y for a one-unit change in the corresponding X_i, $i = 1, 2, \ldots, n$, holding all other X_j constant, where $i \neq j$.

3) **Polynomial Regression:** This is used when the relationship between the independent variable X and dependent variable Y is modeled as nth degree polynomial. The equation can be expressed as:

$$Y = \beta_0 + \beta_1 X + \beta_2 X^2 + \beta_3 X^3 + \ldots + \beta_n X^n + \varepsilon \tag{5.4}$$

Although polynomial regression fits a nonlinear model of X to the data, as a statistical estimation problem, it is linear, in terms of estimated regression coefficients, β_i for $i = 0, 1, \ldots, n$. Therefore, polynomial regression is considered as a special case of polynomial regression. In polynomial regression model, there can be multiple independent variables with both linear and polynomial terms.

4) **Logistic Regression:** In statistics, the logistic model (or logit model) is a statistical model that models the probability of an event taking place. The event is modeled as a binary dependent variable (two classes, coded by an indicator variable, where the two values are labeled "0" and "1"), while the independent variables can each be a binary variable (two classes, coded by an indicator variable) or a continuous variable (any real value). In regression analysis, logistic regression (or logit regression) is estimating the parameters of a logistic model (the coefficients in the linear combination).

Logistic regression is used for binary classification problems; binary values of 1 or 0 indicate two categories, for example, good or bad, pass or failure. A very simple logistic function model is the following:

$$P(Y = 1) = P(x) = \frac{1}{1 + e^{-(\beta_0 + \beta_1 x)}} \tag{5.5}$$

Here,
- $P(Y = 1)$ is the probability of the dependent variable equaling a certain class (usually "1")
- e is the base of natural logarithms
- β_0 and β_1 are the coefficients in the model; they are to be estimated from the data

We can see that probability of Y belonging to a certain class is a function of x; x is a factor that may affect the property of Y.

From Eq. (5.5), it is easy to derive that:

$$\text{Odds} = \frac{P(x)}{1 - P(x)} = e^{\beta_0 + \beta_1 x} \tag{5.6}$$

where Odds (in favor) of an event or a proposition is the ratio of the probability that the event will happen to the probability that the event will not happen. Equation (5.5) can be shown to be equivalent to:

$$\log(\text{Odds}) = \log\left(\frac{P(x)}{1 - P(x)}\right) = \beta_0 + \beta_1 x \tag{5.7}$$

Equation (5.7) actually means that log(Odds) is a linear predictor function.

When there are multiple independent variables that may affect Y's property, then we will have a multiple logistic regression model:

$$\log(\text{Odds}) = \log\left(\frac{P(x)}{1-P(x)}\right) = \beta_0 + \beta_1 x_1 + \beta_2 x_2 + \ldots + \beta_n x_n \tag{5.8}$$

These regression models provide a starting point for building predictive quality models. They are particularly useful when the goal is to understand the effect of different factors on the quality classifier, Y, as the coefficients directly interpret these effects. However, they assume a specific form for the relationship between the independent and dependent variables (linear or polynomial), which may not hold in all scenarios. In such cases, more flexible models like machine learning or deep learning models may be more appropriate.

5.3.2 Time Series Model

In mathematics, a time series is a series of data points indexed (or listed or graphed) in time order. Most commonly, a time series is a sequence taken at successive equally spaced points in time. Thus it is a sequence of discrete-time data.

Time series analysis comprises methods for analyzing time series data in order to extract meaningful statistics and other characteristics of the data. Time series forecasting is the use of a time series model to predict future values based on previously observed values. Time series models can capture the trends, seasonality, and other temporal structures in the data to make future predictions.

Many quality-related parameters are monitored over time, so they are time series data. If we can develop a model that can predict its future values accurately, it will be very useful. Therefore, the time series model is another tool for predictive quality applications. Here are several commonly used time series models:

1) **Autoregressive (AR) Models:** These models use the dependent relationship between an observation and many lagged observations (observations at previous time points). The autoregressive model specifies that the output variable depends linearly on its own previous values and on a stochastic term; thus the model is in the form of a stochastic difference equation. The formula for an AR(p) model is:

$$Y_t = c + \varphi_1 Y_{t-1} + \varphi_2 Y_{t-2} + \ldots + \varphi_p Y_{t-p} + \varepsilon_t \tag{5.9}$$

Here,
- Y_t is the value of Y at time t
- $\varphi_1, \varphi_2, \ldots, \varphi_p$ are the parameters of the model
- c is a constant
- ε_t is the error term at time t
- p is the number of lagged observations in the model

2) **Moving Average (MA) Models:** These models use the dependency between an observation and a residual error from a MA model applied to lagged observations. The formula for an MA(q) model is:

$$Y_t = \mu + \varepsilon_t + \theta_1 \varepsilon_{t-1} + \theta_2 \varepsilon_{t-2} + \ldots + \theta_q \varepsilon_{t-q} \tag{5.10}$$

Here,
- Y_t is the value at time t
- μ is the mean of the series

- ε_t is the error term at time t
- $\theta_1, \theta_2, ..., \theta_q$ are the parameters of the model
- q is the number of lagged observations in the model

3) **Autoregressive Moving Average (ARMA) Models:** These models combine both AR and MA models. Given a time series of data Y_t, the ARMA model is a tool for understanding and predicting future values in this series. The AR part involves regressing the variable on its own lagged (i.e., past) values. The MA part involves modeling the error term as a linear combination of error terms occurring contemporaneously and at various times in the past. The model is usually referred to as the ARMA(p, q) model where p is the order of the AR part and q is the order of the MA part. The formula for an ARMA(p, q) model is:

$$Y_t = c + \varphi_1 Y_{t-1} + ... + \varphi_p Y_{t-p} + \varepsilon_t + \theta_1 \varepsilon_{t-1} + ... + \theta_q \varepsilon_{t-q} \qquad (5.11)$$

Here, the parameters are the same as those defined for AR and MA models.

ARMA models can be estimated by using the Box–Jenkins method [14].

4) **Autoregressive Integrated Moving Average (ARIMA) Models:** These models add a differencing step to the ARMA models to make the time series stationary (i.e., the properties of the series do not depend on the time at which the series is observed). The formula for an ARIMA(p, d, q) model is more complex and is not usually written out in the same way as AR, MA, or ARMA models, but it involves the same components with an additional differencing step [15].

These time series models are particularly effective for forecasting quality parameters, Y_t, when there is a strong temporal component to the data. However, they assume that the same temporal patterns will continue into the future, which may not always be the case, especially in the presence of unexpected events or structural changes.

5.3.3 Machine Learning Model

Machine Learning models can handle complex, nonlinear relationships between variables and are particularly useful when there are many input variables, and the relationships between these variables and the output variable are unknown or too complex to be captured by simpler models.

Here, we will introduce a few commonly used machine learning models for Predictive Quality tasks:

1) **Decision Trees:** Decision trees [16, 17] split the data into branches based on the values of the input variables. Each split is chosen to maximize the separation of the values of the output variable for the resulting subgroups. The process continues until a stopping criterion is met. In the end, a new observation is assigned a value (for regression trees) or a class (for classification trees) based on the terminal node it ends up in. There is no specific formula to represent a decision tree, as it is represented by a series of if-then rules.

2) **Random Forests:** Random forests or random decision forests is an ensemble learning method for classification, regression, and other tasks that operates by constructing a multitude of decision trees at training time. Each tree is built on a bootstrap sample of the data, and at each split, only a random subset of the predictors is considered. For classification tasks, the output of the random forest is the class selected by most trees. For regression tasks, the mean or average prediction of the individual trees is returned. [18, 19] Random forests generally outperform decision trees, frequently used as black box models in businesses, generating reasonable predictions across a wide range of data.

3) **Support Vector Machines (SVM):** Classifying data is a common task in machine learning. Suppose some given data points each belong to one of two classes, and the goal is to decide which class a new data point will be in. In the case of SVMs, a data point is viewed as a p-dimensional vector (a list of p numbers), and we want to know whether we can separate such points with a $(p-1)$-dimensional hyperplane. This is called a linear classifier. There are many hyperplanes that might classify the data. One reasonable choice as the best hyperplane is the one that represents the largest separation. For a linear classifier problem, SVMs find the hyperplane that maximizes the margin between the two classes in a binary classification problem.

Just look at an intuitive example in the following Figure 5.1, assuming that there are two categories: circles and squares. SVM method tries to find a hyperplane (the red line in Figure 5.1) so that margin of separation is maximized.

Mathematically, refers to Figure 5.2, if the two categories are linearly separable (can be cleanly divided by at least one hyperplane). SVM method tries to find a hyperplane so that the margin of separation is maximized, in Figure 5.2, which means that $d = (d+)+(d-)$ is maximized compared to all possible separating hyperplanes.

To find the best separating hyperplane (with the maximum margin), a quadratic program problem must be solved [20]; the equation of hyperplane equation (H_0) in Figure 5.2 is:

$$\mathbf{w}^{\mathrm{T}}\mathbf{x}+b=0 \tag{5.12}$$

Here

- $\mathbf{x}$ is the vector of input variables
- $\mathbf{w}$ is the vector of coefficients
- b is the bias term

$\mathbf{w}$, and b will be solved from the quadratic programming problem based on the data from the training set.

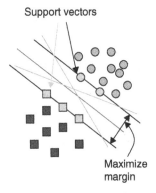

Figure 5.1 Two Categories Classification Problem

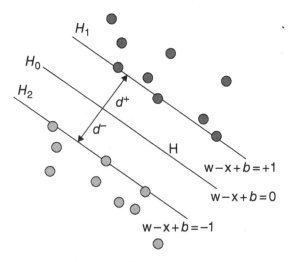

Figure 5.2 Separating Hyperplane and Supporting Vectors

Also, when

$$\mathbf{w}^T\mathbf{x} + b \geq 1, \text{then } y = +1\left(\text{in category} 1\right)$$

$$\mathbf{w}^T\mathbf{x} + b \leq -1, \text{then } y = -1\left(\text{in category} 2\right)$$

H_1 and H_2 are the hyperplanes:

$$\mathbf{w}^T\mathbf{x} + b = +1 \tag{5.13}$$

$$\mathbf{w}^T\mathbf{x} + b = -1 \tag{5.14}$$

Support vectors are the input vectors that just touch the boundaries specified by H_1 and H_2, they are the points touching H_1 and H_2 in Figure 5.2, and they are the elements of the training set that would change the position of the dividing hyperplane if removed.

In some classification problems, there will be no linear hyperplane that can separate different categories. Then we have a nonlinear SVM problem. For nonlinear boundaries, the data is transformed using a kernel function; then they can be separated by maximum margin hyperplanes [21]. There has been tremendous development in this SVM method recently, and it is a very popular machine learning method. SVM finds many applications in the area of text categorization, classification of images, classification of satellite data, hand-writing recognition, and many biological and other science areas.

4) **Gradient Boosting:** Gradient Boosting [22] represents a powerful ensemble method *that* constructs an iterative series of decision trees, each striving to rectify the errors made by its predecessor. Unlike other ensemble methods where trees are built independently, Gradient Boosting establishes a synergistic relationship between the trees, ensuring each new tree focuses on the inaccuracies of the previous one.

The technique adopts a "boosting" approach, which leverages the concept of learning from mistakes. Each decision tree is weak, with minimal depth, meaning it only captures a small portion of the complexity in the data. However, when these trees are combined sequentially, they complement each other, enabling the model to capture more complex patterns and make more accurate predictions.

The final prediction in Gradient Boosting is not a simple average but a weighted sum of the predictions made by all trees. These weights are assigned based on each tree's performance, meaning trees that reduce error more effectively carry greater influence in the final prediction. This distinctive feature of Gradient Boosting ensures a continuous improvement in prediction accuracy as more trees are added to the model, leading to a robust and high-performing predictive tool.

5) **K-Nearest Neighbors (KNN):** The KNN [23, 24] algorithm operates on a fundamental concept of similarity, where a data point's category or value is determined by considering its "k" nearest neighbors in the feature space. The term "k" represents a user-defined constant and signifies the number of neighboring points taken into account when making predictions.

For classification tasks, KNN assigns a new data point to the majority class of its "k" nearest neighbors. It essentially uses voting among the nearest neighbors to classify the new point; the class that receives the most votes becomes the classification of the unknown data point.

In regression tasks, rather than identifying a majority class, KNN calculates the average of the "k" nearest neighbors. This average, which can be a simple arithmetic mean or a weighted average where closer neighbors contribute more significantly, is used as the predicted value for the new data point.

In both scenarios, the determination of "nearness" or "distance" between data points typically involves a distance metric, such as Euclidean or Manhattan distance. One critical aspect to consider in the KNN algorithm is the choice of "k," as it can significantly impact the performance of the model. While smaller "k" values make the model sensitive to noise, larger "k" values make it computationally expensive and potentially biased toward dominant classes.

6) **Neural Networks:** Neural networks [25, 26], foundational to the field of deep learning, are complex systems designed to mimic the human brain's structure and function. They consist of multiple layers of interconnected nodes, known as "neurons," organized in a hierarchical manner.

Figure 5.3 illustrates a basic neural network structure. All the round nodes are neurons. Each neuron receives input from previous layer neurons, with each connection characterized by an associated weight. These weights represent the strength or influence of a particular input on the neuron's output. After receiving the inputs, each neuron performs a weighted sum of these inputs, adding a bias term to the result. This sum is then passed through a nonlinear activation function, which introduces complexity and enables the network to learn and model nonlinear relationships in the data.

The weights and biases within the network are initially set randomly and then iteratively adjusted through a process called backpropagation. Backpropagation, paired with an optimization algorithm like stochastic gradient descent, minimizes the error between the network's predictions and the actual values. This is achieved by propagating the error back through the network and adjusting the weights and biases to reduce the overall error.

Neural networks can have multiple layers (referred to as "deep" networks) and various architectures, depending on the problem at hand. They can model complex patterns and relationships in high-dimensional data, making them versatile and powerful tools for tasks such as image and speech recognition, natural language processing, and more.

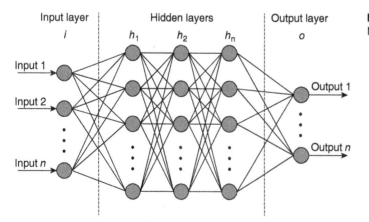

Figure 5.3 A Basic Neural Network Structure

Remember that each machine learning model has its assumptions, strengths, and weaknesses, so the choice of model should depend on the problem at hand, the nature of the data, and the trade-off between interpretability and predictive accuracy.

5.3.4 Deep Learning Models

Deep learning [27, 28], a subset of machine learning, employs algorithms that are based on the structure and function of the human brain called artificial neural networks. Deep learning models are particularly effective at handling high-dimensional data and can automatically extract features from raw data, which sets them apart from traditional machine learning models. Deep learning models are often used in Predictive Quality tasks that involve image, audio, or text data but can also be used for numerical data.

Here, we will introduce a few commonly used deep learning models:

1) **Feedforward Neural Networks (FFNN):** Also known as Multilayer Perceptron (MLP) [29, 30], these are the simplest type of artificial neural network. In an FFNN, information moves in only one direction—forward—through the layers, as shown in Figure 5.4.

   No loops or cycles are involved. Each neuron applies a nonlinear transformation to the weighted sum of its inputs. The weights are learned through a process called backpropagation. The formula for a neuron in a FFNN is:

$$y = f\left(\mathbf{w}^{\mathrm{T}}\mathbf{x} + b\right) \tag{5.15}$$

   Here,
   - y is the output of the neuron
   - f is the activation function
   - $\mathbf{x}$ is the vector of input variables
   - $\mathbf{w}$ is the vector of weights
   - b is the bias term

2) **Convolutional Neural Networks (CNN):** Recognized for their profound usage in image processing tasks, CNNs have carved a niche for themselves in the domain of AI. Their primary

Figure 5.4 Basic Structure of a Feedforward Neural Network

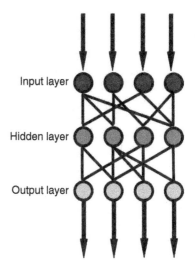

Input layer

Hidden layer

Output layer

strength lies in their exceptional ability to comprehend spatial dependencies within the data, thereby enhancing their suitability for tasks involving images and vision.

Their architecture includes a set of specialized layers known as convolutional layers. These layers are designed to perform a mathematical operation known as convolution, in which they apply a suite of filters to the input data. The convolution operation focuses on smaller regions of the input at a time, enabling the network to recognize complex patterns that would otherwise be missed by regular neural networks.

The salient feature of CNNs is their ability to reduce the dimensions of the input data, a technique known as dimensionality reduction. This process is integral for managing computational complexity, yet it preserves the crucial spatial information within the data. In other words, despite the reduced data size, the network retains the important relationships and patterns within the image's structure, ensuring no critical information is lost.

Moreover, CNNs are equipped with pooling layers, which further condense the data by downsampling the output of the convolutional layers. They serve to decrease the computational load and to prevent overfitting while preserving the most salient features of the data. These pooling layers, alongside convolutional layers, help CNNs achieve translational invariance, which means the network can recognize patterns regardless of their location within the input space.

The prowess of CNNs in handling image-based data stems from their sophisticated architecture that carefully balances computational efficiency with the preservation of spatial relationships, enabling them to effectively identify and understand complex visual patterns. CNNs are mainly used for image processing tasks due to their ability to capture spatial dependencies in the data. They include convolutional layers that apply a set of filters to the input, reducing its dimensions while preserving its spatial information.

This method is especially useful for deep learning of unstructured image data. In Section 5.5.1, I will present a case example of applying CNNs in automotive manufacturing.

3) **Recurrent Neural Networks (RNN):** RNNs are designed to work with sequential data. They include loops that allow information to be carried across different points in the sequence. However, standard RNNs suffer from the vanishing gradient problem, which makes it difficult to learn long-term dependencies.

4) **Long Short-Term Memory (LSTM) Networks:** LSTMs are a special type of RNN that are capable of learning long-term dependencies using a system of gates. This makes them particularly effective for tasks like time series forecasting or natural language processing, where the order of the observations matters.

5) **Autoencoders:** These are used for unsupervised learning tasks. They consist of an encoder, which maps the input data to a lower-dimensional representation, and a decoder, which maps this representation back to the original data. Autoencoders can be used for dimensionality reduction, anomaly detection, or to initialize the weights of a neural network for a subsequent supervised learning task.

Deep learning models have achieved state-of-the-art performance on many Predictive Quality tasks. However, they require a large amount of data and computational resources, and their predictions are often difficult to interpret due to their complexity.

5.4 Performance Metrics in Predictive Modeling

The true value of a predictive model lies in its performance, and understanding how to measure this performance is a vital aspect of Predictive Quality. In order to determine how well a model is doing, we use various metrics that each provide a different perspective on the model's quality.

This is particularly essential when comparing different models or adjusting parameters within the same model to find the best fit. This section will discuss five primary metrics used to assess the quality of predictive models: Accuracy, Precision, Recall, F1 score, and Area Under the Receiver Operating Characteristic Curve (AUC-ROC). Understanding these metrics is crucial to developing and applying effective Predictive Quality models.

5.4.1 Accuracy

Accuracy is one of the most straightforward metrics used in predictive modeling. It is simply the proportion of correct predictions made out of all predictions.

In the fields of science and engineering, the accuracy of a measurement system is the degree of closeness of measurements of a quantity to that quantity's true value.

In a binary classification problem, where the model is predicting one of two possible outcomes (usually labeled as "positive" and "negative"), the predictions can fall into four categories:

1) **True Positives (TP)**: These are positive instances that were correctly predicted as positive.
2) **True Negatives (TN)**: These are negative instances that were correctly predicted as negative.
3) **False Positives (FP)**: These are negative instances that were incorrectly predicted as positive.
4) **False Negatives (FN)**: These are positive instances that were incorrectly predicted as negative.

The formula for accuracy is:

$$\text{Accuracy} = \left(\text{TP} + \text{TN}\right) / \left(\text{TP} + \text{TN} + \text{FP} + \text{FN}\right) \tag{5.16}$$

In multiclass classification problem, the formula for accuracy is:

$$\text{Accuracy} = \frac{\text{Correct Classifications}}{\text{All Classifications}} \tag{5.17}$$

This metric is easy to understand and interpret, which makes it widely used. However, it may not be the best metric in all scenarios. Specifically, if the classes are imbalanced (one class has many more instances than the other), a model that simply predicts the majority class will have high accuracy, but it may not be a good model because it fails to predict the minority class, which is often the class of interest. Therefore, when dealing with imbalanced classes, it is important to also consider other metrics like precision, recall, F1 score, and AUC-ROC.

5.4.2 Precision

Precision, also known as positive predictive value, is a metric used in predictive modeling to measure the exactness or quality of the prediction.

In engineering, precision is usually related to reproducibility and repeatability, the degree to which repeated measurements under unchanged conditions show the same results.

In machine learning practice, in a binary classification problem, it calculates the proportion of TP predictions (correctly predicted positive instances) out of all predicted positive instances.

Using the same terminology as before with TP, TN, FP, and FN, the formula for precision is:

$$\text{Precision} = \text{TP} / \left(\text{TP} + \text{FP}\right) \tag{5.18}$$

This means that precision is concerned with the question: "Of all the instances the model predicted to be positive, how many were actually positive?"

Precision is particularly useful in scenarios where the cost of FPs is high. For example, if an email spam detection model erroneously flags an important email as spam (an FP), the user might miss an important communication.

However, precision alone does not give a complete picture of a model's performance. A model can achieve perfect precision by making a single correct positive prediction, but this is not very useful if there are many other positive instances it fails to detect. Therefore, precision is often used in conjunction with recall (also known as sensitivity or TP rate), which measures the model's ability to correctly identify positive instances.

5.4.3 Recall

Recall, also known as sensitivity or TP rate, is a metric that measures the completeness of the predictions made by a model. In other words, it quantifies how many of the actual positive instances the model is able to identify.

Using the same terminology as before with TP, TN, FP, and FN, the formula for recall is:

$$\text{Recall} = \text{TP} / (\text{TP} + \text{FN}) \tag{5.19}$$

This means that recall is concerned with the question: "Of all the actual positive instances, how many were correctly predicted by the model?"

Recall is particularly useful in scenarios where the cost of FNs is high. For instance, in medical testing for severe disease, failing to identify a positive case (an FN) could have grave consequences, while falsely identifying a negative case as positive (an FP) would mainly lead to further testing.

However, like precision, recall alone does not provide a complete picture of the model's performance. A model could achieve perfect recall by predicting positive for all instances, but this would also result in many FPs. Therefore, recall is often used alongside precision, and the balance between these two metrics is quantified using the F1 score.

5.4.4 F1 Score

The F1 score, also known as the F-Score or F-Measure, is a metric that combines precision and recall into a single number. It is particularly useful in cases where you want to balance the trade-off between precision and recall. The F1 score is the harmonic mean of precision and recall, which gives more weight to lower values. As a result, the F1 score will only be high if both precision and recall are high.

The formula for the F1 score is:

$$\text{F1 score} = 2 \times \frac{\text{Precision} \times \text{Recall}}{\text{Precision} + \text{Recall}} \tag{5.20}$$

In other words, the F1 score is trying to answer the question: "What is the balanced average between precision and recall?"

In cases where you care equally about precision and recall, the F1 score serves as a good single-number summary of the model's performance. However, if you care more about precision than recall, or vice versa, you might want to use a different variant of the F-Score (like the F0.5 Score or

F2 Score) that gives more weight to one of these metrics. Also, it is important to note that the F1 score assumes that you care equally about FPs and FNs, so if that is not the case, it might not be the best metric to use.

5.4.5 AUC-ROC

The AUC-ROC (Area Under the Receiver Operating Characteristic Curve) is a commonly used metric for binary classification problems. Unlike the other metrics we have discussed, which each produce a single value, the ROC curve is a plot that illustrates the TP rate (recall) against the FP rate for various threshold settings, and the AUC measures the entire two-dimensional area underneath this curve.

The ROC curve is created by plotting the TP rate (TPR) against the FP rate (FPR) at various threshold settings:

$$\text{TPR}\left(\text{True Positive Rate}\right) \text{ or Recall} = \text{TP} / \left(\text{TP} + \text{FN}\right) \tag{5.21}$$

$$\text{FPR}\left(\text{False Positive Rate}\right) = \text{FP} / \left(\text{FP} + \text{TN}\right) \tag{5.22}$$

The AUC-ROC ranges from 0 to 1, where a value of 0.5 denotes a model that does no better than random prediction, and a value of 1 denotes a perfect model. A model whose predictions are 100% wrong has an AUC of 0, but in most cases, the AUC lies between 0.5 and 1.

Figure 5.5 is a real AUC-ROC curve that summarizes a research study using several predictive analytic models, such as logistic regression, neural network, SVM, random forest, etc., to predict patient's avoidable readmission based on patients' medical records.

AUC-ROC is beneficial when the classes are imbalanced, and it is generally more robust than accuracy, precision, and recall for comparing different models.

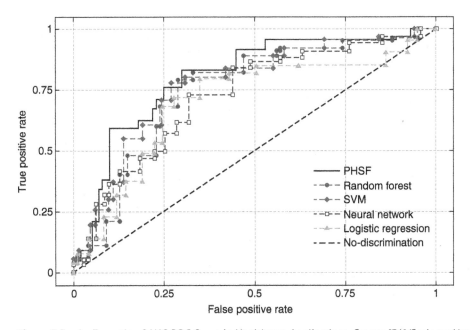

Figure 5.5 An Example of AUC-ROC Curve in Healthcare Applications. *Source:* [31]/Springer Nature

However, AUC-ROC can be overly optimistic when the positive class is rare, or the number of features is very high, leading to overly optimistic performance estimates. This is especially a concern in the context of "big data," where the number of features can easily exceed the number of instances.

In such cases, a precision-recall curve (PRC) and the area under the PRC can be used as an alternative to the ROC curve and AUC-ROC. While ROC curves are a good measure for binary classification problems, precision-recall curves are more suitable for imbalanced datasets.

5.5 Application of Predictive Quality in Various Industries

The value of predictive quality is not restricted to one field but spreads across multiple industries, reflecting its versatile and broad-ranging implications. Predictive quality is a powerful tool that leverages historical data, analytics, and machine learning to anticipate future outcomes. This capability allows businesses and organizations to make proactive decisions, optimize processes, improve customer satisfaction, and ultimately increase profitability. In this section, we will explore the application of predictive quality across five diverse industries: Manufacturing, Healthcare, Retail, Finance, and Information Technology. Each of these industries utilizes predictive quality models differently, based on the unique challenges and goals within their respective fields. Through this exploration, we will gain insights into the vast potential of predictive quality models to transform various aspects of industry operations and strategies.

5.5.1 Manufacturing

Predictive Quality has long played an integral role in the manufacturing industry. The concept of Predictive Quality originated from the need for consistent, high-quality products during mass production, particularly during the industrial revolution when goods were manufactured at an unprecedented scale.

In the early stages, quality control in manufacturing was more reactive, focusing on detecting and correcting defects after production. However, with the advent of statistical process control in the early 20th century, the industry started to shift toward a more proactive approach, monitoring production processes to prevent defects from occurring in the first place.

The application of Predictive Quality has only expanded with the advent of digital technologies. Today, manufacturers in sectors ranging from automotive to electronics, food and beverage, and aerospace, among others, leverage Predictive Quality models to optimize their processes and enhance product quality.

Manufacturers apply Predictive Quality models in several ways:

1) **Predictive Maintenance:** Using historical machine data and advanced algorithms, manufacturers can predict when equipment might fail or require maintenance, reducing downtime and extending the equipment's life.
2) **Process Optimization:** By analyzing production data, manufacturers can identify patterns and correlations that can help optimize production processes, improve efficiency, and reduce waste.
3) **Quality Assurance:** Predictive models can also be used to predict the quality of products at the end of the manufacturing line, enabling early detection of potential quality issues.

A notable success story is that of an international automotive manufacturer that implemented Predictive Quality models to predict the occurrence of defects in their assembly line. By integrating machine learning with real-time data from their manufacturing processes, they were able to identify potential issues before they occurred, reducing defects by 50% and leading to significant cost savings [32].

In another instance, an electronics manufacturer used Predictive Quality to analyze their soldering process. The company utilized a machine learning model to predict soldering defects based on temperature, speed, pressure, material, and placement. This proactive approach allowed them to significantly reduce defects and improve product quality [33].

In the era of Industry 4.0, with the proliferation of IoT devices and increasing digitization, Predictive Quality in manufacturing is set to become even more prevalent, driving the industry toward more efficient and sustainable production methods.

Here are two real case examples of Predictive Quality from General Motors.

Example 5.1   General Motors' Learning Quality Control[1]

Learning Quality Control (LQC) is a system for process monitoring that leverages machine learning and deep learning principles. LQC emphasizes real-time detection or prediction of defects, formulating the task as a binary classification problem. Historical samples, denoted as (X, l), are used to train algorithms that automatically identify patterns linked to defects like anomalies, deviations, or nonconformances. These samples can include process measurements such as features (structured data), images (unstructured data), or signals (unstructured data) denoted by X, along with their associated binary quality label denoted by $l \, \varepsilon$ {good, defective}. Following binary classification notation, a positive label refers to a defective item, and a negative label refers to a good quality, Eq. (5.23):

$$l_i = \begin{cases} 1 & \text{if the } i\text{th item is defective} \, (+) \\ 0 & \text{if the } i\text{th item is good} \, (-) \end{cases} \tag{5.23}$$

The historical samples become the learning data set used by the machine learning algorithms to train a model; learn the coefficients θ of the general form $\hat{y} = f(X; \theta) = p(l = \text{defective})$, where $\hat{y}$ refers to the predictive probability of the positive class (defective). Each predicted probability is then compared to a Classification Threshold (CT) to carry out the classification, as explained in Eq. (5.24):

$$l_i = \begin{cases} 1 & \text{if } \hat{y}_i => \text{CT item is defective} (+) \\ 0 & \text{if } \hat{y}_i < \text{CT item is good} (-) \end{cases} \tag{5.24}$$

The CT is defined based on the business goals considering the costs or implications of the misclassifications (errors). The prediction ability of a classifier is summarized based on four concepts described in the confusion matrix, Table 5.1.

- **True Negative**: it refers to a good item correctly predicted by the classifier.
- **True Positive**: it refers to a defective item correctly predicted by the classifier.

---

1 Example 5.1 is contributed by Dr. Carlos Alberto Escobar Diaz, Dr. Ruben Morales-Menendez, and Dr. José Antonio Cantoral Ceballos, of Tecnológico de Monterrey.

Table 5.1 Confusion Matrix

Actual Quality Status	Predicted Good	Predictive Defective
Good item	True Negative (TN)	False Positive (FP)
Defective item	False Negative (FN)	True Positive (TP)

- **False Positive**: it refers to a good item incorrectly predicted to be defective. From a statistical perspective, the α error (type I) quantifies the FPs.
- **False Negative**: It refers to a defective item incorrectly predicted to be good. The β error (type II) quantifies the FNs.

Regular production often yields imbalanced binary classification of quality datasets, containing many good quality items and a few defects. The primary challenge posed by the Industry 4.0 era is to detect or prevent these few defects. Figures 5.6 and 5.7 visually demonstrate the fundamental concepts of LQC.

LQC's essential concepts include:

1) Offline learning, where historical data containing patterns of concern is used to train machine learning algorithms to develop the classifier, an activity that can last weeks or even months,
2) Online deployment, where the classifier is implemented into production to automatically and in real-time detect patterns associated with quality defects.

Figure 5.6 portrays an LQC monitoring system designed for defect prediction. In this setup, data is gathered early in the process to forecast downstream quality complications. Positive results serve as an alert, prompting the engineering team to troubleshoot the process and implement corrective actions. Since these predictions operate under a veil of uncertainty, FP

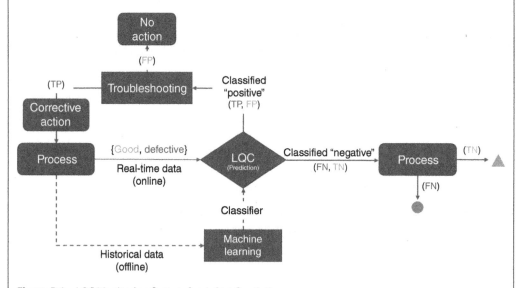

Figure 5.6 LQC Monitoring System for defect Prediction

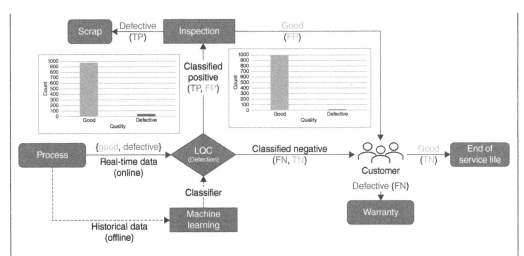

Figure 5.7 LQC Monitoring System for Defect Detection

errors often result in no corrective action, instead leading to a misallocation of the engineering team's time and resources. The FNs, or failures to detect a concerning pattern, result in downstream quality issues.

Figure 5.7 provides a graphical illustration of a general manufacturing process that yields only a handful of defects per million opportunities. The primary aim of LQC in this context is to detect these defective items in real-time. For instance, during a welding process, a signal is generated and passed through a classifier to automatically discern if the weld is good or defective. All positive results or flagged items are subjected to inspection. The TPs are discarded, and the FPs, after a secondary inspection, may continue in the value-adding process.

The primary concern here is the FNs—defective items that slip through undetected and continue in the value-adding process, potentially leading to warranty claims or customer complaints. In this context, while the FPs give rise to the "hidden factory effect" (i.e., inefficiencies), the FNs can cause more grave situations. Thus, efforts to minimize the FNs bear greater importance.

LQC is a progression from Statistical Process Control (SPC). While the latter relies on statistical methods, LQC is based on these statistical methods and augmented with machine learning and deep learning algorithms. This amalgamation solves a broad range of previously intractable engineering problems. These algorithms can effectively learn nonlinear patterns in hyper-dimensional spaces, surpassing traditional statistical SPC methods like control charts. Moreover, deep learning enables the replacement of human-based monitoring systems, an application that was beyond engineering capabilities just a decade ago.

Case Study 1: A Welding Process in Automotive Manufacturing

This case study demonstrates an LQC monitoring system applied to a welding process based on structured data [34]. Here, welding quality is defined by tensile strength, with three features being the most significant predictor variables. Feature 1 represents the maximum recorded welding temperature. Feature 2 signifies the welding machine's elapsed time, and Feature 3 is the sum of the acoustic signal. A virtual dataset (Scaled data of these 3 features) with 10 observed values (samples) of these three features is outlined in Table 5.2 and visually represented in Figure 5.8.

Table 5.2 Feature Values and Quality Status of Each Sample

Index	Feature 1	Feature 2	Feature 3	Quality Status
1	89	91	95	Defective
2	33	20	98	Good
3	95	93	88	Defective
4	40	45	56	Good
5	80	86	92	Defective
6	20	55	12	Good
7	52	60	80	Good
8	90	95	20	Good
9	99	89	60	Good
10	85	82	75	Defective

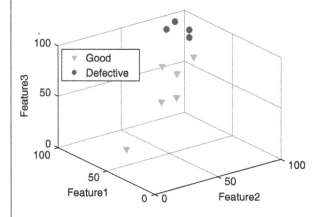

Figure 5.8 3D Patterns for Features and Quality Classification

While a clear separation of quality classes is observable in 3D (with all 3 features), such separation is not present when examining one variable at a time (1D). As shown in the univariate quality classification analysis of Figure 5.9, all the positive (poor quality) classes have high values, but several good-quality items also exhibit high values. A univariate control chart analysis using the I-MR control chart performed on these three features (shown in Figure 5.10) fails to identify poor-quality items.

While the classes do not separate in 2D, as shown in Figure 5.11, a complete separation in the 3D space can be observed. In such cases, an optimal separating hyperplane can be learned by a machine learning algorithm like the SVM. Figure 5.12 demonstrates the separating hyperplane learned by the SVM using a linear kernel; for the code, please refer to https://github.com/Quality40/case-study-0-structured-data-. Other algorithms such as logistic regression or neural networks can also easily separate the classes in this problem.

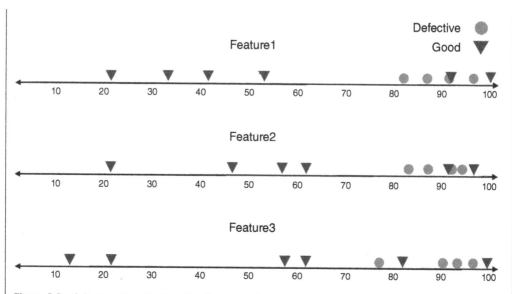

Figure 5.9 1-Feature Quality Classification Analysis

This case study is based on a simple three-features binary quality classification. However, patterns often exist in hyper-dimensional spaces (e.g., 4D or higher) following nonlinear relationships. Machine learning algorithms show a superior capacity to learn these patterns, thereby performing significantly better quality classifications.

Case Study 2: Deep Learning on Unstructured Image Data

This case study is to illustrate how LQC is applied to replace visual-based monitoring systems; a virtual case study with images is presented. The task here is to train a CNN model to automatically monitor the process to detect defects, i.e., significant deviations in the car body. Only 49 pictures were taken from a toy car replica. Figure 5.13 shows the good quality (i.e., negative label), and Figure 5.14 shows cars with significant deviations (i.e., positive labels).

The model contains three convolutional layers, each with convolution kernels of size 3 × 3, stride 1, and padding 1, batch normalization, and ReLU as activation function. The number of filters is 64, 128, and 128 for the first, second, and third convolutional layers, respectively. Finally, the last convolutional activation volume is flattened and passed through a final fully connected layer with two neurons (alternatively, we could only use one neuron and consider binary cross entropy as a cost function). The model was trained using a constant learning rate of 0.00008 for 15 epochs, with mini-batches of two images. The learning rate was found using random exploration, and the model was trained using the Adam optimizer. For full code implementation, refer to https://github.com/Quality40/case-study-2-unstructured-data-.

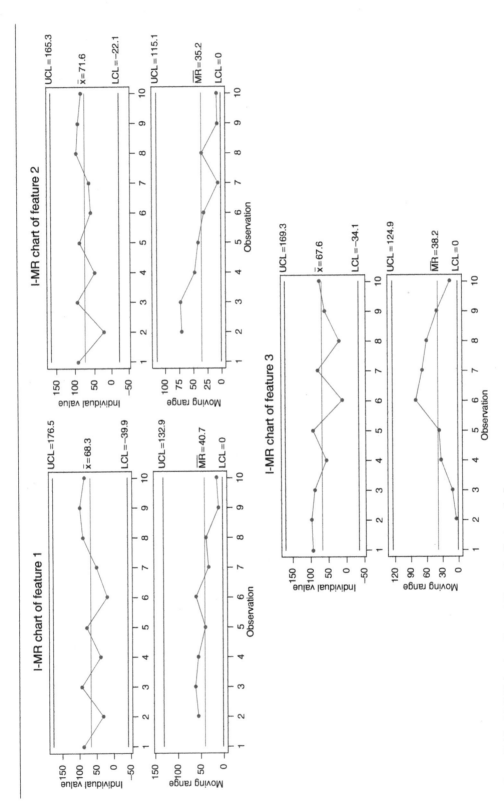

Figure 5.10 I-MR Univariate Control Charts for Features 1–3

Figure 5.11 2D Quality Classification Analysis (Features 1 and 2)

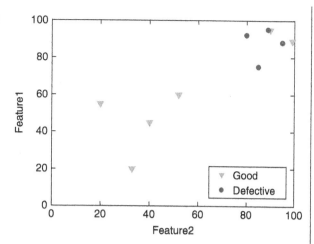

Figure 5.12 Linear Separation Scheme Based on the SVM, Separating Hyperplane Indicated

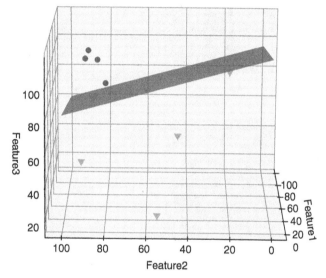

Figure 5.13 Car Body with No Significant Deviations, Good Quality Samples

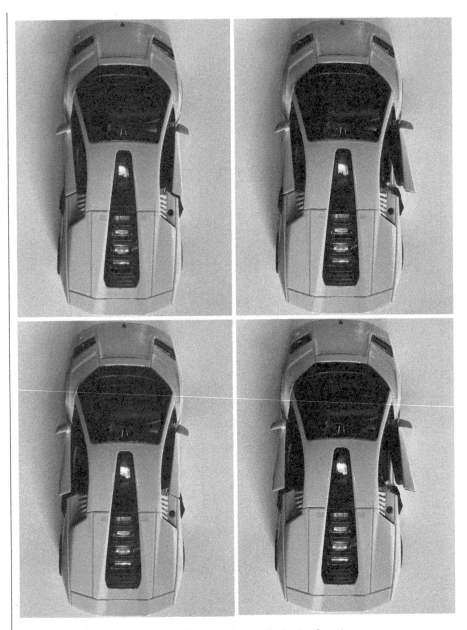

Figure 5.14 Car Bodies with Significant Deviations, Defective Samples

Figure 5.15 shows the confusion matrix of the training, validation, and test sets. The model completely separates the good quality vs. poor quality classes.

This case study demonstrates the application of an LQC for the replacement of a visual monitoring system. The technology is ready; therefore, it is our challenge to find applications and develop innovative solutions. The economic benefit has been widely studied; moreover, eliminating these monotonous tasks allows management to use the power brain of the team to solve other tasks that require higher cognitive levels.

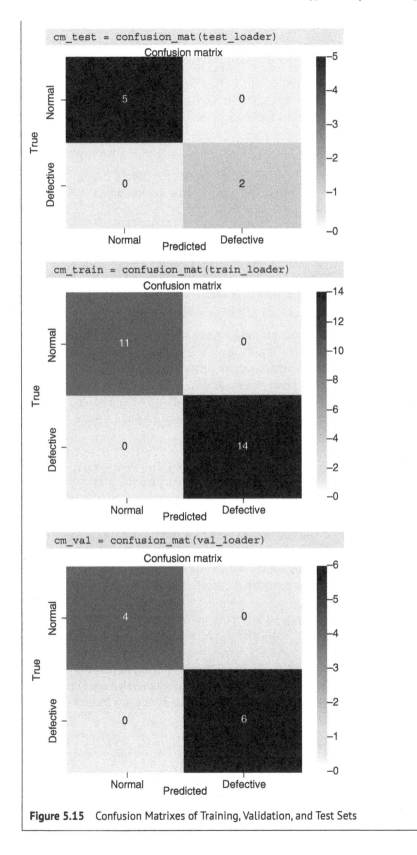

Figure 5.15 Confusion Matrixes of Training, Validation, and Test Sets

5.5.2 Healthcare

The application of predictive quality methods in the healthcare sector has been a transformational aspect of medical advancements in recent decades. Initially, the concept of quality in healthcare revolved around maintaining consistent and high standards of patient care and ensuring that all medical procedures were carried out correctly and safely. However, as medical data grew and technology advanced, the healthcare industry began leveraging predictive quality models to optimize healthcare delivery and patient outcomes.

The use of Predictive Quality in healthcare encompasses a wide range of applications:

1) **Patient Care and Treatment:** Predictive models can forecast a patient's health trajectory based on past medical history, genetic factors, and lifestyle choices, leading to personalized treatment plans.
2) **Disease Surveillance:** By analyzing public health data, predictive models can identify emerging disease trends, helping to mitigate potential health crises.
3) **Resource Management:** Predictive Quality can forecast patient flow, helping healthcare providers optimize staffing, bed allocation, and other critical resources.
4) **Preventive Medicine:** Predictive quality models can forecast the likelihood of an individual developing certain diseases, allowing for early intervention and prevention strategies.

One of the success stories in healthcare is the use of predictive analytics in managing diabetes [35]. By analyzing patient data and using predictive quality models, healthcare providers can predict potential health risks, optimize treatment plans, and improve patient outcomes.

The Mayo Clinic, for instance, developed a predictive model that helps doctors identify individuals at risk of developing heart disease [36]. The model considers a variety of factors such as cholesterol levels, blood pressure, and smoking status and provides personalized preventive measures to those at risk.

Similarly, Google's DeepMind Health project has made significant progress in using AI and predictive analytics for the early detection of diseases. In partnership with the US Department of Veterans Affairs, DeepMind developed an AI model that could predict the onset of acute kidney injury up to 48 hours in advance, which could give doctors a head start in preventing the condition from worsening [37].

This project was publicly announced, and you can find the peer-reviewed research paper titled "A clinically applicable approach to the continuous prediction of future acute kidney injury," published in *Nature* in 2019 [38]. This study showed that their AI model could predict acute kidney injury (AKI) in patients up to two days in advance, with a level of accuracy significantly better than the traditional predictive methods.

However, as with any AI model in healthcare, it is important to note that while the results are promising, the implementation and clinical validation process is complex and requires strict regulatory approval.

It is clear that the application of Predictive Quality in healthcare is bringing about a paradigm shift in how care is delivered. The integration of machine learning, data analysis, and predictive models into healthcare promises a future where medical treatment is not just reactive but also proactive, personalized, and preventive.

5.5.3 Retail

The retail industry has been one of the major sectors to leverage Predictive Quality methods, especially with the advent of e-commerce and digital technologies. In the early stages, quality assurance in retail primarily involved ensuring that products met certain standards and that transactions and customer interactions were handled properly. With the increase in online shopping, data generation, and technological advancements, the retail sector has moved beyond traditional methods to using Predictive Quality models for optimizing processes and enhancing customer satisfaction.

Here are some areas where Predictive Quality models are applied in the retail industry:

1) **Inventory Management:** Predictive models can forecast product demand, helping retailers optimize inventory levels and reduce stockouts or overstocking.
2) **Price Optimization:** Predictive analytics can be used to determine the optimal price for products based on factors like demand, competition, and market trends.
3) **Personalization:** By analyzing customer behavior and purchase history, predictive models can personalize product recommendations, enhancing the customer experience and increasing sales.
4) **Customer Churn:** Predictive models can identify customers who are likely to stop doing business with the retailer, enabling proactive measures to retain these customers.

Amazon, a global e-commerce giant, is an example of successful implementation of predictive models. They use predictive analytics for demand forecasting [39], which helps manage inventory levels efficiently. Their recommendation engine, powered by machine learning algorithms, personalizes the shopping experience for each user based on their browsing and purchasing history, driving increased customer engagement and sales.

Another noteworthy example is Walmart. With their advanced data analytics capabilities, they can analyze and predict consumer buying patterns, optimize inventory, and adjust pricing dynamically [40].

In the fashion industry, companies like Zara have revolutionized supply chain management using predictive models [41]. Zara uses real-time data analytics to track consumer trends and manages to deliver new collections in just weeks, compared to months for other fashion retailers. This quick response to market trends is a significant factor in Zara's global success.

In summary, Predictive Quality significantly transforms the retail industry, driving efficiency, customer satisfaction, and profitability. As technology evolves, the applications of Predictive Quality models in retail are set to increase, shaping the future of the industry.

5.5.4 Finance

The finance industry has a rich history of applying predictive modeling, given the vast amounts of historical data and the need for accurate forecasting. Initially, the finance sector relied heavily on statistical models for risk assessment, portfolio management, and market trend analysis. However, the recent explosion of available data and advancements in technology have led to the adoption of more sophisticated predictive quality models.

Here are a few applications of Predictive Quality in finance:

1) **Risk Assessment:** Predictive models are used to assess the creditworthiness of borrowers, predict loan defaults, and manage financial risk.
2) **Fraud Detection:** Machine learning models can identify unusual patterns and anomalies in financial transactions, helping to detect and prevent fraud.
3) **Algorithmic Trading:** Predictive models can analyze market data and trends to forecast price movements and make trading decisions.
4) **Customer Segmentation:** By analyzing customer behavior and transaction data, predictive models can help in identifying profitable customer segments and personalizing services.

One of the prominent examples of predictive quality in finance is the use of credit scoring models by banks and financial institutions. These models analyze an individual's credit history, income, employment status, and other relevant information to predict their likelihood of repaying a loan. FICO (Fair Isaac Corporation) scores are a well-known example of such predictive models [42].

In the realm of fraud detection, companies like PayPal use machine learning algorithms to analyze transactions in real-time and identify potentially fraudulent activities [43]. These predictive models have significantly reduced the occurrence of financial fraud, saving millions of dollars annually.

Another notable use case is in the area of algorithmic trading. Investment firms and hedge funds use complex predictive models to analyze market trends and make trading decisions. Renaissance Technologies, a famous hedge fund, is known for its heavy use of predictive models and has achieved remarkable success in the financial markets [44].

The advent of AI and machine learning has significantly expanded the possibilities for Predictive Quality in finance. These technologies are enabling more accurate forecasts, better decision-making, and improved financial outcomes. As technology continues to advance, the finance industry is set to benefit significantly from the use of predictive models.

5.5.5 Information Technology

Predictive Quality has been integral to the information technology (IT) industry, with its roots dating back to the earliest stages of software development. Traditional software quality assurance methods relied on a reactive approach, detecting and correcting defects after the software was built. However, as the complexity and scale of software projects grew, this approach became increasingly inadequate.

Predictive Quality in IT involves using models to forecast the quality of software products and services based on data collected during the development process and postdeployment. These predictive models allow for proactive quality assurance, enabling teams to identify and address potential issues before they cause problems.

Here are some key applications of Predictive Quality in the IT sector:

1) **Software Defect Prediction:** Machine learning models can analyze code and development data to predict where defects might occur. This allows teams to focus their testing efforts more effectively and rectify issues earlier in the development cycle.
2) **Service Availability Prediction:** Predictive models can forecast service downtime based on performance data and usage patterns, enabling proactive measures to ensure service availability.
3) **Cybersecurity Threat Prediction:** By analyzing network activity and patterns, predictive models can identify potential cybersecurity threats and breaches, facilitating timely countermeasures.
4) **User Behavior Prediction:** Predictive models can analyze user interaction data to anticipate user behavior and preferences, leading to more user-friendly design and personalized services.

Microsoft provides a noteworthy example of Predictive Quality in IT. They use machine learning models to predict software bugs in the Windows operating system [45]. By analyzing large volumes of code, these models can predict potential problem areas, allowing developers to address these issues proactively.

Another notable example is Google's use of Predictive Quality in its services. For instance, Google uses machine learning models to predict user search queries, enhancing the speed and accuracy of its search engine [46].

In the field of cybersecurity, companies like Darktrace use AI and machine learning to predict and detect cybersecurity threats. Their technology, which they liken to the human immune system, learns what normal network activity looks like and can then detect and respond to anomalies that may signify a threat [47].

The application of Predictive Quality in IT has significantly improved the efficiency and effectiveness of software development and IT service management. As technology continues to evolve, the role of Predictive Quality in the IT sector is expected to grow, providing exciting opportunities for future advancements.

5.6 The Challenges and Limitations of Predictive Quality

While Predictive Quality modeling has shown immense potential and benefits in various sectors, it is not without its challenges and limitations. As organizations embrace these tools to improve quality and efficiency, they must also contend with issues such as data privacy and security, model interpretability, risk of overfitting and underfitting, and the need for high-quality and relevant data. Understanding these challenges can guide better implementation and management of predictive quality models.

5.6.1 Data Privacy and Security Issues

Data is the cornerstone of any predictive model. The more data available, the more accurate the prediction can be. However, collecting, storing, and processing data raises significant concerns about privacy and security. Organizations must ensure they comply with all applicable data protection laws and regulations, such as general data protection regulation (GDPR) in Europe or California Consumer Privacy Act (CCPA) in California. Failure to properly handle sensitive data not only leads to legal consequences but can also damage an organization's reputation. Furthermore, the security of data is paramount, as breaches can lead to significant financial and reputational losses.

5.6.2 Model Interpretability

While predictive models can provide valuable insights, their "black box" nature can be a significant challenge. Complex models like deep learning are often difficult to interpret, making it challenging to understand how the model arrived at a particular prediction. This lack of transparency can be a hurdle, especially in sectors like healthcare or finance, where interpretability is crucial for decision-making and regulatory compliance.

5.6.3 Overfitting and Underfitting

Overfitting occurs when a model learns the training data too well, including its noise and outliers, leading to poor performance on unseen data. On the other hand, underfitting happens when a model is too simplistic and fails to capture underlying patterns in the data. Both scenarios result in poor predictive performance and can be difficult to manage, particularly with complex models.

Careful model selection, cross-validation, and regular performance evaluation are essential to address these issues.

5.6.4 Need for High-Quality and Relevant Data

The quality and relevance of the data used to train predictive models significantly influence their accuracy. Poor quality data, such as data with many missing or erroneous values, can lead to inaccurate predictions. Likewise, if the data is not representative of the problem at hand, the model may fail to generalize well to new data. Therefore, robust data collection, preprocessing methods, and domain expertise are vital for developing effective predictive quality models.

Addressing these challenges is crucial for the successful application of Predictive Quality models. By acknowledging and addressing these limitations, organizations can better harness the power of Predictive Quality to drive improvements and innovation.

5.7 The Future of Predictive Quality

The future of Predictive Quality (PQ) heralds a promising revolution across multiple sectors, propelled by burgeoning digitalization, developments in machine learning (ML) and AI, and the escalating accessibility of data. These factors synergistically enhance the accuracy, efficiency, and broad-based usability of PQ models [48].

PQ is anticipated to be a pivotal element in the global paradigm shift toward proactive and predictive business operations, departing from the more traditional, reactive methodologies [49]. It is predicted that PQ will gain ground in quality assurance processes.

With the evolution of ML interpretability, the mystifying "black box" aspect of intricate models is set to loosen up, fostering improved comprehension and confidence in their predictions [50]. The emergence of more sophisticated, adaptable, and resilient models is also expected to resolve the predicaments of overfitting and underfitting, thereby ensuring reliable and precise predictions [51].

Furthermore, the introduction of advanced data preprocessing and analytical tools will make the predictive model more robust toward imperfect data [52]. This will not only amplify the precision of PQ models but will also extend their usage to scenarios with limited or imperfect data.

In the context of practical applications, PQ models are projected to infiltrate more deeply into industries already employing them and discover fresh applications in sectors that are yet to unlock their full potential. For example, the healthcare sector could witness a shift toward more personalized and proactive care guided by predictions about individual health outcomes [53].

The potentialities are immense, and the influence of PQ in fueling efficiency, profitability, and innovation is poised to grow exponentially. As we progress in technological advancements and innovations, it becomes imperative to critically evaluate the ethical and societal implications of these technologies to ensure their responsible use, fostering benefits for all strata of society.

References

1 Eckerson, W.W. (2007). Predictive analytics. Extending the value of your data warehousing investment. *TDWI Best Practices Report*, 1, pp. 1–36.
2 Kumar, V. and Garg, M.L. (2018). Predictive analytics: a review of trends and techniques. *International Journal of Computer Applications* 182 (1): 31–37.

3 Semssar, S.A. (2022). Predictive quality analytics. Doctoral dissertation. Purdue University Graduate School.

4 Nalbach, O., Linn, C., Derouet, M., and Werth, D. (2018). Predictive quality: towards a new understanding of quality assurance using machine learning tools. In: *Business Information Systems: 21st International Conference, BIS 2018*, Berlin, Germany (18–20 July 2018), Proceedings 21, 30–42. Springer International Publishing.

5 Liu, H. and Motoda, H. (ed.) (2007). *Computational Methods of Feature Selection*. CRC Press.

6 Choudhary, A.K., Harding, J.A., and Tiwari, M.K. (2009). Data mining in manufacturing: a review based on the kind of knowledge. *Journal of Intelligent Manufacturing* 20: 501–521.

7 Wang, J., Ma, Y., Zhang, L. et al. (2018). Deep learning for smart manufacturing: methods and applications. *Journal of Manufacturing Systems* 48: 144–156.

8 Nithya, B. and Ilango, V. (2017). Predictive analytics in health care using machine learning tools and techniques. In: *2017 International Conference on Intelligent Computing and Control Systems (ICICCS)*, 492–499. IEEE.

9 Palaniappan, S. and Awang, R. (2008). Intelligent heart disease prediction system using data mining techniques. In: *2008 IEEE/ACS International Conference on Computer Systems and Applications*, 108–115. IEEE.

10 Malini, N. and Pushpa, M. (2017). Analysis on credit card fraud identification techniques based on KNN and outlier detection. In: *2017 Third International Conference on Advances in Electrical, Electronics, Information, Communication and Bio-Informatics (AEEICB)*, 255–258. IEEE.

11 Broby, D. (2022). The use of predictive analytics in finance. *The Journal of Finance and Data Science* 8: 145–161.

12 Wang, R.Y., Ziad, M., and Lee, Y.W. (2006). *Data Quality*, vol. 23. Springer Science & Business Media.

13 Pipino, L.L., Lee, Y.W., and Wang, R.Y. (2002). Data quality assessment. *Communications of the ACM* 45 (4): 211–218.

14 Box, G.E., Jenkins, G.M., Reinsel, G.C., and Ljung, G.M. (2015). *Time Series Analysis: Forecasting and Control*. Wiley.

15 Asteriou, D. and Hall, S.G. (2011). ARIMA models and the Box–Jenkins methodology. *Applied Econometrics* 2 (2): 265–286.

16 Quinlan, J.R. (1986). Induction of decision trees. *Machine learning* 1: 81–106.

17 Charbuty, B. and Abdulazeez, A. (2021). Classification based on decision tree algorithm for machine learning. *Journal of Applied Science and Technology Trends* 2 (01): 20–28.

18 Ho, T.K. (1995). Random decision forests. In: *Proceedings of 3rd International Conference On Document Analysis and recognition*, vol. 1, 278–282. IEEE.

19 Ho, T.K. (1998). The random subspace method for constructing decision forests. *IEEE Transactions on Pattern Analysis and Machine Intelligence* 20 (8): 832–844.

20 Cortes, C. and Vapnik, V. (1995). Support-vector networks. *Machine Learning* 20: 273–297.

21 Boser, B.E., Guyon, I.M., and Vapnik, V.N. (1992). A training algorithm for optimal margin classifiers. In: *Proceedings of the Fifth Annual Workshop on Computational Learning Theory*, 144–152. ACM (Association for Computing Machinery).

22 Friedman, J.H. (2002). Stochastic gradient boosting. *Computational Statistics & Data Analysis* 38 (4): 367–378.

23 Cover, T. and Hart, P. (1967). Nearest neighbor pattern classification. *IEEE Transactions on Information Theory* 13 (1): 21–27.

24 Deng, Z., Zhu, X., Cheng, D. et al. (2016). Efficient kNN classification algorithm for big data. *Neurocomputing* 195: 143–148.

25 Anderson, J.A. (1995). *An Introduction to Neural Networks*. MIT Press.

26 Hassoun, M.H. (1995). *Fundamentals of Artificial Neural Networks*. MIT Press.

27 LeCun, Y., Bengio, Y., and Hinton, G. (2015). Deep learning. *Nature* 521 (7553): 436–444.

28 Bengio, Y., Goodfellow, I., and Courville, A. (2017). *Deep Learning*, vol. 1. Cambridge, MA: MIT Press.

29 Svozil, D., Kvasnicka, V., and Pospichal, J. (1997). Introduction to multi-layer feed-forward neural networks. *Chemometrics and Intelligent Laboratory Systems* 39 (1): 43–62.

30 Sazli, M.H. (2006). A brief review of feed-forward neural networks. *Communications Faculty of Sciences University of Ankara Series A2-A3 Physical Sciences and Engineering* 50 (01): 11–17.

31 Shams, I., Ajorlou, S., and Yang, K. (2015). A predictive analytics approach to reducing 30-day avoidable readmissions among patients with heart failure, acute myocardial infarction, pneumonia, or COPD. *Health Care Management Science* 18: 19–34.

32 Escobar, C.A. and Morales-Menendez, R. (2018). Machine learning techniques for quality control in high conformance manufacturing environment. *Advances in Mechanical Engineering* 10 (2), 1687814018755519.

33 Tsai, T.N. (2012). Development of a soldering quality classifier system using a hybrid data mining approach. *Expert Systems with Applications* 39 (5): 5727–5738.

34 Escobar, C.A., Macias-Arregoyta, D., and Morales-Menendez, R. (2023). The decay of Six Sigma and the rise of Quality 4.0 in manufacturing innovation. *Quality Engineering* 1–20. https://doi.org/10.1080/08982112.2023.2206679.

35 Ellahham, S. (2020). Artificial intelligence: the future for diabetes care. *The American Journal of Medicine* 133 (8): 895–900.

36 Lopez-Jimenez, F., Attia, Z., Arruda-Olson, A.M. et al. (2020). Artificial intelligence in cardiology: present and future. In: *Mayo Clinic Proceedings* (Vol. 95, No. 5, 1015–1039. Elsevier.

37 VA, DeepMind develop machine learning system to predict life-threatening disease before it appears, VA New Release (31 July 2019). https://news.va.gov/press-room/va-deepmind-develop-machine-learning-system-to-predict-life-threatening-disease-before-it-appears (accessed 28 July 2023).

38 Tomašev, N., Glorot, X., Rae, J.W. et al. (2019). A clinically applicable approach to continuous prediction of future acute kidney injury. *Nature* 572 (7767): 116–119.

39 Rastogi, R. (2018). Machine learning@ amazon. In: *The 41st International ACM SIGIR Conference on Research & Development in Information Retrieval*, 1337–1338. ACM (Association for Computing Machinery).

40 Marr, B. (2017). Really big data at Walmart: real-time insights from their 40+ petabyte data cloud. *Forbes* (23 January). https://www.forbes.com/sites/bernardmarr/2017/01/23/really-big-data-at-walmart-real-time-insights-from-their-40-petabyte-data-cloud/?sh=a9144436c105.

41 Uberoi, R. (2017) ZARA: achieving the "Fast" in fast fashion through analytics. *Digital Innovation and Transformation*. Harvard University (5 April). https://d3.harvard.edu/platform-digit/submission/zara-achieving-the-fast-in-fast-fashion-through-analytics.

42 Arya, S., Eckel, C., and Wichman, C. (2013). Anatomy of the credit score. *Journal of Economic Behavior & Organization* 95: 175–185.

43 Knorr, E. (2015). Ahead of the Curve, How PayPal beats the bad guys with machine learning. *InfoWorld* (13 April). https://www.infoworld.com/article/2907877/how-paypal-reduces-fraud-with-machine-learning.html.

44 Peng, N.Y. (2020). How renaissance beat the markets with machine learning. *Towards Data Science* (2 January). https://towardsdatascience.com/how-renaissance-beat-the-markets-with-machine-learning-606b17577797.

45 Warren, T. (2020). How Microsoft tackles the 30,000 bugs its 47,000 developers generate each month. *The Verge* (22 April). https://www.theverge.com/2020/4/22/21230816/microsoft-developers-bugs-machine-learning-numbers-statistics.

46 Wiggers, K. (2020). Google details how it's using AI and machine learning to improve search. *VentureBeat* (15 October). https://venturebeat.com/ai/google-details-how-its-using-ai-and-machine-learning-to-improve-search

47 Thiyagarajan, P. (2020). A review on cyber security mechanisms using machine and deep learning algorithms. In: *Handbook of Research on Machine and Deep Learning Applications for Cyber Security* (ed. P. Thiyagarajan), 23–41. IGI Global.

48 Chui, M., Manyika, J., and Miremadi, M. (2016). Where machines could replace humans-and where they can't (yet). *The McKinsey Quarterly*, pp. 1–12. https://www.mckinsey.com/capabilities/mckinsey-digital/our-insights/where-machines-could-replace-humans-and-where-they-cant-yet (accessed 28 August 2023).

49 Kiron, D., Prentice, P.K., and Ferguson, R.B. (2014). The analytics mandate. *MIT Sloan Management Review* 55 (4): 1.

50 Doshi-Velez, F. and Kim, B. (2017). Towards a rigorous science of interpretable machine learning. *arXiv preprint* arXiv:1702.08608. https://arxiv.org/abs/1702.08608 (accessed 28 August 2023).

51 Goodfellow, I., Bengio, Y., and Courville, A. (2016). *Deep Learning*. MIT Press.

52 Fan, C., Chen, M., Wang, X. et al. (2021). A review on data preprocessing techniques toward efficient and reliable knowledge discovery from building operational data. *Frontiers in Energy Research* 9: 652801.

53 Chen, L. (2020). Overview of clinical prediction models. *Annals of Translational Medicine* 8 (4): 71.

6

Data Quality

6.1 Introduction

As we traverse the landscape of data-driven decision-making, the old adage "Garbage In, Garbage Out" rings more true than ever. The caliber of data leveraged to guide decisions, sculpt strategies, and drive operations remains a prime consideration for contemporary organizations. Hence, the understanding and enhancement of data quality form the bedrock of successful data management, emerging as a focal point for quality professionals.

Fundamentally, data quality gauges the state of data based on criteria such as accuracy, completeness, consistency, reliability, and relevancy. High-quality data should not only be precise but also pertinent for its intended purpose, and it needs to be delivered promptly and in an intelligible format.

For quality professionals, safeguarding data quality is central to their duties. Superior quality data lays a robust foundation for processes and systems, curtails operational inefficiencies, and bolsters the reliability of outcomes. These professionals shoulder crucial responsibilities in defining data governance policies, setting standards for data collection, and deploying practices to monitor and enhance data quality. Armed with high-quality data, quality professionals can produce more accurate analyses, make well-informed decisions, and heighten overall operational efficacy.

The dawn of the digital age has amplified the significance of data quality exponentially. The rise of big data, artificial intelligence, and advanced analytics has thrust data into the limelight, transforming it into a potent force of innovation and competitive advantage. However, to unlock the full potential of these cutting-edge technologies, it is vital that the underlying data maintains a high-quality standard. Substandard data quality can skew insights, lead to incorrect conclusions, and, ultimately, give rise to misguided strategies and actions. In an epoch characterized by digital transformation, preserving high data quality is not merely a best practice—it is a business imperative.

Historically, the importance of data quality began to surge with the advent of computer systems in the 1960s and 1970s [1, 2]. As more business processes were automated, the impact of poor data quality on operations and decision-making became increasingly apparent [3, 4]. Over the years, as businesses started relying more on data, there was a growing realization of the need for dedicated resources and processes to ensure and improve data quality [5, 6].

As we venture further into this chapter, we will delve into the nuances of data quality, its repercussions on various sectors, and tactics to ascertain and elevate it. We'll also examine the challenges thrown up by subpar data quality and the potential cost organizations may incur due to its oversight. Through this journey, we aim to emphasize the pivotal role of data quality in harnessing the true power of data in today's digital epoch.

Quality in the Era of Industry 4.0: Integrating Tradition and Innovation in the Age of Data and AI,
First Edition. Kai Yang.

This chapter comprises the following sections:

Section 6.2 will spotlight the types of data susceptible to quality issues, common reasons for poor data quality, and the potential costs and impacts of low-quality data.

Section 6.3 will discuss data quality dimensions and measurable metrics employed by information management professionals to gauge and improve the quality of data in comparison to a well-defined standard.

Sections 6.4 and 6.5 will delve into data quality management and governance, exploring strategies, tools, and techniques such as data governance policies, data cleansing, and data auditing, which are instrumental in managing and enhancing data quality.

Section 6.6 will underscore the role of quality professionals in data quality, encompassing setting data standards, monitoring data quality, and training staff on data management best practices.

Finally, Section 6.7 will look ahead to future trends in data quality, including the impact of emerging technologies, regulatory alterations, and shifts in business practices.

6.2 Data and Data Quality

6.2.1 Overview

As we delve into the complex world of data quality, it is essential to first ground ourselves in understanding the foundational concepts. Data quality is not a one-dimensional concept, nor is it a static one. It encompasses a range of characteristics, extends over different types of data, and its requirements can shift depending on its use. In this section, we begin our exploration by defining the types of data involved in data quality study and providing a comprehensive definition of data quality itself.

6.2.1.1 Data Involved in Data Quality Study

Data quality studies encompass various types of data depending on the specific context and requirements of the study. Data can be broadly categorized into Metadata, Transactional Data, Master Data, Reference Data, Operational Data, Analytical Data, Unstructured Data, Semi-structured Data, Big Data, Real-time Data, Historical Data, and Sensitive Data [7, 8]. Each of these categories has unique characteristics and applications, and thus, their quality needs to be assessed in accordance with their intended purpose and use. These data categories will be described in detail later.

Metadata, for instance, provides information about other data, such as when and by whom the data was collected and for what purpose. Transactional data, on the other hand, involves the data generated during transactions like customer purchase history or web browsing history. Master data refers to the core data about an organization's business entities, while reference data is used to categorize other data. Each of these categories, and others not listed here, are integral to the overall data ecosystem of an organization and have specific data quality requirements that need to be addressed.

6.2.1.2 Definition of Data Quality

Data quality refers to the degree to which a set of data fulfills the requirements of its intended use. This often encompasses factors such as accuracy, completeness, consistency, reliability, timeliness, and relevancy. Accuracy refers to whether the data is correct and free from errors. Completeness is about whether all necessary data is present. Consistency pertains to whether data is uniform across different data sets, while reliability considers whether the data is trustworthy and dependable. Timeliness addresses whether the data is up-to-date and available when needed, and relevancy examines whether the data fits the needs of the current context or task. In Section 6.3, I will give some detailed information about how to measure the data quality quantitatively, given the actual data set.

High-quality data is not just about having error-free data; it is about having the right data in the right form at the right time for the right purpose. Ensuring data quality, therefore, is a multi-faceted process requiring continuous monitoring and improvement to adapt to changing needs and circumstances.

6.2.2 Categories of Data

Data quality study generally encompasses various types of data. Here is a list of key data categories that can be involved:

1) **Metadata**: This includes data that provides information about other data. For instance, it might indicate when and by whom a particular set of data was collected, and for what purpose.

Example 6.1   Metadata

Metadata is often referred to as "data about data." It provides information about a certain set of data including its means of creation, purpose, time and date of creation, creator or author, location on a computer network where the data was created, and standards used.

For instance, consider a photograph taken with a digital camera. The photograph itself (the digital file) is the primary data, but the digital camera also records metadata about the photograph. Here is an example of such metadata (Table 6.1):

Table 6.1   Metadata Example

Metadata Type	Example Metadata
Date and time of creation	2023-05-20 14:45
Camera make and model	Canon EOS 5D Mark IV
Aperture	f/2.8
Shutter speed	1/200 s
ISO speed	ISO-400
Focal length	70 mm
Flash used	No
Orientation	Landscape
GPS (Location where the photograph was taken)	40.7128° N, 74.0060° W
File size	5.6 MB
File format	JPEG

In this example, each row of the table is a type of metadata about the photograph. This metadata can be helpful for various purposes, such as organizing photographs, searching for a specific photograph, validating the authenticity of the photograph, or adjusting the camera settings for future photographs. Some software even allows users to edit the metadata, though doing so can potentially undermine its value as a record of the photograph's history and original conditions.

2) **Transactional Data**: These are the data generated during transactions. They could include customer purchase history, web browsing history, and so on. Transactional data is usually collected during business operations and is generally unique to each transaction or event.

Example 6.2 Transactional Data

Consider an example from an online e-commerce website. Every time a customer places an order, the system captures several pieces of transactional data related to that order. Here is how that data might look:

Table 6.2 Example of Transactional Data

Order ID	Customer ID	Product ID	Product Name	Quantity	Price per Item	Total Price	Transaction Date
1001	C123	P567	Running shoes	1	$120.00	$120.00	2023-05-01
1002	C456	P890	Yoga mat	2	$30.00	$60.00	2023-05-02
1003	C789	P321	Sports bottle	1	$15.00	$15.00	2023-05-03

In this example, each row of the Table 6.2 represents a unique transaction (an order), and the columns provide different types of transactional data about the order. This includes data such as the order ID, customer ID, product ID, product name, quantity ordered, price per item, total price, and the date of the transaction. This kind of transactional data is crucial for keeping track of sales, inventory management, understanding customer behavior, financial reporting, and many other aspects of business operations.

---

3) **Master Data**: This is the consistent and uniform set of identifiers and extended attributes that describe the core entities of the enterprise, such as customers, suppliers, products, sales, etc.

Example 6.3 Master Data

Consider a simplified example of master data from a business perspective, focusing on customers and products (Tables 6.3a and 6.3b):

Table 6.3a Customer Master Data

Customer ID	First Name	Last Name	Email Address	Phone Number	Billing Address
C123	John	Doe	john.doe@example.com	555-123-4567	123 Main St, NY
C456	Jane	Smith	jane.smith@example.com	555-765-4321	456 Elm St, CA
C789	Bob	Johnson	bob.johnson@example.com	555-987-6543	789 Pine St, TX

Table 6.3b Product Master Data

Product ID	Product Name	Category	Unit Price	Manufacturer
P567	Running shoes	Footwear	$120.00	Footwear Inc.
P890	Yoga mat	Fitness	$30.00	Fitness Co.
P321	Sports bottle	Beverage	$15.00	Beverage Ltd.

In these examples, the Customer Master Data table includes information about the custom-ers of the business. This could include data fields such as Customer ID, First Name, Last Name, Email Address, Phone Number, and Billing Address.

Similarly, the Product Master Data table includes information about the products that the business sells. This could include fields such as Product ID, Product Name, Category, Unit Price, and Manufacturer.

This master data remains relatively constant over time, and it is crucial to various business operations and decision-making processes. It is important to note that maintaining the quality of this data is vital, as errors can propagate across different systems and processes, leading to widespread issues and inaccuracies.

4) **Reference Data**: This is a subset of master data used to classify or categorize other data, used to facilitate the identification, retrieval, or use of data. It generally remains static and is non-transactional in nature.

Example 6.4   Reference Data

Consider an example from a global business perspective. In international business operations, we often need to deal with various types of currencies. To handle these operations, we would need a standard set of currency codes that can be used across the organization. This could serve as our reference data (Table 6.4).

Table 6.4   Reference Data Example

Currency Code	Currency Name	Country
USD	United States Dollar	United States
EUR	Euro	European Union
JPY	Japanese Yen	Japan
GBP	British Pound	United Kingdom
AUD	Australian Dollar	Australia
CAD	Canadian Dollar	Canada
CHF	Swiss Franc	Switzerland
CNY	Chinese Yuan	China
SEK	Swedish Krona	Sweden
NZD	New Zealand Dollar	New Zealand

In this table, each row represents a type of currency, and the columns provide the standard currency code, the name of the currency, and the country or region where it is used.

This kind of reference data ensures consistency across the organization and makes it pos-sible to easily categorize and analyze data from different systems or regions. For example, if a company is analyzing its global sales data, it could use this reference data to convert all sales figures into a common currency for comparison purposes.

5) **Operational Data**: This type of data supports the operation of a particular process or system. It could include system logs, real-time process data, etc.

Example 6.5   Operational Data

Consider a logistics company that needs to keep track of all the shipments it is currently handling. The operational data for this company might look something like this (Table 6.5):

Table 6.5   Operational Data Example

Shipment ID	Vehicle ID	Driver ID	Origin	Destination	Departure Time	Expected Arrival	Current Status
S001	V101	D345	London	Paris	15 2023-05-15 09:00	2023-05-15 18:00	En route
S002	V102	D678	Berlin	Madrid	2023-05-15 08:00	2023-05-15 20:00	Delayed
S003	V103	D912	Rome	Munich	2023-05-15 10:00	2023-05-15 17:00	Delivered

In this example, each row of the table represents a shipment, and the columns provide various pieces of operational data about each shipment: the shipment ID, vehicle ID, driver ID, origin and destination of the shipment, departure time, expected arrival time, and the current status of the shipment.

This operational data is crucial for the company's operations, as it allows them to track shipments in real-time, anticipate and resolve delays, manage their fleet of vehicles, and ensure that their service is running smoothly. It also provides a valuable source of information for analysis and decision-making, such as identifying operational bottlenecks or optimizing delivery routes.

---

6) **Analytical Data**: This type of data is used for supporting decision-making processes, predictions, and insights. For example, data used in business intelligence tools, predictive analytics, and machine learning models.

7) **Unstructured Data**: This is data that does not have a predefined data model or is not organized in a predefined manner. It includes data from emails, word-processing documents, videos, photos, social media posts, etc.

8) **Semi-structured Data**: This type of data does not conform to the formal structure of data models, but it contains tags or other markers to enforce hierarchy and relationships among elements. Examples include XML, JSON data, etc.

9) **Big Data**: This is characterized by volume, variety, and velocity. It can include anything from logs to social media data to the Internet of Things (IoT) data.

10) **Real-time Data**: This is data that is created, processed, stored, and accessed in real-time. It often informs real-time decision-making and responses.

11) **Historical Data**: Data about past events and changes in the systems. It is often used for trend analysis, forecasting, and other types of statistical analysis.

12) **Sensitive Data**: This includes data that, if disclosed, could harm individuals or organizations. This could include personal data like Social Security numbers, confidential corporate data, etc.

The relevance of these categories in a data quality study depends on the specific context and requirements of the study.

6.2.3 Causes of Poor Data Quality

Poor data quality can be caused by a wide range of factors, including:

1) **Human Error**: This is one of the most common causes of poor data quality. Mistakes made during data entry, misunderstanding of data fields, or typos can all introduce errors.
2) **Inconsistent Data Entry Standards**: If data entry standards are not clearly defined or consistently applied, discrepancies can occur. For example, date formats might vary, or different abbreviations might be used for the same term.
3) **Lack of Data Governance**: Without strong data governance policies and procedures, data quality can quickly deteriorate. This includes defining who is responsible for maintaining data quality, setting data standards, and monitoring compliance with these standards.
4) **System or Software Errors**: Errors in the systems used to collect, process, store, and retrieve data can also contribute to poor data quality. This might include bugs in the software or failures in the hardware.
5) **Data Integration Issues**: When integrating data from different sources, there may be inconsistencies and discrepancies if different data standards or formats are used. For instance, one system might use different naming conventions or coding schemes from another.
6) **Incomplete Data**: If data is missing or incomplete, this can also lead to poor data quality. This might occur if fields are not filled out, if data collection is interrupted, or if certain data is not collected at all.
7) **Obsolete or Outdated Data**: Data that is not regularly updated can become outdated, which can also contribute to poor data quality.
8) **Lack of Staff Training**: If staff members do not have adequate training on data collection, data entry, and data management practices, this can lead to errors and inconsistencies in the data.
9) **Lack of Quality Assurance Processes**: Without processes in place to regularly check, clean, and validate data, errors, and inaccuracies can go unnoticed and uncorrected.
10) **Poor Data Architecture or Database Design**: If the underlying data architecture or database design is flawed, this can lead to issues with data quality. For example, if the design does not accommodate all the necessary data fields, or if it does not properly link related records, this can introduce errors and inconsistencies.

Addressing these issues typically involves developing a comprehensive data management strategy that includes clear data governance policies, consistent data standards, regular data quality checks, and ongoing staff training.

6.2.4 Cost of Poor Data Quality

Poor data quality can have significant costs, both direct and indirect, for organizations. Here are a few ways in which these costs can manifest:

1) **Operational Inefficiencies**: Poor data quality can lead to errors and inefficiencies in operations, slowing down processes and requiring time and resources to correct. This also includes the time spent identifying and resolving data quality issues.
2) **Poor Decision-Making**: Data is crucial for informed decision-making. If the data is of poor quality, the decisions based on that data will likely be flawed, potentially leading to lost opportunities, inefficient resource allocation, and other negative outcomes.

3) **Reputational Damage**: If poor data quality leads to errors that affect customers or other stakeholders, this can harm the organization's reputation. In some cases, this could also lead to loss of customer trust and potentially loss of business.

4) **Regulatory Fines**: Particularly in regulated industries, poor data quality can lead to noncompliance with laws and regulations, resulting in hefty fines or other penalties.

5) **Increased Costs**: Poor data quality can lead to redundant work, such as the need to re-enter or clean data, and increased costs related to manual data fixing efforts.

6) **Customer Dissatisfaction**: If poor data quality impacts customer service—for example, by leading to errors in billing or failures in service delivery—this can lead to customer dissatisfaction and churn.

7) **Inability to Leverage Data**: Organizations with poor data quality may find it more challenging to leverage their data for beneficial purposes, such as analytics, forecasting, and machine learning, limiting their ability to derive valuable insights and benefits from their data.

8) **Loss of Competitive Advantage**: In today's data-driven business environment, organizations that are unable to maintain high-quality data may find themselves at a competitive disadvantage.

The exact cost of poor data quality can vary widely depending on the specific circumstances and the size and nature of the organization. However, some studies suggest that the cost can be significant. For example, a report from IBM estimated that the annual cost of poor data quality in the US alone could be up to $3.1 trillion [9].

6.3 Data Quality Dimensions and Measurement

In our journey through the complex terrain of data quality, it is paramount to comprehend its multi-dimensional character. Data quality is far from being a singular, binary attribute; rather, it comprises several dimensions, each contributing to the overall utility and credibility of the data. These dimensions—such as accuracy, completeness, consistency, timeliness, and relevancy—underpin the essence of data quality, offering a structured framework for its understanding, assessment, and enhancement.

However, merely understanding these dimensions is not sufficient; their effective measurement is equally crucial. Quantitative evaluation of data quality dimensions sheds light on the data's health, providing vital insights that drive informed decision-making and strategic planning.

In this section, we will delve into these various data quality dimensions and examine the methodologies employed to measure them [10–12], offering an all-encompassing perspective on the data quality landscape.

6.3.1 Data Quality Dimensions

Data Quality Dimensions are characteristics or aspects used to measure or describe the quality of data. They provide a way to understand and evaluate how well a dataset fits the purposes for which it is intended. Here are some of the key dimensions often used in assessing data quality:

1) **Accuracy**: This refers to how well the data reflects the real-world entities or phenomena it is supposed to represent. Accurate data closely matches the actual values.

2) **Completeness**: This refers to the extent that all required data is present in the dataset. Missing data can affect the reliability of conclusions drawn from the dataset.

3) **Consistency**: This refers to how well data values are maintained across the dataset and over time. Inconsistent data can lead to conflicting reports and analytics.

4) **Timeliness**: This refers to whether data is up-to-date and available when needed. Timely data allows for real-time decision-making and responsive business practices.

5) **Validity**: This refers to whether the data conforms to the required syntax (format, type, and range of values). Invalid data can lead to errors in data processing and analytics.

6) **Uniqueness**: This refers to the absence of unnecessary duplicate entries in the dataset. Too many duplicates can misrepresent the data and waste storage resources.

7) **Integrity**: This refers to the structural accuracy and consistency of the dataset. For instance, does the data in one field logically correspond to the data in another?

8) **Relevance**: This refers to the usefulness of the data in relation to the purpose for which it is collected and used. Irrelevant data can cause confusion and waste resources.

9) **Reliability**: This refers to the degree to which data remains accurate and consistent over its lifecycle.

Understanding and managing these data quality dimensions is crucial for any organization that uses data to inform decision-making, optimize operations, or provide services. Each dimension provides a different lens through which to assess the fitness and reliability of the data.

6.3.2 Measurement of Data Quality

The measurement of data quality through its various dimensions is a pivotal aspect of data management. This process serves as a diagnostic tool, offering detailed insights into the state of data and illuminating potential areas for improvement. However, assessing data quality is not a one-size-fits-all task. Different dimensions require unique metrics and methods of measurement. From evaluating the accuracy of data entries, gauging the completeness of data sets, ensuring consistency across different data sources, to verifying the timeliness and relevancy of data, each dimension offers a specific lens to view and measure the quality of data. This journey into measuring data quality dimensions is an essential step toward fostering trust in data and making it a reliable cornerstone for decision-making and strategy formulation. I will go over a step-by-step procedure on how to measure data quality through these dimensions.

6.3.2.1 Measuring Accuracy in Data Quality

1) **Define**: The first step in measuring accuracy is to define what constitutes an accurate data point in the context of the specific data set you are dealing with. This will be the "gold standard" against which you will compare your data. For example, let us consider a simple data set that contains customer information, including customer names and email addresses. An accurate data point, in this case, would be a correctly spelled name and a valid email address.

2) **Identify and Isolate**: Next, identify and isolate the data points you want to measure for accuracy. Using our customer information example, you might measure the accuracy of the customer names or email addresses.

3) **Compare**: Compare each data point against the "gold standard" you defined in the first step. This might involve checking each customer's name for spelling errors or verifying that each email address is in a valid format.

4) **Calculate Accuracy Rate:** The accuracy rate can be calculated as the number of accurate data points divided by the total number of data points. Multiply this by 100 to get a percentage. For example, if you have a data set of 100 customer email addresses and 95 are in the correct format, your accuracy rate would be $(95/100)* 100 = 95\%$.

5) **Address Inaccuracies:** Once you have measured the accuracy of your data, the next step is to address any inaccuracies you have found. This could involve correcting spelling errors in customer names or fixing improperly formatted email addresses.

6) **Continuous Monitoring**: Keep in mind that data accuracy is not a one-time process. It requires continuous monitoring and validation to maintain high-quality data.

It is also important to note that while accuracy is a key data quality dimension, it is not the only one. A data set might be highly accurate but still have issues with completeness, consistency, timeliness, or relevancy. Therefore, it is important to measure and monitor all data quality dimensions regularly.

6.3.2.2 Measure Completeness in Data Quality

The procedure for Measuring Completeness in Data Quality is as follows:

1) **Define**: Completeness refers to the absence of null or empty values in a data set. The first step in measuring completeness is to identify the data fields that should always contain data. For instance, in a dataset of customer details, fields such as "Customer ID," "First Name," "Last Name," and "Email Address" are typically expected to always have values.
2) **Identify and Isolate**: After defining what completeness means for your data set, identify the fields that need to be checked. For our example, these would be the aforementioned fields in the customer details data set.
3) **Count Empty Values**: Count the number of empty or null values in each of these fields.
4) **Calculate Completeness**: Completeness can be calculated by subtracting the number of empty values from the total number of values, dividing by the total number of values, and then multiplying by 100 to get a percentage.
5) **Address Incompleteness**: After measuring completeness, take steps to fill in the missing values where possible. This might involve reaching out to the customers to update their details or checking other sources for missing information.
6) **Continuous Monitoring**: As with accuracy, completeness should be continuously monitored and validated to ensure data quality over time.

As always, keep in mind that completeness is just one dimension of data quality. It is important to measure and monitor all aspects of data quality to get a holistic view of the health of your data.

Example 6.6   Measuring Completeness for an E-commerce Data Set

Here is a dataset from an e-commerce company. This dataset includes customer orders with fields like "Order ID," "Customer ID," "Product ID," "Quantity," "Price," "Shipping Address," and "Shipping Method."

Suppose we exam a segment of the data set as follows (Table 6.6):

Table 6.6   Example of Measuring Completeness

Order ID	Customer ID	Product ID	Quantity	Price	Shipping Address	Shipping Method
O001	C123	P567	2	$240	123 Main St, NY	Standard
O002	C456	P890	1	$30	"n/a"	Express
O003	C789	P321	3	$45	789 Pine St, TX	"n/a"
O004	C123	P567	"n/a"	$120	123 Main St, NY	Standard
O005	C456	P890	1	$30	456 Elm St, CA	Express

In this case, completeness implies that all fields for each order should be populated with valid data. We will check the completeness of the "Quantity," "Shipping Address," and "Shipping Method" fields. Then we count the number of empty or "n/a" values in each of these fields. Here, "Quantity" has 1, "Shipping Address" has 1, and "Shipping Method" has 1 missing values in Table 6.6. If we are looking at the completeness for the entire dataset, we would add up the total missing values (3 in this case), and use the total number of values (15 in this case, as we have 5 rows and 3 columns we are checking). So the completeness would be ((15 − 3)/15)*100 = 80%.

6.3.2.3 Measure Consistency in Data Quality

Procedure for Measuring Consistency in Data Quality is as follows:

1) **Define**: Consistency in data quality refers to the absence of conflict within a data set or between multiple data sets. The first step is to define the standards or rules that your data should adhere to. For instance, in a customer database, one rule might be that a single customer ID should always refer to the same individual.
2) **Identify and Isolate**: Identify the data points that need to be checked for consistency. For a customer database, this might involve checking customer IDs, email addresses, and phone numbers.
3) **Check for Conflicts**: You will need to compare data points to each other to find any conflicts. This could involve checking if a single customer ID is linked to different names, addresses, or other details.
4) **Calculate Consistency**: Calculate the percentage of data points that are consistent based on the total number of data points. For example, if there are 5 inconsistencies out of 100 customer IDs, the consistency rate would be 95%.
5) **Address Inconsistencies**: Once inconsistencies are identified, you will need to take steps to resolve them. This might involve deduplication or correction of conflicting data points.
6) **Continuous Monitoring:** Data consistency should be regularly monitored to ensure that the database remains reliable over time.

Example 6.7   Consistency in Data Quality

This example involves an organization's customer database. Suppose we have two different systems that keep track of customer data and want to ensure consistency between them (Tables 6.7a and 6.7b).

Table 6.7a   Customer Database – System A

Customer ID	Name	Email	Phone Number
C001	Alice	alice@email.com	1234567890
C002	Bob	bob@email.com	2345678901
C003	Carol	carol@email.com	3456789012

Table 6.7b   Customer Database – System B

Customer ID	Name	Email	Phone Number
C001	Alice	alice@email.com	1234567890
C002	Bob	bob123@email.com	2345678901
C003	Carol	different@email.com	4567890123

In this example, there is a difference in Bob's email and Carol's email and phone number across the two systems. This inconsistency needs to be addressed for reliable customer communication and accurate analysis of customer data.

6.3.2.4 Measure Timeliness in Data Quality

Procedure for Measuring Timeliness in Data Quality

1) **Define**: Timeliness in data quality refers to the degree to which the data is up-to-date and available when required. It is important to define what is considered "timely" for your data. For instance, in a sales database, a new sale might be considered timely if it is recorded within 24 hours of the transaction.
2) **Identify and Isolate**: Identify the data points that need to be checked for timeliness. For a sales database, you might check the "transaction date" field.
3) **Check for Delays**: Look for any data points that were not recorded within the defined timely period. For instance, find any sales that were not recorded within 24 hours of the transaction.
4) **Calculate Timeliness**: Calculate the percentage of data points that are timely based on the total number of data points. For example, if there are 15 late entries out of 200 transactions, the timeliness rate would be 92.5%.
5) **Address Delays**: Once you have identified any delays in data recording, you will need to investigate the cause and take steps to prevent future delays. This might involve changes to data collection processes or systems.
6) **Continuous Monitoring**: Data timeliness should be regularly monitored to ensure that the database remains up-to-date and reliable.

Example 6.8   Measure Timeliness in Data Quality

This example involves a sales database for an online store. Suppose the database has fields like "Transaction ID," "Customer ID," "Transaction Time," and "Recorded Time."
Here is a simplified representation of our data (Table 6.8):

Table 6.8   Example of Timelines in Data

Transaction ID	Customer ID	Transaction Time	Recorded Time
T001	C123	2023-06-01 10:15:00	2023-06-01 11:30:00
T002	C456	2023-06-01 10:30:00	2023-06-01 15:45:00
T003	C789	2023-06-01 10:45:00	2023-06-01 10:55:00
T004	C123	2023-06-01 11:00:00	2023-06-02 10:00:00
T005	C456	2023-06-01 11:15:00	2023-06-01 11:25:00

If we set our timeliness standard as recording transactions within four hours, we see that Transaction T002 was recorded within that time frame, but Transaction T004 was not. We would then calculate the timeliness as 80% and would need to investigate why T004 was recorded late.

6.3.2.5 Measure Validity in Data Quality

Procedure for Measuring Validity in Data Quality is as follows:

1) **Define**: Validity in data quality refers to the degree to which the data conforms to the defined business rules or constraints. The first step in measuring validity is to establish these rules or constraints. For instance, in a patient dataset, a valid age might be defined as any integer between 0 and 120.
2) **Identify and Isolate**: Identify the data fields that need to be checked for validity. For our patient dataset, this would be the "age" field.
3) **Check for Invalid Data**: Compare each data point to your established rules. In our case, we would find any ages that are less than 0 or greater than 120, or that are not integers.
4) **Calculate Validity**: Calculate the percentage of data points that are valid. For instance, if there are 2 invalid ages out of 100 patients, the validity would be 98%.
5) **Address Invalid Data**: After identifying invalid data points, take steps to correct them where possible. This could involve checking the original source for errors, or asking the provider to confirm the information.
6) **Continuous Monitoring**: Regularly monitor the validity of data to maintain the overall quality of the database.

Example 6.9   Measure Validity in Data Quality

Consider an example involving a dataset from a hospital that includes "Patient ID," "Patient Name," "Age," "Height," and "Weight."
   Here is a simplified representation of our data (Table 6.9):

Table 6.9   Example of Validity in Data

Patient ID	Patient Name	Age	Height (cm)	Weight (kg)
P001	John	25	180	75
P002	Sarah	32	165	65
P003	Tim	−3	150	50
P004	Mary	30	"n/a"	55
P005	Paul	130	170	70

For this dataset, we have established that valid ages are integers between 0 and 120, valid heights are between 50 and 250 cm, and valid weights are between 10 and 200 kg. Looking at the data, we see that Tim's age and Mary's height are invalid, and Paul's age is also invalid. We can calculate the validity for each field and then take steps to correct the invalid entries.

6.3.2.6 Measure Uniqueness in Data Quality

Procedure for measuring uniqueness is as follows:

1) **Define**: Uniqueness in data quality refers to the absence of duplicate entries in a data set. For instance, in a customer database, uniqueness might be defined as having only one record per customer.
2) **Identify and Isolate**: Identify the fields that should be unique. For a customer database, this might be the "Customer ID" or "Email Address" field.

3) **Check for Duplicates**: Look for any records that have the same data in the fields that should be unique. In our example, you might find two records that have the same customer ID or email address.

4) **Calculate Uniqueness**: Calculate the percentage of unique data points. If there are 5 duplicate records out of 100 customers, the uniqueness would be 95%.

5) **Address Duplicates**: After identifying duplicate records, take steps to merge or delete them. This might involve creating a single record that combines data from the duplicates, or keeping the most recent record and deleting the others.

6) **Continuous Monitoring**: Regularly monitor the uniqueness of data to maintain the overall quality of the database.

Example 6.10   Measure Uniqueness in Data Quality

Let's consider a customer database with fields like "Customer ID," "Name," "Email," and "Phone Number."

Here is a simplified representation of our data (Table 6.10):

Table 6.10   Example of Uniqueness in Data

Customer ID	Name	Email	Phone Number
C001	Alice	alice@email.com	1234567890
C002	Bob	bob@email.com	2345678901
C003	Carol	carol@email.com	3456789012
C004	Alice	alice@email.com	1234567890
C005	Dave	dave@email.com	4567890123

From the data, we see that Customer ID C001 and C004 have the same name, email, and phone number, suggesting that these records are duplicates. We would calculate the uniqueness as 80% and then need to decide how to handle the duplicate records, whether by merging them or keeping the most recent.

6.3.2.7   Measuring Integrity in Data Quality

The procedure for measuring integrity in data quality is as follows:

1) **Define**: Integrity in data quality refers to the maintenance of consistency, accuracy, and trustworthiness of data over its entire life cycle. It is important to define the rules that uphold the integrity of your data. For instance, in a relational database of orders and products, one rule might be that every order must be linked to an existing product in the product table.

2) **Identify and Isolate**: Identify the data points that need to be checked for integrity. For an order database, you might check the "product_id" field in the order table against the "product_id" field in the product table.

3) **Check for Integrity Violations**: Look for any data points that violate your defined integrity rules. For instance, find any orders that are linked to a nonexistent product.

4) **Calculate Integrity**: Calculate the percentage of data points that uphold the integrity rules based on the total number of data points. For example, if there are 5 orders linked to nonexistent products out of 100 orders, the integrity would be 95%.

5) **Address Integrity Violations**: Once integrity violations are identified, you will need to take steps to correct them. This might involve checking the original data source for errors, updating the data, or even changing the integrity rules if they are found to be flawed.
6) **Continuous Monitoring**: Data integrity should be regularly monitored to ensure that the database remains consistent and trustworthy over time.

Example 6.11 Measuring Integrity in Data Quality

For this example, consider a relational database with two tables: one for products and one for orders (Table 6.11).

Table 6.11 Example of Integrity in Data

(a) Product Table

Product ID	Product Name	Price
P001	Book	10
P002	Pen	1
P003	Notebook	5

(b) Order Table

Order ID	Product ID	Quantity
O001	P001	2
O002	P003	1
O003	P004	3
O004	P001	1
O005	P002	5

In this case, we see that Order ID O003 is linked to a nonexistent Product ID P004. This is an integrity violation. To maintain high data quality, we would need to find the source of this error and correct it, perhaps by referencing the original order or contacting the person who placed the order to confirm the product ID.

6.3.2.8 Measuring Relevance

The procedure for measuring relevance in data quality is as follows:

1) **Define**: Relevance in data quality refers to the applicability and usefulness of the data in relation to the context in which it is used. For instance, in a market research project, relevant data might include customer demographics, purchase history, and product preferences, while data about the customers' favorite colors might be irrelevant.
2) **Identify and Isolate**: Identify the data fields that are relevant to your specific use case or project. For a market research project, this might include fields like "Age," "Gender," "Location," "Purchases," and "Product Preferences."
3) **Check for Irrelevant Data**: Identify any data fields that do not contribute to your use case or project. For our market research example, this might include fields like "Favorite Color" or "Pet's Name."

4) **Calculate Relevance**: Calculate the percentage of data fields that are relevant based on the total number of fields. For instance, if 2 out of 10 fields are irrelevant, the relevance would be 80%.
5) **Address Irrelevant Data**: Once irrelevant data fields are identified, you can choose to ignore them during analysis, or even remove them from the dataset if they are not being used for other projects or use cases.
6) **Continuous Monitoring**: Regularly reassess the relevance of your data as your project progresses or your use case evolves. Relevance can change over time as the context changes.

Example 6.12   Measuring Relevance in Data Quality

In this example, let's consider a customer database for an online bookstore. The database includes fields like "Customer ID," "Name," "Age," "Gender," "Location," "Favorite Color," "Favorite Genre," "Purchase History," "Pet's Name," and "Product Reviews."
  Here is a simplified representation of our data (Table 6.12):

Table 6.12   Example of Relevance in Data

Customer ID	Name	Age	Gender	Location	Favorite Color	Favorite Genre	Purchase History	Pet's Name	Product Reviews
C001	Amy	35	F	NY	Blue	Mystery	[P001, P002]	Fluffy	[Good, Excellent]
C002	Bob	45	M	CA	Green	Sci-Fi	[P003]	Coco	[Average]

  In the context of a market research project aiming to improve book recommendations, fields like "Favorite Color" and "Pet's Name" are irrelevant. Therefore, the relevance of this data set would be 80% (8 out of 10 fields are relevant). To maintain a high quality of data for this specific use case, these irrelevant fields can be ignored or even removed.

6.3.2.9   Measuring Reliability

The procedure for measuring reliability in data quality is as follows:

1) **Define**: Reliability in data quality refers to the extent to which data remains consistent over time. It reflects the level of trustworthiness and dependability of the data source. For instance, in a climate study, reliability might be determined by the consistency of temperature readings over time.
2) **Identify and Isolate**: Identify the data fields that need to be checked for reliability. For a climate study, you might check the "Temperature" field.
3) **Check for Inconsistencies**: Look for any data points that are inconsistent over time. For example, you might find that the temperature readings fluctuate wildly from day to day without any corresponding change in weather conditions.
4) **Calculate Reliability**: Calculate the percentage of data points that are consistent over time. If there are five inconsistent readings out of 100 days, the reliability would be 95%.
5) **Address Inconsistencies**: After identifying inconsistencies, take steps to correct them. This might involve checking the original data source for errors, recalibrating the data collection equipment, or identifying other factors that might be causing the inconsistencies.
6) **Continuous Monitoring**: Regularly monitor the reliability of data to maintain its overall quality.

Example 6.13   Measuring Reliability in Data Quality

In this example, let's consider a climate study that includes "Date," "Location," "Temperature," "Humidity," and "Precipitation."

Here is a simplified representation of our data (Table 6.13):

Table 6.13   Example of Reliability in Data

Date	Location	Temperature	Humidity	Precipitation
01/01/2023	NY	20 °C	70%	5 mm
01/02/2023	NY	21 °C	72%	6 mm
01/03/2023	NY	15 °C	70%	4 mm
01/04/2023	NY	55 °C	68%	7 mm
01/05/2023	NY	20 °C	69%	5 mm

From the data, we see that the temperature on 01-04-2023 is inconsistent with the other readings. This indicates a potential reliability issue. To maintain high data quality, we would need to investigate the source of this inconsistency and correct it. This could involve checking the temperature sensor for malfunctions, verifying the data against other sources, or identifying other factors that might have led to the unusual reading.

Example 6.14   Evaluate Data Quality in Multiple Dimensions

Let's consider a scenario where we have data for a customer database at a hypothetical online retail store.

The data set is as follows (Table 6.14):

Here is how we can measure different dimensions of data quality using this data set:

Table 6.14   Data quality in multiple dimensions

Customer ID	Name	Email	Age	Location	Gender	Last Purchase Date
101	John Doe	johndoe@example.com	28	NY	M	2022-05-01
102	Jane Doe	jane.doe@example	32	TX	F	2022-03-15
103	Mary Sue	mary.sue@example.com	NA	FL	F	NA
104	Jim Bean	jimbean@exampel.com	45	CA	M	2022-07-10
105	NULL		39	IL	M	2022-04-20
106	Mike Lee	mike.lee@example.com	51	TX	NA	2022-06-15

1) **Completeness**: We can identify missing data in our data set. From a glance, we see that "Customer ID" 105 does not have a name, and "Customer ID" 106 does not have a gender specified. Similarly, "Customer ID" 103 doesn't have a "Last Purchase Date." We can calculate completeness by counting the number of filled data points divided by the total data points.

For example, in the "Name" field, we have five out of six entries filled, giving a completeness of around 83%.

2) **Accuracy**: This involves checking the data points for errors. The email address for "Customer ID" 104 is spelled incorrectly ("exampel" instead of "example"). Depending on the data we have, we might need additional steps to verify the accuracy of data, such as cross-checking with another reliable data source.

3) **Validity**: Here we check if the data follows the rules we have defined. In our case, we could have rules that the "Age" must be a positive number and "Last Purchase Date" should be a date in the past. The age for "Customer ID" 103 is "NA," which is not a valid entry. Similarly, we need to check all the purchase dates are in the past.

4) **Consistency**: We need to ensure data across all records is consistently formatted. An inconsistency can be seen in the "Email" field, where "Customer ID" 102 has the email "jane.doe@example," missing the ".com" at the end.

5) **Uniqueness**: This refers to the absence of unnecessary duplicate entries in the data set. In our small data set, it seems there are no duplicates, but in larger data sets, we could use algorithms to check for duplicate entries.

After identifying the issues with data quality, we would proceed to clean and process the data, potentially involving steps like filling missing values, correcting inaccuracies, standardizing and validating formats, and removing duplicates.

After we discussed data quality dimensions and their measurements, those are basic data quality metrics and their quantitative assessment values in an era where data forms the backbone of decision-making processes and strategic planning, ensuring high data quality has become essential for businesses across all industries. The task of maintaining and improving data quality lies in robust data quality management and governance.

Data Quality Management (DQM) and Data Governance (DG) are two interrelated disciplines that provide the foundation for reliable, accurate, and meaningful data in an organization. DQM involves the execution of policies, practices, and procedures that ensure data is fit for its intended uses in operations, decision-making, and planning. On the other hand, DG refers to the overall management of the availability, usability, integrity, and security of the data employed in an enterprise. It provides a framework for data stewardship responsibilities, ensuring data meets the organization's quality requirements and complies with regulations and business rules.

The objective of DQM and governance is to create and uphold high-quality data that improves operational efficiency, enhances decision-making, and boosts customer satisfaction. Through these practices, organizations can avoid the risks and costs associated with poor data quality, such as inaccurate reports, misguided decisions, regulatory noncompliance, and decreased customer trust.

I will discuss DQM and governess in the next two sections in detail.

6.4 Data Quality Management

DQM [13–15] represents a critical discipline within any data-reliant organization. As businesses increasingly harness data for decision-making, strategic planning, and operational management, safeguarding data quality emerges as an imperative. DQM incorporates a range of processes, policies, and methodologies devised to ensure data is accurate, consistent, reliable, timely, and apt for its intended purposes. By maintaining top-tier data quality, organizations can enhance operational efficiency, refine decision-making, and boost customer satisfaction. On the flip side, lackluster attention to DQM can result in costly errors, misdirected strategies, and dwindling trust among stakeholders and customers.

A rudimentary framework for DQM and governance incorporates the delineation of the organization's data quality objectives, the establishment of policies and procedures to accomplish these goals, the assignment of data stewardship roles and responsibilities, the implementation of data quality measures and tools, and the periodic auditing and refinement of the data quality program. By adopting such a framework, organizations can preserve high-quality data, propelling their goals, and driving their success.

In the subsequent subsections, we delve into the intricacies of DQM, shedding light on their significance, methodologies, and the guiding principles that steer these practices.

6.4.1 Reactive Versus Proactive Data Quality Management

Reactive and proactive approaches represent two different strategies in DQM [16, 17]. Both have their merits and are used in different scenarios depending on the organization's resources, requirements, and existing data quality.

Reactive Data Quality Management: This approach involves responding to data quality issues as they occur. In this case, data errors are discovered during data usage, such as in reporting or analysis, and then resolved. Reactive DQM is often seen in organizations that lack a formalized DG process, or where resources for proactive data management are not available. While this approach can resolve data errors, it does not prevent them and can often lead to repeated issues.

Proactive Data Quality Management: This strategy focuses on preventing data quality issues before they occur. Proactive DQM involves setting up systems, procedures, and governance to ensure data is captured accurately, consistently, and completely right from the source. This approach requires an understanding of data quality dimensions, potential sources of errors, and an ongoing commitment to maintaining and improving data quality.

Comparison of Reactive and Proactive DQM

1) **Effort and Resource Allocation**: Reactive DQM can be less resource-intensive in the short term as issues are addressed only when they arise. However, in the long run, this could lead to more resources spent on fixing recurring issues. Proactive DQM requires an initial investment in setting up robust DG and quality measures but can save resources in the long run by preventing issues.

2) **Impact on Business Operations**: Errors in data can disrupt business operations and decision-making. Reactive DQM addresses issues after they have already had an impact, while proactive DQM aims to prevent such disruptions from occurring in the first place.

3) **Quality of Insights**: The quality of insights derived from data analysis depends on the quality of data used. Proactive DQM ensures high-quality data is available consistently for analysis, leading to more reliable insights. Reactive DQM, however, runs the risk of using low-quality data for analysis before errors are detected and fixed.

4) **Scalability**: As organizations grow and their data becomes more complex, maintaining data quality becomes more challenging. Reactive DQM can become increasingly time-consuming and less effective at scale. Proactive DQM, with its emphasis on processes and systems, can better adapt to larger and more complex data environments.

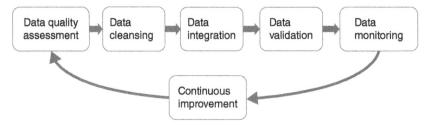

Figure 6.1 Proactive Data Quality Management

5) **Data Culture**: Proactive DQM promotes a culture of data quality awareness throughout the organization. Reactive DQM, on the other hand, may result in a culture of "fire-fighting" data issues.

In conclusion, while both approaches have their uses, most organizations find that a proactive approach to DQM offers the best long-term benefits. It encourages prevention over cure, reduces long-term costs, and supports more effective and reliable use of data across the organization.

A proactive DQM process can be illustrated by the following flow chart (Figure 6.1)

Here I will discuss each step in detail.

6.4.2 Data Quality Assessment

Data Quality Assessment (DQA) is the process of examining data systematically to determine if it meets the necessary quality standards for its intended purpose. This can be an ongoing process, but it is particularly crucial when first starting a DQM program. The assessment should consider key dimensions of data quality such as accuracy, completeness, consistency, timeliness, validity, and relevance.

Here are the typical steps involved in a DQA:

1) **Identify Data to Assess**: The first step is to identify the data that you want to assess. This could be a particular dataset, a type of data (e.g., customer data), or data stored in a particular system.
2) **Define Data Quality Criteria**: Define what constitutes quality for the selected data. These criteria should be based on the data quality dimensions that are most relevant to your situation. For example, if timeliness is crucial for your operations, one of your criteria might be that all data is up-to-date.
3) **Data Profiling**: Data profiling is the process of examining the data available in an existing database and collecting statistics and information about that data. The goal of data profiling is to uncover inconsistencies, anomalies, and patterns in the data. Data profiling tools can provide insights such as:
   - **Descriptive Statistics**: These are basic statistics like min, max, average, standard deviation, and others that give you a summary of your data.
   - **Frequency Distributions**: This tells you how often each value appears in the data. It can help identify outliers and common values.
   - **Data Types**: Understanding the types of data you have (e.g., numeric, categorical, and date/time) can help in assessing the suitability of the data for its intended use.
4) **Assess Data Quality**: This is the actual assessment, where you evaluate the selected data against your data quality criteria. You might do this by using a data quality tool, running queries against the data, or examining the data manually.

5) **Report on Data Quality**: Once the assessment is complete, the findings are typically documented in a data quality report. The report should outline the findings of the assessment, including any data quality issues that were identified.
6) **Develop a Data Quality Improvement Plan**: If the data does not meet the necessary quality standards, the next step is to create a plan for improving data quality. This might include steps for data cleansing, setting up new data governance procedures, or training for staff who enter data.

In the end, the aim of DQA is to understand the condition of your data, identify any issues that may affect its usability, and to plan for how to address these issues. It is the first step toward managing and improving data quality.

In step 5, a DQA report is a document that presents the findings from a DQA. It should be detailed, informative, and tailored to the needs of its intended audience. The specifics of a Data Quality Report can vary significantly depending on the nature of the data, the specific quality dimensions being assessed, and the business context.

A sample DQA report is shown in Table 6.15.

6.4.3 Data Cleansing

Data Cleansing (or Data Cleaning) is the process of detecting and correcting or removing corrupt, inaccurate, or irrelevant parts of data within a dataset. This process can involve multiple stages and techniques, depending on the nature and scale of the data quality issues.

Here are the typical steps involved in a data cleansing process:

1) **Data Audit**: Similar to a DQA, the first step is to examine the dataset to understand its structure, the types of data it contains, and the scale and nature of any quality issues. This might involve techniques like data profiling or exploratory data analysis.
2) **Workflow Specification**: Based on the data audit, you should specify a workflow for the data cleansing process. This might include defining rules for handling specific types of errors or inconsistencies, outlining the steps for cleaning the data, and specifying the order in which these steps will be carried out.

Table 6.15 A sample data quality assessment report

Data Quality Dimension	Assessment Method	Findings	Identified Issues
Accuracy	Automated verification against source documents	95% of records were accurate	5% of records had incorrect values
Completeness	Missing values analysis	10% of records were missing one or more values	Missing values in key fields
Consistency	Cross-dataset comparison	Inconsistent formats found for date fields across different datasets	Inconsistency in date formats
Timeliness	Date of last update versus record date	On average, records were updated 2 d after the event	Delayed updates
Relevance	Manual review of data fields versus business requirements	All fields were relevant to the business requirements	None

3) **Workflow Execution**: Next, you execute the data cleansing workflow. This could involve running scripts or using data cleansing tools to apply the defined rules to the data. Here are some common data cleansing techniques:
   - **Data Transformation**: You might need to change the format or structure of the data to make it more consistent or easier to work with. For example, you could convert all dates to a standard format, or transform categorical data into numerical data.
   - **Data Imputation**: Missing data can be filled in using statistical methods. For example, you could fill in missing values with the mean or median of the remaining values in the column.
   - **Deduplication**: Duplicate entries in the data should be identified and removed to ensure that each record is unique.
   - **Error Correction**: Spelling errors, typos, or inconsistent naming conventions can be corrected using rules or algorithms. For example, you could use a text-matching algorithm to identify and correct spelling errors.
4) **Postprocessing and Checking**: After the initial cleansing, you should review the cleaned data to verify that the cleaning has been successful and has not introduced new issues. You might also run additional checks or validations to ensure that the data now meets the required quality standards.
5) **Reporting**: Finally, document the data cleaning process and its results, similar to the DQA. This could include a description of the cleansing workflow, the techniques used, any challenges encountered, and the final state of the data. This report can help inform future data-cleaning efforts and contribute to overall data governance within the organization.

Data cleansing is an iterative process. It might take several rounds of cleaning and checking to address all data quality issues, particularly for large or complex datasets. And like all aspects of DQM, data cleansing should be an ongoing effort, rather than a one-time task.

6.4.4 Data Integration

Data Integration is the process of combining data from different sources and providing users with a unified view of these data. This process becomes significant in a variety of situations, which include both commercial (when two similar companies need to merge their databases) and scientific (combining research findings from different bioinformatics repositories, for example) domains.

Data integration appears with increasing frequency as the volume and the need to share existing data increases. It has become the focus of extensive theoretical work, and numerous open problems remain unsolved.

Here are the typical steps involved in a data integration process:

1) **Data Identification**: The first step is to identify the data sources that need to be integrated. This could include databases, data files, data streams, APIs, or any other data sources.
2) **Schema Mapping**: In this step, you define how the schemas (or structures) of the different data sources map to each other. This can be a complex task, particularly if the data sources have different structures or use different naming conventions. Techniques used in schema mapping include:
   - **Manual Mapping**: Experts with knowledge of both schemas manually specify the mappings.
   - **Automated Mapping**: Machine learning or rule-based techniques are used to automatically infer the mappings.
3) **Data Transformation**: The data from the different sources is then transformed based on the schema mapping. This could involve changing the format or structure of the data, converting data types, or resolving naming inconsistencies.

4) **Data Fusion**: The transformed data is then combined or "fused" into a single dataset. Techniques used in data fusion include:
   - **Record Linkage**: Identifying records that represent the same entity across different data sources. This might involve techniques like fuzzy matching or deduplication.
   - **Data Consolidation**: Merging records from different data sources into a single, "golden" record. This could involve choosing the most recent or the most reliable data when there are conflicts.
5) **Data Loading**: The integrated data is then loaded into a data warehouse or another storage system where it can be accessed by users.
6) **Data Validation**: Finally, the integrated data is checked to ensure that it meets the necessary quality standards. This could involve data profiling, data quality checks, or user validation.

The ultimate goal of data integration is to provide a unified view of data from different sources, which can improve the completeness, consistency, and usefulness of the data. However, it is a complex process that requires careful planning and execution, as well as ongoing management to handle changes to the source data or the integration requirements.

6.4.5 Data Validation

Data Validation is a process that ensures the data in question maintains a high degree of quality, correctness, and usefulness. It involves a series of checks and procedures to ensure the data meet specified criteria in terms of format, content, and structure. This process can significantly enhance the reliability of your datasets and, by extension, any insights derived from them.

Here are the typical steps involved in a data validation process:

1) **Define Validation Rules**: Before the validation process starts, the first step is to define the validation rules. These rules are derived from the specific requirements for the data. For example, a validation rule might specify that a particular field must always contain a numerical value, or that a date must always be in a specific format.
2) **Apply Validation Rules**: The next step is to apply these rules to the dataset. This can be done through software programs that scan through the data, or it can be done manually if the dataset is small enough.
3) **Identify and Rectify Errors**: During this stage, the data that does not comply with the validation rules are identified. Depending on the nature of the error, it may be possible to rectify the error automatically. For example, a common error is that dates are entered in an inconsistent format. In such a case, it may be possible to write a script that automatically converts all dates to a consistent format. In other cases, it might be necessary to manually rectify the error or even source the original data again.
4) **Repeat Validation**: After the errors have been rectified, the validation process is repeated to ensure that no errors remain. It is quite common for the validation process to be an iterative one, with each iteration identifying fewer and fewer errors until none remain.
5) **Verification**: Finally, a subset of the data is often checked manually in order to verify that the validation process has been successful. This typically involves selecting a random sample of data and checking each piece of data against the original source document or system.

Common validation techniques include:

- **Range Check**: This checks that the value is within a specified range. For example, the age of a person should be between 0 and 120.

- **Type Check**: This checks that the value is of the correct type. For example, an email field should contain a string.
- **Format Check**: This checks that the value follows a specified format. For example, a date might need to be in the format "YYYY-MM-DD."
- **Uniqueness Check**: This checks that the value is unique across the dataset. For example, a user ID field should contain a unique value for each user.
- **Consistency Check**: This checks that the data is consistent across the dataset. For example, the same product should not have different prices in different parts of the dataset.

Data validation is an ongoing process that should be built into your regular data management workflows. By maintaining high data quality, you will ensure that your data is reliable, actionable, and ready for further analysis or processing.

6.4.6 Data Monitoring

Data Monitoring is a critical aspect of DQM. It involves the continuous tracking and review of data to ensure consistency, accuracy, and overall quality over time. By continuously monitoring your data, you can detect and correct quality issues promptly, minimizing their impact on your business or research outcomes.

Here are the typical steps involved in a data monitoring process:

1) **Establish Monitoring Metrics**: To monitor your data, you first need to define what you will monitor. This involves establishing metrics related to the quality dimensions important to your data, such as accuracy, completeness, timeliness, and consistency. These metrics serve as indicators of your data's health and should align with your data quality goals and business objectives.

2) **Set Up Monitoring Tools**: Once you have defined your metrics, the next step is setting up tools and systems to track these metrics. This could involve using data quality software, custom scripts, or built-in features of your database system. The tools you choose should enable automatic and continuous monitoring of your data.

3) **Monitor Data**: With your metrics and tools in place, you can begin monitoring your data. This involves regularly collecting data quality metrics and reviewing them for potential issues or trends. If your tools support it, you can also set up alerts to notify you when metrics fall outside acceptable ranges.

4) **Analyze and Report**: Data monitoring is not just about collecting metrics; it is also about analyzing them to gain insights into your data quality. This might involve trend analysis, root cause analysis of quality issues, or evaluating the impact of recent changes to your data or systems. You should also report on your monitoring results, sharing insights and issues with stakeholders as necessary.

5) **Take Action**: Finally, if your monitoring efforts uncover quality issues, you need to take action to resolve them. This could involve data cleansing, adjusting your data processes, or even reevaluating your data quality goals or metrics. The actions you take should be documented and communicated to ensure everyone is on the same page.

There are various techniques and tools used in data monitoring, including:

- **Data Profiling**: This involves reviewing the data to gain insights into its structure, content, and quality. Data profiling tools can help automate this process, identifying potential issues such as missing data, duplicate entries, or inconsistencies.

- **Data Quality Dashboards**: These provide a visual representation of data quality metrics, enabling you to quickly assess the state of your data and identify trends or issues.
- **Automated Alerts**: These can notify you when data quality metrics fall outside acceptable ranges, allowing you to quickly respond to potential issues.

Data monitoring is a proactive process that should be built into your ongoing data management workflows. By continuously monitoring your data, you can maintain high data quality and ensure your data remains reliable and actionable.

6.4.7 Technology, Tools, and Software on Data Quality Management

The landscape of technology, tools, and software for DQM is broad and rapidly evolving. The digital age has ushered in a new era of data generation, collection, and processing. As the volume of data multiplies exponentially, ensuring its quality has become a challenge and a necessity. High-quality data fuels informed decision-making, drives operational efficiency, and enhances customer satisfaction. Consequently, the demand for robust and effective DQM tools has escalated in various sectors, such as healthcare, finance, retail, manufacturing, and information technology. These tools assist in ensuring data integrity, accuracy, consistency, timeliness, and relevance, thereby enhancing the overall trustworthiness of data. This section explores the diverse array of technologies, tools, and software designed to manage data quality, from traditional database management systems to sophisticated AI-powered data quality platforms [18, 19]. It will address the features, benefits, limitations, and application scenarios of these tools, providing a comprehensive view of the present and future prospects of technology in DQM.

6.4.7.1 Technologies and Tools

In DQM, various technologies have emerged that contribute to the maintenance, control, and improvement of data quality. Some of these technologies are:

1) **Data Profiling Tools**: These are used to assess the current state of data quality within a dataset. Data profiling tools generally offer insights into various aspects of data, such as completeness, accuracy, consistency, and uniqueness. They perform functions such as identifying missing or null values, detecting duplicates, and exploring data distributions. Some popular data profiling tools include Informatica Data Quality, Talend Open Studio, and IBM Infosphere Information Analyzer.

2) **Data Cleansing Tools**: These tools are used to fix detected issues in datasets. They can help to correct spelling errors, remove duplicates, standardize formatting, and fill in missing values, among other tasks. They often use techniques such as matching algorithms, parsing, and standardization to clean data. Examples of such tools include Trifacta, Data Ladder, and WinPure.

3) **Data Integration Tools**: These technologies are used to combine data from different sources while ensuring that the integrated data maintains high quality. They perform tasks such as data transformation, data deduplication, and data mapping. They may also use techniques such as data warehousing, data virtualization, and data federation. Prominent data integration tools include Oracle Data Integrator, Microsoft SQL Server Integration Services (SSIS), and SAP Data Services.

4) **Master Data Management (MDM) Tools**: MDM tools help manage, govern, and distribute master data across an organization. Master data is critical data about customers, products, suppliers, and other business entities that are used in various business processes. MDM tools ensure consistency, accuracy, and control of this data. Some widely used MDM tools are Informatica MDM, SAP Master DG, and IBM InfoSphere MDM.

5) **Data Governance Tools**: These tools help implement processes, policies, standards, and metrics for DQM. They enable organizations to manage data quality issues, comply with regulations, and support business objectives. Tools like Collibra, Alation, and Informatica Axon are known for providing such services.

6) **Artificial Intelligence (AI) and Machine Learning (ML) Tools**: More recently, AI and ML are being leveraged for DQM. ML algorithms can learn from the data to identify patterns, correlations, and anomalies that may not be easily detected by traditional data quality tools. AI-powered data quality tools can predict and prevent potential data quality issues and continuously improve their performance by learning from past errors. Tools like TIBCO, Databricks, and Google AutoML are making headway in this space.

The selection of a DQM technology or tool should be based on the specific needs, resources, and objectives of an organization. The above categories are not mutually exclusive and often work together to provide a comprehensive DQM solution. Each category has its strengths and weaknesses, and a combination of tools and technologies is often the most effective approach to DQM.

6.4.7.2 Data Quality Management Software

Here are some well-known DQM software solutions, with a brief description of their features, pros and cons, and suitable applications:

1) **Informatica Data Quality (IDQ)**:
   - **Functions and Features**: IDQ provides robust data profiling, standardization, matching, and enrichment capabilities. It offers a unified environment that enables you to discover, profile, cleanse, and monitor data quality.
   - **Pros**: It provides a flexible, scalable solution for DQM and is capable of processing large volumes of data efficiently. It also supports a broad range of data sources.
   - **Cons**: The user interface may not be very intuitive, and the learning curve for new users can be steep.
   - **Fitted Applications**: IDQ is well-suited for large enterprises dealing with complex data integration and quality challenges, particularly when the organization already uses other Informatica products.

2) **IBM InfoSphere QualityStage**:
   - **Functions and Features**: IBM QualityStage is a part of IBM's InfoSphere Information Server. It enables users to create and monitor data quality rules, cleanse data, and build and maintain customer views. It supports profiling, standardizing, matching, and enriching data.
   - **Pros**: It integrates well with other IBM products and supports a variety of data sources. It also offers strong capabilities for handling complex data quality scenarios.
   - **Cons**: The user interface is not very modern or user-friendly. Some users have reported that the software requires significant technical expertise to utilize effectively.
   - **Fitted Applications**: IBM QualityStage is ideal for organizations that already use IBM's platform and have complex, enterprise-level data quality needs.

3) **Talend Data Quality**:
   - **Functions and Features**: Talend Data Quality offers data profiling, cleansing, and enrichment capabilities. It also enables users to monitor data quality over time. It can integrate data across various sources and supports ML capabilities.
   - **Pros**: Talend is known for its intuitive interface and ease of use. It also offers a free, open-source version, which makes it a cost-effective solution for some businesses.

- **Cons**: While the open-source version offers solid capabilities, advanced features are only available in the paid version. Some users have reported that customer service could be improved.
- **Fitted Applications**: Talend Data Quality is a good fit for mid-sized to large businesses, especially those looking to start with an open-source solution and scale up as their needs grow.

4) **SAS Data Quality**:
- **Functions and Features**: Statistical Analysis System (SAS) Data Quality provides data profiling, standardization, cleansing, and matching capabilities. It also allows users to establish data quality rules and monitor data quality over time.
- **Pros**: SAS provides a comprehensive and flexible suite of data management tools, including strong capabilities for statistical analysis.
- **Cons**: SAS software can be expensive, and the complexity of the software means it often requires significant training or technical expertise.
- **Fitted Applications**: SAS Data Quality is suitable for large organizations with complex data quality needs, particularly those that need advanced statistical capabilities or are already invested in the SAS ecosystem.

The best data quality software for a given organization depends on a variety of factors, including the specific data quality challenges the organization faces, the complexity and volume of the data, the organization's technical capabilities, and budget constraints.

6.5 Data Governess

Data governance, an essential element of any comprehensive data strategy, serves as the custodian of an enterprise's data. It encompasses the systematic management of data availability, usability, integrity, and security within a given organization. This complex process requires establishing well-defined rules, policies, and procedures to guide how data is managed across an organization. It addresses integral aspects such as data quality, data privacy, security, lifecycle management, and regulatory compliance. Furthermore, it clarifies the allocation of authority within the organization regarding decision-making about various data types. With clearly delineated guidelines and policies, DG ensures that data remains consistent, reliable, and immune to misuse. It sets a robust framework delineating who can perform specific actions on certain data under different circumstances.

The principal objectives of DG involve guaranteeing data quality, preserving data privacy and security, achieving regulatory compliance, and optimizing data value for business decision-making. Additionally, it plays a crucial role in facilitating data integration and interoperability across an organization.

In the current data-driven era, the significance of data governance is paramount. Effective data governance minimizes the risk of data breaches, bolsters operational efficiency, augments customer satisfaction, and drives revenue growth by fostering informed decision-making. The evolution of DG has mirrored the broader changes in data management and technology. In the early 1990s, with the birth of data warehousing, organizations began to understand the importance of data consistency and quality for decision-making [20].

The DG responsibility generally lies with a dedicated DG team or a data steward. They collaborate with a diverse array of stakeholders—including IT professionals, data scientists, business analysts, and executives—to devise and enforce DG policies and procedures. In some organizations, a DG council may be formed, or a Chief Data Officer (CDO) may be appointed, to oversee these activities.

Thomas, 2006, highlights the rise of the CDO role, emphasizing its significance in aligning data management with organizational strategy [21].

This section will delve into the fundamentals, strategies, and best practices of DG, thereby elucidating its role in attaining data excellence. The journey from DG being a mere concept in the 1990s to its prevalent role today illustrates how organizations increasingly understand the strategic importance of DG in their operations [22, 23].

6.5.1 Data Governance Strategy

DG strategy forms the cornerstone of an organization's efforts to ensure the quality, consistency, and security of its data. It provides a framework for managing the company's data assets and making informed decisions about them.

6.5.1.1 Fundamentals

A good DG strategy starts with recognizing data as a valuable asset. The strategy should be built around the following fundamentals:

- **Data Stewardship**: Assigning clear roles and responsibilities to manage and oversee data. This could be a DG team, a data steward, or a CDO.
- **Data Quality**: Ensuring that the data is accurate, consistent, and reliable. This includes data cleaning, validation, and quality assurance processes.
- **Data Security and Privacy**: Protecting data from unauthorized access and complying with relevant data protection regulations.
- **Data Architecture and Integration**: Structuring data in a way that is useful and accessible for decision-making and ensuring it can be integrated across systems.
- **Data Lifecycle Management**: Managing the entire lifecycle of data, from its creation and storage to its disposal.

6.5.1.2 Objectives

The objectives of a DG strategy should align with the organization's overall business goals. Some common objectives include:

- Improving data quality
- Ensuring compliance with data protection laws
- Protecting the company's data assets
- Enhancing the value derived from data
- Increasing efficiency and productivity by reducing data-related errors and issues

6.5.1.3 Winning Strategy

A winning DG strategy is one that successfully achieves its objectives and delivers value to the organization. Some key elements of a successful strategy include:

- **Alignment with Business Goals**: The strategy should support the organization's overall business goals and strategy.
- **Stakeholder Engagement**: All relevant stakeholders, from executives to IT staff, should be involved in the strategy. This ensures buy-in and support from all levels of the organization.
- **Clear Policies and Procedures**: The strategy should establish clear policies and procedures for managing and using data. This includes data quality standards, data security policies, and data usage guidelines.

- **Continuous Improvement**: The strategy should be flexible and adaptable, with regular reviews and updates to ensure it remains effective and relevant.

In conclusion, a winning DG strategy is not just about managing data—it is about transforming data into a strategic asset that drives business value. It requires a comprehensive, organization-wide approach and ongoing commitment to data excellence.

6.5.2 Data Governance Framework

A DG framework is a structured approach to managing, processing and utilizing data. It defines the principles, policies, roles and responsibilities, and processes that guide data management activities within an organization. The framework is designed to ensure the quality, security, and compliance of data, as well as its effective and efficient use for decision-making and business operations.

Major components of a DG framework include:

1) **Rules and Policies**: These are the guidelines that dictate how data should be handled, maintained, accessed, and used within the organization. They typically include data quality standards, data privacy and security rules, data usage guidelines, and data lifecycle management policies.

   Specifically, Here is a deeper dive into some of the key categories of rules and policies in a DG framework:

   - **Data Quality Standards**: These policies define what constitutes "good" data within the organization. They set out the criteria for accuracy, completeness, consistency, timeliness, and relevance. These standards guide the processes of data cleaning, validation, and quality assurance.
   - **Data Privacy and Security Rules**: These rules aim to protect data from unauthorized access or misuse. They align with local and international data protection regulations (such as GDPR, HIPAA, etc.) and include guidelines on data encryption, access controls, data anonymization, and incident response.
   - **Data Usage Guidelines**: These policies dictate how data can be used within the organization. They might outline acceptable uses of data, detail procedures for sharing data both internally and externally, and provide guidance on ethical data practices.
   - **Data Lifecycle Management Policies**: These rules guide the management of data throughout its entire lifecycle—from its initial creation or collection, through its active use, to its eventual archiving or disposal. They address areas like data retention, archival, and disposal to ensure compliance and mitigate risks.
   - **Metadata Management Policy**: These guidelines refer to the handling of metadata (data about data). It includes how metadata should be created, stored, and maintained to ensure easy data discovery and understanding.
   - **Data Ownership Rules**: These policies define who within the organization is responsible for various types of data. Clear data ownership is essential for accountability and effective decision-making about data.

2) **Processes**: These are the procedures and workflows that define how DG activities should be carried out. Specifically, processes in a DG framework refer to the defined set of procedures or activities that guide how data is handled throughout its lifecycle in an organization. These processes enable organizations to manage data effectively, improve data quality, and ensure compliance with relevant regulations.

Key processes in a DG framework might include:

- **Data Collection**: This process involves the acquisition of data from various sources. It also includes determining what data to collect, how often, and in what format.
- **Data Classification and Categorization**: Once data is collected, it needs to be classified and categorized based on various attributes such as data type, sensitivity level, or business function. This process helps in understanding the data and implementing appropriate controls based on its classification.
- **Data Storage and Retention**: This process guides how and where data is stored and for how long. The data storage process must align with data privacy and security policies and consider aspects such as data redundancy, backup, and disaster recovery. The data retention process, on the other hand, determines how long data should be kept based on regulatory requirements and business needs.
- **Data Access and Security**: This involves defining who has access to what data, how data is secured, and how data privacy is ensured. It includes setting up user roles and permissions, implementing data encryption, and monitoring data access.
- **Data Quality Management**: This process focuses on ensuring the accuracy, completeness, consistency, timeliness, and relevance of data. It involves data validation, cleaning, transformation, and ongoing quality checks.
- **Data Integration**: This process is about consolidating data from different sources and providing a unified view. It involves data mapping, transformation, and loading.
- **Data Usage and Distribution**: This process outlines how data can be used, shared, and distributed both internally and externally. It includes aspects like data sharing agreements, data anonymization for privacy, and defining how data should be used for decision-making.
- **Data Archiving and Disposal**: This involves the safe archival and disposal of data that is no longer required. This process needs to comply with data retention policies and regulations.

3) **People, Roles, and Responsibilities**: This involves all the individuals who have a role in DG. Specifically, in a data governess context, these individuals, each with their distinct roles and responsibilities, are crucial for the successful implementation and execution of DG initiatives. Roles and responsibilities clarify who is responsible for what in the DG process. Here are the key roles and people who are responsible for them:

- **DG Council or Steering Committee**: This is usually a group of high-level stakeholders who provide guidance, oversight, and direction for the DG program. They define the strategic vision for DG and make key decisions regarding policies, standards, and procedures.
- **Data Owners**: Data owners are usually senior management personnel who have ultimate responsibility for specific datasets. They make decisions about what data to collect, how it is used and secured, and who can access it. They are accountable for the data quality and compliance of their data.
- **Data Stewards**: Data stewards are responsible for the operational execution of DG initiatives. They work to ensure data quality, implement data policies, manage data issues, and promote DG's best practices throughout the organization. Data stewards often come from business units and have a deep understanding of the data needs and processes of their departments.
- **Data Custodians**: Data custodians manage the technical aspects of storing, securing, and maintaining data. They often work in IT departments and are responsible for the physical and technical storage and security of data.
- **Data Users**: Data users are those who use the data in their day-to-day work. This group can include everyone from business analysts and data scientists to customer service representatives and marketing professionals.

- **Data Scientists/Analysts**: These individuals work directly with data to generate insights. They ensure that data is correctly interpreted and utilized.
- **Data Privacy and Compliance Officers**: These roles are responsible for ensuring the organization's data practices comply with relevant laws and regulations. They also handle data privacy and protection issues.

  The exact roles and responsibilities can vary between organizations, depending on the size, industry, and specific needs of the organization. What is most important is to clearly define and communicate these roles, ensuring each person understands their responsibilities in supporting the organization's DG objectives. This creates a culture of data accountability and responsibility, which is critical to the success of DG efforts.

4) **Technology**: This refers to the tools and systems used to support DG. This includes databases, data warehouses, data integration tools, data quality tools, data security tools, and DG platforms.

  These technologies facilitate data collection, storage, management, analysis, and security and help automate many aspects of DG to increase efficiency and reduce the risk of human error. Specifically, the technology component should include the following items:

- **Data Storage and Management Systems**: These systems provide the infrastructure for storing, organizing, and managing data. This category includes databases, data warehouses, and data lakes, which store structured and unstructured data, respectively.
- **Data Integration Tools**: These tools help consolidate data from different sources and formats, providing a unified and consistent view of data. They include Extract, Transform, and Load (ETL) tools, which are used to collect data from various sources, convert it into a suitable format, and load it into a database or data warehouse.
- **Data Quality Tools**: These tools help maintain the quality of data by identifying and correcting errors, inconsistencies, and duplicates in the data. They can automate tasks such as data cleaning, validation, and profiling, making it easier to ensure data is accurate, complete, and consistent.
- **Data Security Technologies**: These technologies help protect data from unauthorized access and breaches. They include encryption tools, access control systems, and data masking tools, which help ensure data privacy and compliance with data protection regulations.
- **Data Cataloging Tools**: These tools help create a metadata repository or a data catalog, which provides information about the data's origin, usage, relationships, and quality. This promotes better data understanding and discovery.
- **DG Platforms**: These are comprehensive solutions that support various aspects of DG, including policy management, data stewardship, data quality, and compliance. They provide a central hub for DG activities and often come with features for workflow management, reporting, and collaboration.
- **Business Intelligence and Data Analytics Tools**: These tools help analyze data and derive insights, which support decision-making. They allow users to visualize data in the form of dashboards, reports, and charts.

  While technology is a crucial component of DG, it is not a silver bullet. It must be accompanied by robust DG policies, processes, and people, all working together toward the organization's DG objectives. The right technology should enhance and enable these other components, rather than being seen as a standalone solution.

5) **Structure of the Framework**:

A well-structured DG framework typically starts with the organization's DG vision and objectives. It then outlines the DG strategy and principles, followed by the policies and procedures for managing data.

Next, it defines the roles and responsibilities of the various individuals involved in DG. This includes the DG council or team, data owners, data stewards, and data users.

Following this, the framework lays out the processes for data collection, storage, access, modification, and disposal. It also specifies the tools and systems that will be used to support these processes.

Lastly, the framework includes a plan for monitoring and improving DG. This involves regularly reviewing the effectiveness of the DG activities, tracking performance against the objectives, and making necessary adjustments.

In conclusion, a DG framework provides a holistic, structured approach to managing data. It not only helps ensure data quality, security, and compliance, but also maximizes the value of data for the organization.

6.5.3   Data Stewardship

Data stewardship, is a critical function within a DG program. It refers to the management and oversight of an organization's data assets to provide business users with high-quality data that is easily accessible in a consistent manner. The main goal of data stewardship is to ensure that data is treated as a valuable business asset.

Data stewards are the individuals or groups who carry out this role. They serve as the link between IT and business operations and are responsible for maintaining the quality, usability, and security of data. Data stewards ensure that data management processes are properly implemented and adhered to, and they assist in resolving data-related issues.

Here are some best practices for effective data stewardship:

1) **Define Clear Roles and Responsibilities**: The role of data steward often falls into several categories, such as data quality steward, data definition steward, or data privacy steward. Clearly defining these roles and their responsibilities is crucial to ensure effective data management.
2) **Establish Data Standards and Policies**: Data stewards should help establish and enforce data standards and policies, including data naming conventions, data classification schemas, and metadata standards. This ensures consistency and quality in the data.
3) **Ensure Data Quality**: One of the main responsibilities of data stewards is to ensure the accuracy, completeness, consistency, and reliability of data. This involves conducting regular data quality checks, resolving data quality issues, and promoting data quality awareness within the organization.
4) **Promote DG Education**: Data stewards should promote DG awareness and education throughout the organization. They can organize training sessions and workshops to educate employees about the importance of DG and their roles in ensuring good DG.
5) **Implement Data Security Measures**: Data stewards should work with the IT department to ensure data is protected and that data privacy and compliance regulations are met.
6) **Utilize DG Tools**: There are many tools available that can assist data stewards in their role, such as data catalogs, data quality tools, and DG platforms. These tools can help automate many data stewardship tasks and improve efficiency.

An effective data stewardship is a team effort that involves cooperation between data stewards, data owners, and data users. It requires a culture that values data and understands its importance in driving business decisions and outcomes.

6.5.4 Data Life Cycle Management

Data lifecycle management (DLM) refers to the process of managing the flow of data throughout its lifecycle, from creation and initial storage to the time when it becomes obsolete and is deleted. It involves the policies, rules, and strategies used to effectively manage and protect data, ensuring it remains useful and accessible throughout its lifecycle. The main goal of DLM is to improve efficiency, ensure compliance with data-related regulations, and optimize the value of data.

Here is a general framework and process for managing the data lifecycle:

1) **Data Creation**: This is the first stage of the data lifecycle. Data can be created in various ways, such as data entry, data acquisition from sensors, or data generation through applications and systems.
2) **Data Storage and Retention**: Once data is created, it is stored in a database, data warehouse, or other storage systems. The data retention policy should dictate how long data should be kept based on its type, purpose, and the legal or regulatory requirements.
3) **Data Use**: Data is then used for various purposes like analysis, decision-making, or process execution. This stage includes querying, aggregating, and reporting of data.
4) **Data Sharing and Distribution**: In this stage, data is shared or distributed across different departments, individuals, or other systems as per need. This stage needs careful access control and security measures.
5) **Data Archiving**: When data is no longer actively used but needs to be kept for compliance or future reference, it is archived. Archived data is moved to a lower-cost storage but can still be accessed when needed.
6) **Data Deletion or Destruction**: The end of the data lifecycle is the disposal stage. When data is no longer needed, and the required retention period has passed, it should be securely deleted or destroyed.

Each of these stages requires careful management and consideration of data quality, data privacy, and data security. Organizations should set up policies and practices to guide each stage of the data lifecycle and regularly review these policies to ensure they stay current with changing business needs and regulatory requirements.

Effective DLM is about more than just technology; it requires a combination of people, processes, and technology to ensure data is effectively managed from creation to disposal.

6.5.5 Data Governess Tools and Technology

DG tools and technologies support the implementation of DG initiatives, offering functionalities to manage, maintain, and monitor the quality and security of data across an organization. Here are some noteworthy DG tools available [24].

1) **IBM InfoSphere Information Server**: This platform offers data integration, data quality, and DG functionalities. It helps organizations cleanse, monitor, transform, and deliver data, while collaborating to bridge the gap between business and IT.
2) **Informatica Axon**: A DG tool that helps businesses engage all constituencies, from business users to data scientists, to effectively discover, understand, govern, and steward data.
3) **Collibra**: This is a robust platform that provides wide-ranging DG capabilities. It offers functionalities such as metadata management, data cataloging, data privacy and risk management, and data quality.

4) **SAP Master DG**: SAP MDG is a MDM solution that provides comprehensive functionality for DQM, process management, and DG.
5) **Talend**: Known for its data integration capabilities, Talend also offers data quality and governance features, such as data preparation, stewardship, and cataloging, making it a versatile tool for managing the data lifecycle.
6) **Alteryx**: It provides a user-friendly, self-service DG model that empowers business users and data analysts to discover and catalog data assets, fostering collaboration between business and IT.
7) **SAS Data Management**: This is an enterprise-level DG tool offering functionalities like data integration, data quality, data preparation, and MDM.

The right tool depends on the specific needs and objectives of your organization's DG program. Consider factors like usability, scalability, integration capabilities, security, and compliance features when choosing a DG tool. Also, remember to review recent updates, as new features and functionalities are regularly added to these tools.

6.6 The Role of Quality Professionals

As data continues to fuel digital transformation and decision-making processes across all industries, the role of quality professionals in data quality has become more critical than ever. In this section, we will explore the role of these key individuals, their evolving responsibilities, and the significant impact they can have on an organization's data quality initiatives. Quality professionals, with their deep understanding of quality management principles and processes, are ideally positioned to champion data quality, turning raw data into a strategic asset that drives business improvement. They play a significant role in establishing DG strategies, implementing data quality standards, and leading data stewardship activities. This section underscores why quality professionals are essential in managing, monitoring, and improving data quality, and how they can contribute to fostering a data-driven culture within an organization.

Quality professionals have a critical role in ensuring data quality within an organization. As experts in quality management principles, they bring a unique perspective to data management that can significantly improve the integrity, reliability, and usability of data. Here are some roles and responsibilities of quality professionals in data quality:

1) **Establish Data Quality Standards**: Quality professionals are responsible for establishing data quality standards that align with the business objectives and regulatory requirements of the organization. They ensure that these standards promote accuracy, consistency, completeness, and relevance in the data.
2) **Implement Data Quality Measures**: They play a pivotal role in implementing measures to assess and monitor data quality. This involves the use of various data quality dimensions such as accuracy, completeness, consistency, validity, timeliness, and relevance.
3) **Develop Data Quality Management Strategies**: Quality professionals often lead the development of DQM strategies. This includes determining the methodologies, technologies, and processes to be used to ensure and improve data quality.
4) **Conduct Data Quality Audits**: Quality professionals are typically responsible for conducting regular data quality audits. These audits help identify any data issues, assess the effectiveness of the current data management practices, and determine areas of improvement.
5) **Collaborate with IT and Business Units**: They act as a bridge between IT and business units to ensure that data quality initiatives are understood and implemented effectively across the

organization. They also facilitate communication between these units, fostering a shared understanding and commitment to data quality.

6) **Promote DG**: Quality professionals play an important role in promoting DG within the organization. They help establish DG policies and practices and promote adherence to these standards.

7) **Data Stewardship**: In many organizations, quality professionals take on the role of data stewards. They ensure that data is used and managed according to established policies and guidelines, and they often oversee the resolution of data quality issues.

The reason why these roles and responsibilities fall within the purview of quality professionals is because of their expertise in quality management principles and processes. They understand the importance of quality control and assurance, and they are experienced in identifying and resolving quality issues. With the increasing importance of data in today's digital age, these skills are essential to ensure that data is reliable, accurate, and fit for purpose.

To successfully play the role of a data quality professional, it is important to gain new skills and knowledge that align with the changing data landscape. Here are some key areas for quality professionals to consider:

1) **Understanding of Data Governance and Data Management**: Quality professionals should understand the principles of DG and data management, which include data quality, metadata management, data privacy, and data security. This will help them design and implement effective data quality strategies and programs.

2) **Data Analysis Skills**: Ability to work with data, including data cleaning, manipulation, analysis, and interpretation, is crucial. This includes knowledge of data analysis tools and programming languages such as SQL, Python, or R.

3) **Knowledge of Data Quality Tools and Technologies**: There are many tools and technologies used in DQM, including data profiling, data quality rules implementation, data cleansing, matching, and monitoring tools. Familiarity with these tools will help quality professionals execute their roles more effectively.

4) **Understanding of Data-related Regulations**: With the growing concern around data privacy and security, it is crucial to be familiar with relevant data protection laws and regulations, such as GDPR, HIPAA, or CCPA.

5) **Project Management Skills**: Implementing data quality initiatives often requires a project management approach. Therefore, skills like planning, coordinating, and monitoring project activities are essential.

6) **Communication and Change Management Skills**: Ensuring data quality is not just a technical task, but also requires cultural change within the organization. Being able to communicate the importance of data quality, and manage the change associated with implementing new data quality initiatives, is a critical skill.

7) **Continual Learning**: Given the rapidly evolving nature of the data field, a commitment to continual learning and staying updated with the latest trends, technologies, and best practices in data management is vital.

By acquiring these skills and knowledge, quality professionals will be better equipped to ensure and enhance the quality of data in their organizations, thereby driving more effective decision-making and operational efficiencies.

Quality professionals, data analysts/scientists, IT professionals, and other stakeholders must work collaboratively in an organization to ensure data quality. Here is how these roles can effectively work together:

1) **Collaboration with Data Analysts/Scientists**: Quality professionals need to work with data analysts/scientists to understand the data requirements, ensure that the data is of high quality, and suitable for the intended analysis. They can also learn from data analysts/scientists about the specific problems they encounter with data quality during their analysis work. Furthermore, quality professionals can help analysts/scientists understand the importance of maintaining data quality and guide them on best practices.

2) **Collaboration with IT Professionals**: Quality professionals should work closely with IT professionals as they are typically the ones who manage the databases and systems where data is stored. IT can provide support with data management tools and technologies and help enforce data security measures. Together, they can implement data quality rules, standards, and validation checks within the organization's IT systems. Moreover, quality professionals can aid IT by defining business rules for data handling.

3) **Collaboration with Other Stakeholders**: Quality professionals need to work with all stakeholders to foster a data quality culture. This includes leadership, who need to understand the value and importance of data quality to support the necessary investments, as well as staff who need to be trained and motivated to follow data quality procedures. They can also work with stakeholders to understand their data needs, get their buy-in for data quality initiatives, and involve them in defining data quality requirements.

4) **Data Governance**: Quality professionals should be involved in DG bodies or committees within the organization, which often include representatives from various functions. These bodies oversee data standards, policies, and procedures, ensure alignment with business strategy, and resolve data-related issues.

By facilitating effective communication and collaboration between these groups, quality professionals can ensure that data quality initiatives are comprehensive, well-coordinated, and aligned with the needs and goals of the organization.

6.7 Future Trends in Data Quality

The rapid progression of technology perpetually transforms the data quality landscape, posing both hurdles, and possibilities for organizations across every sector. This section endeavors to provide a glimpse into the imminent trends in data quality, exploring the evolution of notions, the advent of innovative technologies, and the progress of methodologies that will shape the forthcoming generation of DQM. With the escalating volume and complexity of data, the urgency to maintain and enhance data quality has become paramount. By staying informed of these trends, organizations can better prepare to effectively harness their data, thereby catalyzing improved decision-making, boosting operational efficiency, and securing a competitive advantage. This section aims to offer a futuristic viewpoint on data quality, emphasizing the prospective strategies, tools, and skills essential to traverse this dynamic landscape in the upcoming years.

Here are potential future trends in data quality:

1) **Enhanced Automation in Data Quality Management**: Automation technologies, like AI and ML, are predicted to play a pivotal role in data quality tasks such as data cleaning, validation, and monitoring. Their utilization will enable organizations to handle more extensive data volumes efficiently, reduce human errors, and free up resources for tasks of higher value [25].

2) **Real-time Data Quality**: As real-time data processing and decision-making gain momentum, the demand for instantaneous data quality checks also intensifies. This can yield more immediate and relevant insights, thereby improving decision-making speed and accuracy [26].

3) **Data Privacy and Security**: With tightening data privacy regulations globally, ensuring data quality also signifies assuring its privacy and security. This involves validating personal data accuracy, safeguarding its usage and storage, and providing rectification or erasure when needed [27].

4) **Data Quality in the Cloud**: With the increasing shift of data to the cloud, ensuring data quality across distributed systems presents a notable challenge. Understanding the impact of cloud technologies on data quality becomes crucial [28].

5) **Integration of Multiple Data Sources**: As data sources diversify and expand, the need to integrate these sources while preserving data quality grows. This necessitates a solid understanding of how to manage and integrate various data sources.

6) **Adoption of Data Quality Standards**: With heightened emphasis on data quality, more organizations are expected to adopt data quality standards and frameworks, thereby enhancing consistency and reliability [11].

7) **Emphasis on Data Literacy**: As data becomes a more critical factor in decision-making, there is a growing focus on data literacy—the ability to read, understand, generate, and communicate data effectively. This includes understanding data quality concepts and their importance.

To remain current with these trends, it is vital to keep informed about the latest advancements in the data field. This could involve tracking relevant news, blogs, and social media, attending webinars, conferences, and training courses, and participating in professional networks and forums. Pursuing further studies or certifications in data management or related fields to enhance your knowledge and skills is also advisable.

References

1 Ivanov, K. (1972). Quality-control of information: on the concept of accuracy of information in data-banks and in management information systems. Doctoral dissertation. KTH Royal Institute of Technology.

2 Hansen, M.D. (1991). Zero defect data. Doctoral dissertation. Massachusetts Institute of Technology.

3 Kahn, B.K., Strong, D.M., and Wang, R.Y. (2002). Information quality benchmarks: product and service performance. *Communications of the ACM* 45 (4): 184–192.

4 Wand, Y. and Wang, R.Y. (1996). Anchoring data quality dimensions in ontological foundations. *Communications of the ACM* 39 (11): 86–95.

5 Wang, R.Y. (1998). A product perspective on total data quality management. *Communications of the ACM* 41 (2): 58–65.

6 Weidema, B.P. and Wesnæs, M.S. (1996). Data quality management for life cycle inventories—an example of using data quality indicators. *Journal of Cleaner Production* 4 (3–4): 167–174.

7 Batini, C., Cappiello, C., Francalanci, C., and Maurino, A. (2009). Methodologies for data quality assessment and improvement. *ACM Computing Surveys (CSUR)* 41 (3): 1–52.

8 Chapman, A.D. (2005). *Principles of Data Quality*. GBIF.

9 Côrte-Real, N., Ruivo, P., and Oliveira, T. (2020). Leveraging internet of things and big data analytics initiatives in European and American firms: is data quality a way to extract business value? *Information and Management* 57 (1): 103141.

10 Wang, R.Y. and Strong, D.M. (1996). Beyond accuracy: what data quality means to data consumers. *Journal of Management Information Systems* 12 (4): 5–33.

11 Pipino, L.L., Lee, Y.W., and Wang, R.Y. (2002). Data quality assessment. *Communications of the ACM* 45 (4): 211–218.

12 Batini, C. and Scannapieco, M. (2016). *Data and Information Quality*. Cham, Switzerland: Springer International Publishing.

13 Redman, T.C. (1997). *Data Quality for the Information Age*. Artech House, Inc.

14 Davenport, T.H. and Prusak, L. (1997). *Information Ecology: Mastering the Information and Knowledge Environment*. Oxford University Press on Demand.

15 English, L.P. (1999). *Improving Data Warehouse and Business Information Quality: Methods for Reducing Costs and Increasing Profits*. Wiley.

16 Hüner, K.M., Ofner, M., and Otto, B. (2009). Towards a maturity model for corporate data quality management. In: *Proceedings of the 2009 ACM symposium on Applied Computing*, 231–238. ACM (Association for Computing Machinery).

17 Lucas, A. (2010). Corporate data quality management: From theory to practice. In: *5th Iberian Conference on Information Systems and Technologies*, 1–7. IEEE.

18 Ehrlinger, L. and Wöß, W. (2022). A survey of data quality measurement and monitoring tools. *Frontiers in Big Data* 28.

19 Chien, M. and Jain, A. (2022). Critical capabilities for data quality solutions. Gartner Research. https://www.gartner.com/en/documents/4020778 (accessed 2 November 2022).

20 Redman, T.C. (1992). *Data Quality: Management and Technology*. Bantam Books, Inc.

21 Thomas, G. (2006). *The DGI Data Governance Framework*, 20. Orlando, FL, USA: Data Gov. Institute.

22 Weber, K., Otto, B., and Österle, H. (2009). One size does not fit all—a contingency approach to data governance. *Journal of Data and Information Quality (JDIQ)* 1 (1): 1–27.

23 Al-Ruithe, M., Benkhelifa, E., and Hameed, K. (2019). A systematic literature review of data governance and cloud data governance. *Personal and Ubiquitous Computing* 23: 839–859.

24 Gartner, Inc. (2022). A Guide to Gartner Data Governance Research—Market Guides, Hype Cycles, and Peer Reviews. https://atlan.com/gartner-data-governance/ (accessed 4 April 2023).

25 Williams, D. and Tang, H. (2020). Data quality management for industry 4.0: a survey. *Software Quality Professional* 22 (2): 26–35.

26 Woodall, P., Oberhofer, M., and Borek, A. (2014). A classification of data quality assessment and improvement methods. *International Journal of Information Quality 16* 3 (4): 298–321.

27 Singh, D. (2020). Towards data privacy and security framework in big data governance. *International Journal of Software Engineering and Computer Systems* 6 (1): 41–51.

28 Fan, W. and Bifet, A. (2013). Mining big data: current status, and forecast to the future. *ACM SIGKDD Explorations Newsletter* 14 (2): 1–5.

7

Risk Management in the 21st Century

7.1 Introduction

In the fast-paced, interconnected world of the 21st century, organizations confront a spectrum of risks that could potentially influence their operations, reputation, and overall success. Risk management has, therefore, evolved as an indispensable discipline that aids in the navigation of uncertainties and proactive confrontation of possible threats. This chapter delves deep into the domain of risk management within the context of the 21st century, exploring the mutating risk landscape, methodologies for risk identification, assessment, and mitigation, as well as the emerging trends that are set to shape the future of risk management [1–4]. A growing body of literature on risk management underscores its growing importance in a volatile, uncertain, complex, and ambiguous (VUCA) world [5]. Borge [6] presents a comprehensive guide on risk management, examining different risks and corresponding mitigation strategies. He argues that understanding risks is crucial for strategic planning and decision-making in organizations.

The chapter initially builds a foundational understanding by defining risk and inspecting its various manifestations. It delves into a wide range of risks that have arisen in the 21st century, including technological advancements, environmental issues, socioeconomic shifts, geopolitical dynamics, and health crises. By understanding the nature and breadth of these nascent risks, organizations are better equipped to face the challenges of an evolving risk landscape.

Effective risk management necessitates robust frameworks to guide organizations' efforts. This chapter scrutinizes both traditional and modern risk management approaches, underscoring the necessity to integrate risk management with strategic planning. It delves into the crucial components of risk identification and assessment, showcasing diverse techniques to recognize and evaluate risks. Emphasis is placed on the comprehensive understanding of risks via both quantitative and qualitative analyses.

Risk mitigation and control are fundamental in effective risk management. This chapter investigates risk response strategies, including risk transfer and insurance, risk avoidance and reduction, and constant risk monitoring and control. It also highlights the significance of fostering a risk-aware organizational culture and emphasizes the crucial roles leadership, communication, and stakeholder engagement play in successful risk management.

Moreover, the chapter acknowledges the revolutionary impact of technology on risk management practices. It explores the role of automation, artificial intelligence (AI), and data analytics in

Quality in the Era of Industry 4.0: Integrating Tradition and Innovation in the Age of Data and AI,
First Edition. Kai Yang.
© 2024 John Wiley & Sons, Inc. Published 2024 by John Wiley & Sons, Inc.

improving risk assessment and prediction. It also underscores the criticality of resilience and business continuity planning in enabling organizations to withstand and recover from adverse events.

Illustrated through engaging case studies and best practices, this chapter provides real-world exemplars of effective risk management. It concludes by predicting future trends in risk management, considering the impact of technology innovations, climate change, and socioeconomic shifts on the risk landscape. By staying attuned to these trends, organizations can proactively adapt their risk management strategies to thrive in an ever-evolving world.

In summation, this chapter's objective is to furnish readers with a comprehensive understanding of risk management in the 21st century. By traversing various facets of risk management, from risk identification to mitigation, and by inspecting the crucial role of organizational culture, technology, ethics, and resilience, this chapter offers insights and guidance to traverse the complex risk landscape of the modern era.

The role of quality professionals in the 21st century has expanded beyond ensuring product and process excellence. They now need to master risk management to fulfill their responsibilities in a changing world. Quality professionals have expertise in identifying and addressing sources of variability and nonconformities, making them well-suited for contributing to risk management. By incorporating risk management principles, they can anticipate and address risks, protecting the delivery of high-quality products, organizational reputation, and customer satisfaction.

In today's complex business environment, risks are interconnected and multifaceted. Quality professionals, with their understanding of processes, supply chains, and customer expectations, can identify risks affecting product quality, regulatory compliance, and operational effectiveness. By integrating risk management into their systems, they can assess, prioritize, and mitigate risks, minimizing disruptions, product failures, and noncompliance.

The 21st century introduces new and emerging risks that demand the attention of quality professionals. Technological advancements, environmental concerns, and societal changes introduce novel risks requiring adaptive strategies. Quality professionals can identify these risks and develop effective mitigation strategies, contributing to organizational resilience and sustainability.

Mastering risk management empowers quality professionals to take on leadership roles. By advocating for risk management practices and integrating them into decision-making processes, they shape a proactive and risk-aware organizational culture. Analytical skills enable quality professionals to utilize data for risk assessment and prediction, identifying patterns and indicators to inform decision-making and targeted mitigation strategies.

In conclusion, mastering risk management is crucial for quality professionals in the 21st century. By embracing risk management principles, they contribute to organizational success and resilience. Their ability to identify, assess, and mitigate risks positions them as key players in navigating the complexities of the business landscape, ensuring high-quality products, safeguarding reputation, customer satisfaction, and long-term viability.

7.1.1 Overview of Risk Management

Risk management signifies a proactive, systematic procedure for the identification, evaluation, prioritization, and mitigation of risks that aims to accomplish organizational objectives and augment decision-making. It entails the acknowledgment of potential uncertainties, both positive and negative, while strategizing actions to curtail threats and exploit opportunities. In the current intricate and interconnected world, risk management has evolved into a crucial discipline spanning across industries and sectors.

The fundamental objective of risk management is to facilitate organizations to proficiently navigate through uncertainties and safeguard their interests. It necessitates understanding the nature and extent of risks that an organization may encounter, whether originating from internal factors, external factors, or a combination of both. By comprehensively identifying risks, organizations can cultivate a holistic understanding of the potential impact and the probability of occurrence, thus enabling them to allocate resources and devise suitable risk mitigation strategies.

Risk management encapsulates a series of activities and processes collaboratively working to address risks. These comprise risk assessment, which involves evaluating the significance and potential repercussions of identified risks, and risk analysis, which delves into examining the root causes and contributing factors of the risks. Additionally, risk management includes the selection and execution of risk response strategies, such as risk transfer, risk avoidance, risk reduction, or acceptance, contingent on the organization's risk appetite and capacity.

A robust risk management framework offers a structured and consistent method for managing risks across an organization. It delineates clear roles, responsibilities, and processes for identifying, evaluating, and responding to risks. Risk management frameworks also contemplate the context in which an organization operates, considering factors such as legal and regulatory requirements, industry standards, and stakeholder expectations.

Organizational culture assumes a pivotal role in risk management. A risk-aware culture promotes open communication, transparency, and accountability concerning risks. It encourages a proactive mindset, where individuals across all organizational levels actively partake in risk identification and mitigation. Leadership bears an integral role in advocating and integrating risk management practices, setting the tone for risk awareness and accountability throughout the organization.

In the contemporary world, technology is progressively significant in risk management. Automation, data analytics, and AI tools can augment risk assessment, prediction, and monitoring. These tools empower organizations to amass and analyze vast amounts of data, recognize patterns and trends, and make informed decisions regarding risk management strategies.

In conclusion, risk management is an essential discipline that enables organizations to navigate through uncertainties and make knowledgeable decisions in the pursuit of their objectives. It involves systematically identifying and assessing risks, implementing suitable risk response strategies, and fostering a risk-aware culture. By incorporating risk management into their operations, organizations can proactively address potential threats and seize opportunities, ensuring long-term resilience, success, and sustainability.

7.1.2 Redefining Risk Management in the 21st Century

As the 21st century unfolds, we observe a marked evolution in risk management catalyzed by an intricate, globally interconnected environment. Traditional paradigms, once centered on financial and operational risks, have expanded to a broader and anticipatory perspective reflective of today's complexities.

Swift advancements in technology serve as a pivotal factor in this transformation. The rise of digitalization, automation, and interconnected systems has engendered novel risks that mandate astute navigation by organizations. Cybersecurity threats, data breaches, and privacy issues are now key concerns in our digital epoch, necessitating adjustments in risk management to incorporate robust measures that safeguard critical information and thwart cyberattacks.

In parallel, environmental risks have escalated in significance throughout the 21st century. Climate change, natural disasters, and resource scarcity pose grave threats to diverse sectors.

Consequently, risk management has integrated sustainability, formulating strategies to mitigate environmental risks, fortify resilience, and promote responsible practices. The effects of environmental risks extend beyond the planet, impacting an organization's operations, reputation, and bottom line.

Additionally, socioeconomic risks have gained prominence. Societal trends, fueled by rapid population growth, income inequality, and political instability, coupled with trade disruptions, present formidable challenges. Risk management has evolved to encompass these complexities, engaging with stakeholders and assessing the potential impact of unforeseen events.

Global health crises, such as the COVID-19 pandemic, have further sculpted the contours of risk management. Emphasizing preparedness and response to health-related risks, risk management practices now include pandemic planning, business continuity strategies, and supply chain resilience [7].

Risk management's integration into strategic planning heralds another significant shift. No longer viewed merely as a compliance measure, risk management is acknowledged as a strategic imperative [8], weaving risk considerations into the decision-making fabric to ensure careful evaluation of risks and opportunities.

In conclusion, technological progress, environmental concerns, socioeconomic shifts, geopolitical dynamics, and health crises have sculpted the risk management landscape in the 21st century. Beyond traditional confines, risk management now adopts a more holistic, proactive approach, enabling organizations to effectively navigate the labyrinth of modern risks and thrive amidst uncertainty.

7.1.3 The Paramountcy of Risk Management in the Contemporary Context

In our dynamic, interconnected world, risk management has emerged as a critical discipline. The evolving business landscape, with its technological advancements, geopolitical shifts, environmental concerns, and global uncertainties, calls for a proactive and strategic approach to risk management. Understanding the importance of risk management in today's context allows organizations to navigate uncertainty effectively and safeguard their interests.

Increasing complexity and interconnectedness of risks [9, 10] necessitate a comprehensive approach to risk management. It provides a structured framework to identify, assess, and prioritize risks, allowing organizations to allocate resources effectively and develop targeted mitigation strategies.

In the age of rapid technological advancements, risk management's significance amplifies. The dual nature of technology, bringing opportunities alongside risks such as cyber threats and data breaches, necessitates an integrated risk management approach [11, 12]. It enables organizations to anticipate and mitigate technology-related risks while harnessing technology's benefits responsibly.

Environmental risks demand an effective risk management strategy to understand and mitigate their potential impact on operations, supply chains, and long-term sustainability [13, 14]. Risk management fosters proactive responses to environmental risks, reducing ecological footprints and seizing opportunities arising from sustainability practices.

Risk management is also instrumental in navigating socioeconomic risks, helping organizations anticipate and respond to political instability, trade disruptions, social unrest, and economic fluctuations. It enhances their resilience and adaptability in an ever-changing business environment.

Further, risk management serves as a crucial element of corporate governance and compliance. Regulatory bodies and stakeholders expect robust risk management frameworks and effective risk management practices meet these requirements while enhancing reputations and fostering stakeholder trust.

In summary, risk management's importance lies in helping organizations navigate complexity, uncertainty, and interconnected risks. A proactive and strategic approach to risk management fosters resilience, sustainability, and stakeholder trust in an increasingly uncertain and challenging business landscape. It not only safeguards organizations from potential threats but also leverages opportunities, ensuring business continuity and protecting organizational interests.

7.2 Deciphering the Nature of Risk

Risk, an omnipresent facet of our lives, shapes the decisions we make and the outcomes we encounter. As our world grows increasingly complex and dynamic, understanding risk is crucial for individuals, organizations, and societies. Risk extends beyond mere uncertainty; it encapsulates potential hazards, threats, and opportunities impacting our goals, well-being, and overall success. By comprehending risk's multifaceted nature, we can navigate uncertainty more effectively, make informed decisions, and proactively manage potential consequences.

At its essence, risk can be conceptualized as the probability of an event occurring and its potential impact [15]. Risks can germinate from a plethora of sources—natural disasters, technological disruptions, financial volatility, legal complexities, and human behavior. Each risk presents unique characteristics like likelihood, severity, timing, and interconnectedness with other risks [16]. Understanding these attributes is fundamental to grasping the true nature and magnitude of the risks we face.

Effective risk understanding necessitates robust methodologies for risk assessment and analysis. Both quantitative and qualitative approaches facilitate risk evaluation through the analysis of historical data, statistical models, expert opinions, and scenario planning [17]. Systematic risk assessment provides insights into the potential consequences of various risks, aiding in prioritizing and effective resource allocation.

Yet, it is essential to perceive risk as more than a negative concept. Risks also herald opportunities for growth, innovation, and advancement [18]. Embracing and harnessing risks can yield breakthrough discoveries, competitive advantages, and organizational resilience. Understanding risk entails recognizing the potential benefits of calculated risks and weighing them against potential downsides.

Understanding risk extends beyond individual decision-making. Organizations and societies grapple with risks on a larger scale, with impacts that reverberate through economies and communities. Risk management practices offer a structured framework for identifying, assessing, and mitigating risks [19]. By integrating risk management into strategic planning, organizations can enhance their uncertainty navigation ability, protect their interests, and seize opportunities.

As we explore risk, we will delve into various risk types that individuals and organizations encounter, each with unique challenges and considerations. From financial and operational risks to technological, environmental, and societal risks, understanding these diverse risk dimensions allows us to develop tailored approaches for risk mitigation and response [20].

As we commence this journey to understand risk, our aim is to equip ourselves with the knowledge, tools, and insights for effective uncertainty navigation. Embracing risk as an integral part of our decision-making processes enables informed choices, opportunities, and enhanced adaptability in an ever-changing world. Through this exploration of risk, we can unlock the potential for growth, innovation, and resilience in both personal and organizational contexts.

7.2.1 Definition of Risk

Risk management has appeared in scientific and management literature since the 1920s. It became a formal science in the 1950s [2]. Most of the research was initially related to finance and insurance.

Risk, in a scientific context, is defined as the probability or likelihood of a negative event or outcome occurring due to a particular action, decision, or event, combined with the magnitude of the consequences should the event occur.

It is typically represented quantitatively as a combination of the magnitude of potential loss (consequence), and the probability (likelihood) that the loss will occur. It can be expressed mathematically as:

$$\text{Risk} = \text{Probability}(\text{of an event}) \times \text{Impact}(\text{of the event}) \tag{7.1}$$

In this formula:

- "Probability" refers to the likelihood of an event occurring. It is often represented as a value between 0 and 1, where 0 indicates that the event will not occur and 1 indicates that the event will certainly occur.
- "Impact" refers to the magnitude or severity of the consequences if the event does occur. The impact can be quantified in various ways depending on the nature of the risk, such as financial loss, number of people affected, or extent of environmental damage.

Here, the event refers to any particular situation that leads to a negative consequence, and the impact refers to the severity of the negative consequence. The probability is a measure of the frequency of occurrence of the negative event.

In broader perspectives, risk can also encompass uncertainty, where the outcomes may be unknown, and the probabilities may not be well-defined.

In various fields such as finance, engineering, health, environment, and others, different specific definitions of risk may be used, but the underlying concept of weighing the likelihood and impact of negative outcomes remains the same.

Since 1990s, people in risk management community, such as ISO, Institute of Risk Management (IRM), realized that risk should not be perceived solely as a negative concept. Risks also bring opportunities for growth, innovation, and advancement. According to the ISO 31000 Risk Management standard [21]. The ISO defines risk as the "effect of uncertainty on objectives." This definition expands the common perception of risk to incorporate any type of uncertainty that could affect the achievement of objectives, not only those with negative outcomes.

To break down this definition further:

1) **"Effect"**: This refers to a deviation from the expected, which can be positive or negative.
2) **"Uncertainty"**: This encompasses the lack or insufficiency of information, understanding or knowledge about an event, its consequences, or its likelihood.
3) **"Objectives"**: These could be related to various contexts, such as financial, health, safety, and environmental goals, and can apply at different levels such as strategic, organization-wide, project, product, and process.

So, according to this definition, a risk does not necessarily always have a negative impact. Instead, it can result in opportunities (positive impacts) when the uncertainty has an effect that may benefit the objectives. This highlights the importance of the process of risk management in identifying and harnessing potential opportunities while mitigating negative outcomes.

7.2.2 Types of Risks

If we exam the risk from mathematical viewpoint, we can classify the risks into the following three categories based on the numerical ranges of the impact, assuming negative means loss or undesirable, and positive means benefits or opportunities:

1) **Hazard (or Pure) Risk:** This type of risk arises from situations where there are only possibilities for loss, and no opportunities for gain, the impact is negative. These risks are typically insurable. Examples can include natural disasters such as earthquakes, hurricanes, and floods, or accidents, illnesses, and death. Hazard risks are often managed through risk transfer mechanisms, such as insurance, and by implementing safety and prevention measures.

2) **Control (or Uncertainty) Risk:** This type of risk involves situations where the outcome, or impact is uncertain, but the range of outcomes is known. These risks can be managed but not completely controlled, hence the name "control risk." For instance, changes in market conditions, such as fluctuations in prices, interest rates, or exchange rates, fall into this category. These risks are usually managed using techniques such as hedging and diversification.

3) **Opportunity (or Speculative) Risk:** These are risks that involve both the possibility of loss and the potential for gain, so the impact can be both positive or negative. Opportunity risks occur whenever there is uncertainty associated with a decision that includes a trade-off between potential outcomes. Examples could include launching a new product, entering a new market, or making an investment. While there is a chance these actions could result in losses, they also have the potential to result in gains. The management of these risks often involves a strategic balance between risk and reward and may include techniques such as scenario planning, market research, and cost–benefit analysis.

Risks can also be categorized in the field of study or industry; here are some common types of risks:

1) **Strategic Risk:** These risks stem from a company's business strategy and strategic objectives. They could include risks from mergers and acquisitions, partnerships, changes in management, or competition.

2) **Operational Risk:** This includes risks that arise from daily business operations such as system failures, process failures, fraud, theft, and other disruptions to the normal functioning of a business.

3) **Financial Risk:** Financial risks relate to a company's financial health and management. This could include currency risk, liquidity risk, credit risk, interest rate risk, and others. Financial risks could significantly impact a company's ability to meet its financial obligations.

4) **Compliance Risk:** These are risks related to legal and regulatory compliance. They include risks associated with failing to follow laws, regulations, and standards that govern the industry or sector in which a business operates.

5) **Reputational Risk:** This type of risk pertains to damage to a business's reputation or standing due to a negative event, scandal, or any form of negative publicity.

6) **Market Risk:** This refers to the risk of losses in positions arising from movements in market prices. It is often associated with investments and is subdivided into four standard types: interest rate risk, equity risk, currency risk, and commodity risk.

7) **Cybersecurity Risk:** These are risks associated with cyber threats or data breaches that could compromise a business's data integrity or security.

8) **Environmental Risk:** These risks involve potential damage to the physical environment due to a company's operations. It also includes the potential impact of climate change and natural disasters on a business's operations.

9) **Health and Safety Risk:** These risks are associated with the potential harm to the health and safety of employees, customers, or the public caused by the company's operations or products.
10) **Project Risk:** These are potential problems that could impact the successful completion of a project, including schedule delays, budget overruns, resource shortages, and technical risks.

Each of these types of risks requires specific strategies for identification, assessment, and mitigation. It is also important to note that these categories often overlap, and a single event could potentially cause multiple types of risks. Therefore, comprehensive and integrated risk management strategies are essential.

7.2.3 Risk Assessment and Analysis

Risk assessment and analysis are crucial steps in the risk management process. They involve identifying potential risks, estimating the likelihood of those risks occurring, and calculating the potential impact of these risks [22, 23]. The goal of risk assessment and analysis is to provide a basis for risk mitigation, prevention, and management decisions.

Here is a more detailed overview:

1) **Risk Identification:** The first step in the process involves identifying potential risks. This can be accomplished through various methods, including brainstorming sessions, expert consultations, historical data analysis, risk checklists, and scenario analysis, among others. The result is a list of potential risk events that could negatively affect the project, business, or other endeavors.
2) **Risk Analysis:** Once the risks are identified, the next step is to analyze each risk. This usually involves two key components:
   • **Likelihood Estimation:** This involves determining the probability of each risk event occurring. Likelihood can be estimated using historical data, expert judgment, or statistical techniques.
   • **Impact Assessment:** This involves determining the potential consequences if the risk event were to occur. Impacts can be quantified in various ways, such as cost overruns, delays, lost revenue, damage to reputation, or harm to individuals or the environment.
3) **Risk Evaluation:** After analysis, the risks are then evaluated. This typically involves comparing the level of risk against predefined criteria or thresholds. The purpose of this step is to prioritize risks and to make decisions about which risks need to be treated and in what order.
4) **Risk Quantification:** Risk quantification involves assigning a numerical value to the risk. This is usually calculated as the product of likelihood and impact (Risk = Probability * Impact). The resulting risk score allows for risks to be ranked and compared.

The results of risk assessment and analysis form the basis for the next steps in the risk management process, which are risk treatment and monitoring. Risk treatment involves deciding how to respond to the risks—whether to accept, avoid, transfer, or mitigate them. Risk monitoring involves tracking identified risks, regularly reviewing the risk environment, and updating risk assessments as necessary.

It is important to note that risk assessment and analysis should be a continuous and iterative process, as the risk environment can change over time due to various factors, including changes in the external environment, the progress of the project, or other changes in the organization.

In actual risk assessments and analysis, we often use well-defined scales to identify and assess the likelihood of risk (probability) and impact. It helps to standardize the assessment and makes the process more transparent and consistent. In the following example, we assume that all the impact is negative (bad). Table 7.1a provides an example of such a scale.

Table 7.1a   Risk Likelihood Scale

Rating	Definition
1	**Very Unlikely**: The risk event may only occur in exceptional circumstances
2–4	**Unlikely**: The risk event could occur at some point
5–7	**Likely**: The risk event is expected to occur in most circumstances
8–9	**Very Likely**: The risk event is expected to occur frequently
10	**Certain**: The risk event is expected to occur in most circumstances, almost a certainty

Table 7.1b   Risk Impact Scale

Rating	Definition
1	**Minimal**: The risk event would cause a minor inconvenience with no significant delay or cost increase
2–4	**Minor**: The risk event would cause a slight increase in cost or delay, but would not significantly affect the overall project
5–7	**Moderate**: The risk event would cause a noticeable increase in cost or delay, possibly affecting the overall project
8–9	**Major**: The risk event would cause a significant increase in cost or delay and would likely affect the overall project success
10	**Catastrophic**: The risk event would cause such a massive increase in cost or delay that it could jeopardize the entire project

The scales listed in Tables 7.1a and 7.1b are just examples, and the actual definitions may vary depending on the nature of the project, the industry, and other factors. The scales should be defined in a way that makes sense for the particular context and provides useful input for the risk assessment and decision-making process.

Example 7.1   Risk Assessment and Analysis

Let us consider a hypothetical project with certain identified risks. We will quantify these risks using likelihood, impact and calculate the risk score. We'll make a Risk Assessment Table for better illustration, as in Table 7.2.

Table 7.2   Sample Risk Assessment Table

Risk	Likelihood (1–10)	Impact (1–10)	Risk Score (Likelihood × Impact)
Equipment failure	4	7	28
Budget overrun	6	8	48
Delay in supply chain	8	6	48
Regulatory changes	2	9	18
Key personnel leaves	3	8	24

In this table:

- Likelihood is a rating from 1 to 10, where 1 means the risk is unlikely to occur, and 10 means it is almost certain to occur.
- Impact is also rated from 1 to 10, where 1 means the impact will be minor if the risk occurs, and 10 means the impact will be very significant.
- Risk Score is the product of Likelihood and Impact. It gives a numerical measure of the overall risk level.

Here is a detailed analysis of the risks:

1) **Equipment Failure:** The likelihood is estimated as 4, as the equipment is relatively new and well-maintained. However, if it were to fail, the impact would be significant (7), as it would cause project delays and increase costs. The overall risk score is 28.
2) **Budget Overrun:** The likelihood of budget overrun is estimated as 6 due to the complex nature of the project and the uncertainty in cost estimates. The impact is rated as 8, as it would directly affect the project's financial performance and potentially its feasibility. The overall risk score is 48.
3) **Delay in Supply Chain:** With recent disruptions in global logistics, the likelihood of this risk is relatively high (8). While it would cause delays, it is something the project has some flexibility to accommodate; hence, the impact is rated as 6. The overall risk score is 48.
4) **Regulatory Changes:** Changes in regulations related to this project are currently considered unlikely (2). However, if they were to occur, they could have a significant impact on the project, requiring major changes to project plans or even jeopardizing the project's viability (9). The overall risk score is 18.
5) **Key Personnel Leaves:** The likelihood of this risk is estimated as 3, as key personnel are well-compensated and have shown high levels of job satisfaction. However, if key personnel were to leave, it would be quite disruptive to the project (8). The overall risk score is 24.

Based on this analysis, "Budget Overrun" and "Delay in Supply Chain" are the highest-scoring risks and might be the first ones to be addressed in risk mitigation efforts.

It is important to note that this is a simplified example. Real-world risk assessments can be much more complex and may involve more detailed and sophisticated analysis methods.

7.3 Risk Management Frameworks

Risk management plays a crucial role in any organization, ensuring that potential threats and uncertainties are systematically identified, analyzed, and mitigated. Given the dynamism and complexity of the contemporary business environment, a robust risk management framework is more critical than ever. This framework serves as a structured and coherent approach to managing uncertainties that may impact an organization's objectives. This section delves into the salient aspects of risk management, offering an exploration of its evolutionary trajectory and how it can be tightly woven into the fabric of strategic planning. We begin by revisiting Traditional Risk Management Approaches that laid the foundation for how we perceive and handle risks. We then transition to scrutinizing contemporary risk management models, which accommodate the sophisticated and interconnected nature of modern risks. Lastly, we emphasize the importance of Integrating Risk Management with Strategic Planning, underscoring the strategic role of risk management in navigating toward success amidst uncertainties.

7.3.1 Traditional Risk Management Approaches

The genesis of traditional risk management approaches primarily traces back to the mid-20th century. These approaches primarily homed in on insurable risks such as the damage or loss of physical assets, liability risks, or personnel risks. The maturation of risk management during this epoch was heavily swayed by industries like insurance, finance, and engineering, culminating in a predominantly technical and compartmentalized outlook on risk management [24].

The post-World War II era heralded an amplified focus on risk management, reflecting the escalating complexities faced by burgeoning organizations. The emphasis predominantly gravitated toward hazard risks, which could be transferred utilizing insurance mechanisms. Risk was conceptually viewed as a phenomenon to be minimized or evaded. The adoption of quantitative risk assessment methodologies started to permeate, particularly in industries like nuclear power and aerospace, where considerable safety concerns prevailed [3].

The twilight of the 20th century observed a gradual metamorphosis from these traditional methodologies to more integrated and strategic risk management approaches. This transition was propelled by myriad factors, including advancements in information technology, burgeoning regulatory requirements, and high-profile corporate failures, all of which spotlighted the limitations of traditional risk management approaches [18].

Traditional Risk Management Approaches usually involve the following steps:

7.3.1.1 Risk Identification

This step is to identify the potential risks that could affect the organization. It is the first step in the risk management process and plays a vital role in establishing a robust risk management framework. It involves recognizing and documenting potential events or circumstances that could negatively or positively affect the achievement of an organization's objectives. Here are several components and methods of the risk identification process:

- **Information Gathering Techniques**: A variety of techniques can be used to collect information about potential risks, including document reviews, interviews, surveys, and brainstorming sessions. It is often useful to involve people from different parts of the organization and with different types of expertise in this process.
- **Risk Identification Tools and Techniques**: These might include:
  1) **Checklists**: Comprehensive lists of common risks by category or by industry can be helpful, although they must be used carefully to avoid missing out on unique risks relevant to the particular context.
  2) **SWOT Analysis**: Strengths, Weaknesses, Opportunities, and Threats (SWOT) analysis can be used to identify both internal and external risks.
  3) **Scenario Analysis**: This involves imagining different scenarios that could occur in the future and identifying the risks associated with each scenario.
  4) **Risk Breakdown Structure (RBS)**: This is a hierarchical decomposition of risks, starting from high-level risk categories and moving down to detailed risk events.
  5) **Cause and Effect Diagrams**: These can help identify the root causes of risks.

6) **Process Flow Analysis**: This involves mapping out a process and identifying where risks may occur at each step.

- **Risk Categories**: Risks are typically categorized to provide structure to the risk identification process. Common risk categories might include strategic risks, operational risks, financial risks, regulatory risks, technological risks, environmental risks, and reputational risks. The exact categories used will depend on the nature of the organization and its context.

- **Risk Register**: Identified risks are documented in a risk register. For each risk, the register might also include information about its nature, potential causes, potential impacts, risk owner, and any current risk responses.

Effective risk identification requires a systematic approach and a proactive mindset. It is also an ongoing process, as new risks can emerge over time due to changes in the organization's internal or external environment. It is worth noting that while the focus is often on identifying negative risks (threats), it is also important to identify positive risks (opportunities) that could benefit the organization.

7.3.1.2 Risk Analysis

Risk Analysis is to evaluate the likelihood and impact of the identified risks. It is the process that comes after risk identification in the risk management process. It involves comprehending the nature of the risk, the level of risk, and determining appropriate management strategies. This process is vital as it provides a way for organizations to prioritize risks based on their potential impact and likelihood.

Risk analysis can be broken down into the following components:

- **Risk Evaluation:**
  This involves understanding the risk in terms of its potential impact and the likelihood of occurrence.
  1) **Impact Assessment:** Evaluating the potential impact involves understanding the possible consequences if the risk were to materialize. Impacts could be financial, reputational, operational, or strategic, among others. Impact can be assessed qualitatively (low, medium, high) or quantitatively (in terms of financial loss, time delay, etc.).
  2) **Likelihood Assessment:** This involves estimating the probability of the risk event occurring. It can be based on historical data, predictive models, expert judgment, or a combination of these.

- **Risk Quantification:**
  For some types of risks, it is possible to quantify the risk in terms of monetary value. This is often done using techniques like:
  1) **Expected Monetary Value (EMV):** This involves multiplying the potential impact of the risk by the likelihood of it occurring to give a monetary value.
  2) **Sensitivity Analysis:** This examines how different values of an independent variable impact a particular dependent variable under a given set of assumptions. This technique is used within specific boundaries that depend on one or more input variables.
  3) **Monte Carlo Simulation**: This is a computational algorithm that relies on repeated random sampling to obtain numerical results. This method allows for modeling complex situations where many random variables are involved and assessing the impact of risk.

- **Risk Prioritization:**
  After evaluating and quantifying risks, they need to be prioritized. This usually involves ranking them based on their potential impact and likelihood of occurrence. A common way to do this is

by using a risk matrix, plotting risks on a two-dimensional matrix with impact on one axis and likelihood on the other.

- **Risk Tolerance/Acceptance:**
Organizations also need to define their risk tolerance or acceptance levels, i.e., the amount of risk they are willing to accept. This can be different for different types of risk and can help in deciding which risks to treat and how.

Effective risk analysis is not a one-time activity. It needs to be an ongoing part of the risk management process, with regular updates as conditions change or new information becomes available. It is also important to note that while risk analysis can help to inform decision-making, it cannot eliminate all uncertainty. The aim is to reduce uncertainty to a manageable level and make informed decisions about risk management strategies.

7.3.1.3 Risk Treatment

This step is to develop strategies to manage the risks, often through risk transfer mechanisms like insurance. Risk treatment is the process of selecting and implementing measures to modify risk. This process involves the development of a set of actions to enhance opportunities and reduce threats to the project's objectives. It is an iterative process that includes the selection of one or more options for modifying risks, and implementing those options, and then reviewing the outcomes.

Risk treatment can be achieved through several strategies:

- **Risk Avoidance**: This involves taking actions that aim to prevent the risk from occurring or being expressed. For example, this could mean not proceeding with an activity or project that could potentially bring about the risk.
- **Risk Reduction or Mitigation**: This strategy involves taking steps to reduce the likelihood or impact of a risk. For instance, implementing safety procedures, installing redundancies, conducting training, or investing in security technologies can reduce various types of risks.
- **Risk Transfer**: This involves transferring or sharing the risk with other parties. This could be achieved through various means such as insurance, contractual agreements, outsourcing, or partnerships.
- **Risk Acceptance**: This occurs when the organization decides to accept the risk without trying to avoid, mitigate, or transfer it. This might be appropriate for risks that have a low likelihood and impact, or where the cost of other risk treatment strategies outweighs the potential benefit.
- **Risk Exploitation**: In case of positive risks (opportunities), strategies might involve exploiting the risk to ensure the opportunity is realized. This could mean allocating more resources to a certain area, adopting a new approach, or making changes in order to capitalize on the opportunity.

The chosen approach or combination of approaches becomes a part of the organization's risk management plan. Each risk should have an owner responsible for implementing the risk treatment plan and monitoring the risk over time.

The selection of an appropriate risk treatment strategy depends on various factors, including the nature of the risk, the potential impact and likelihood of the risk, the effectiveness of potential control measures, the costs and benefits of potential treatment options, and the organization's risk tolerance levels. It is important to note that not all risks can or should be fully eliminated—often, the goal is to bring the risk to a level that is acceptable given the organization's risk tolerance levels. Risk treatment should also be a dynamic and ongoing process, adjusting as the risk environment changes or as new risks emerge.

7.3.1.4 Risk Monitoring

This step is to conduct regular reviews and update the risk assessment and management strategies. Risk monitoring, often referred to as risk review or risk control, is an ongoing part of the risk management process. It involves regular tracking, evaluating, and communicating risks and the effectiveness of risk responses over time.

Risk monitoring is crucial because the risk landscape is dynamic, with new risks emerging and existing risks changing as the organization's internal and external environment evolves. Additionally, the effectiveness of risk responses can change over time and may need to be adjusted.

Key activities in the risk monitoring process include:

- **Tracking Identified Risks**: This involves keeping an eye on the identified risks to see if their likelihood or impact changes over time. Changes in the risk profile might necessitate a change in the risk response.
- **Monitoring Risk Metrics and Indicators**: Key Risk Indicators (KRIs) are metrics that provide an early signal of increasing risk exposure in various areas of the organization. Monitoring KRIs can help to spot trends and trigger a response before a risk becomes a significant issue.
- **Auditing and Reviewing Risk Responses**: It is important to regularly review and audit the implemented risk responses to ensure they are working as intended and are effective in managing the risks. If a risk response is not working effectively, it may need to be adjusted.
- **Reporting and Communicating about Risks**: Regular reporting on risks is crucial to keep all stakeholders informed about the risk landscape and the organization's risk management activities. This can include reporting on changes in significant risks, emerging risks, and the effectiveness of risk responses.
- **Updating the Risk Register**: The risk register should be regularly updated to reflect the current understanding of risks and the status of risk responses. This includes adding new risks as they are identified and removing risks that are no longer relevant.
- **Lessons Learned**: It is important to document what has been learned from successful and unsuccessful risk management and apply these lessons to future risk management activities.

The frequency of risk monitoring activities can vary depending on the nature of the risk and the rate of change in the organization's environment. Some risks might require daily monitoring, while others might only need to be reviewed monthly or annually.

In conclusion, risk monitoring ensures that an organization's risk management activities are effective and relevant and supports decision-making and continuous improvement in risk management. These steps are usually applied linearly and often siloed, with each department or function in an organization managing its own risks.

7.3.1.5 Pros and Cons of Traditional Risk Management Approaches

Pros

- **Simplicity**: Traditional methods are often simpler and easier to implement than more complex and integrated risk management methods.
- **Quantitative Focus**: These approaches often use quantitative methods, which can provide clear and objective assessments of risks.
- **Well Established**: These approaches are well established and widely understood, with a substantial body of knowledge and experience to draw on.

Cons

- **Limited Scope**: Traditional approaches often focus on insurable or hazard risks and may not adequately address other types of risks, such as strategic or operational risks.

- **Lack of Integration**: Risks are often managed in silos within an organization, which can lead to gaps and inconsistencies in risk management.
- **Reactive Nature**: These approaches are often reactive, focusing on dealing with risks after they have been identified rather than proactively seeking to anticipate and manage risks.

In conclusion, while traditional risk management approaches have contributed significantly to the field, they also have limitations that have led to the development of more advanced and integrated risk management models. The evolving nature of risks in the 21st century, influenced by technology, globalization, and societal changes, necessitates a more comprehensive and strategic approach to risk management.

7.3.2 Contemporary Risk Management Models

Contemporary Risk Management Models have evolved as a response to the limitations of traditional risk management approaches and the changes in the global business environment. Traditional methods, while effective for managing certain types of risks, often lack the ability to address more complex and interrelated risks. Contemporary models, on the other hand, take a more holistic, integrated, and strategic approach to risk management.

The evolution from the traditional approaches toward contemporary risk management models has been driven by several factors:

1) **Increased Complexity and Interdependence:** Modern organizations operate in a complex, interconnected world where risks can quickly propagate across organizational boundaries. This requires a more integrated approach to risk management.
2) **Change in the Nature of Risks:** With the rise of technology, globalization, and regulatory changes, the nature of risks has also changed, necessitating more advanced risk management approaches.
3) **Demand for Better Corporate Governance:** High-profile corporate failures have highlighted the importance of risk management as a part of good corporate governance.
4) **Strategic Focus:** There is a growing recognition that risk management needs to be aligned with an organization's strategy and objectives, rather than being treated as a separate function.

Here are several popular contemporary risk management models:

7.3.2.1 Enterprise Risk Management (ERM)

ERM [25, 26] is a strategic business discipline that supports the achievement of an organization's objectives by addressing the full spectrum of its risks and managing the combined impact of those risks as an interrelated risk portfolio. As a holistic approach to risk management, ERM involves assessing and addressing risks from all sources within an enterprise, both internal and external, to ensure that the organization's strategic goals are met.

The key components of an ERM process, often following guidelines of established frameworks like COSO ERM or ISO 31000, usually include:

1) **Risk Management Culture and Governance:** This involves establishing a risk-aware culture within the organization, where everyone understands the importance of managing risk. The organization's leadership plays a crucial role in establishing and enforcing this culture. This also involves defining the organization's risk appetite and risk tolerance levels.
2) **Risk Identification:** This involves identifying potential risks that could impact the organization's ability to achieve its strategic objectives. It could involve methods like brainstorming sessions, interviews, document reviews, or utilizing technology to detect potential risks.

3) **Risk Assessment:** After identifying risks, they are assessed in terms of their potential impact and likelihood of occurrence. The assessment can be both qualitative (e.g., high, medium, and low) or quantitative (e.g., potential financial loss).

4) **Risk Mitigation:** This involves developing and implementing strategies to manage the identified risks. These strategies can involve risk avoidance, reduction, sharing, or acceptance.

5) **Risk Reporting and Monitoring:** This involves regularly monitoring the identified risks and the effectiveness of the risk mitigation strategies. It also involves reporting on risks to stakeholders to ensure transparency and informed decision-making.

The benefits of ERM include:

- **A Holistic View of Risk**: Instead of addressing risks in silos, ERM provides a comprehensive view of risks across the organization.
- **Strategic Alignment**: ERM aligns risk management with the organization's strategic objectives, helping to ensure that risks are managed in a way that supports the achievement of these objectives.
- **Improved Decision-Making**: By providing a comprehensive understanding of risks, ERM can support more informed and effective decision-making.
- **Enhanced Risk Response**: ERM can improve the organization's ability to anticipate and respond to risks, enhancing its resilience.

However, implementing ERM can also have its challenges:

- **Requires Culture Change**: Implementing ERM often requires a shift in organizational culture toward a more risk-aware mindset.
- **Can Be Complex and Resource-Intensive**: Implementing ERM can be a complex process that requires significant resources and expertise.
- **Potential for Complacency**: There is a risk that the organization may become over-reliant on the ERM process and fail to identify new or emerging risks.

Despite these challenges, ERM has become an essential approach in today's complex and volatile business environment, allowing organizations to manage their risk portfolio more effectively and achieve their strategic objectives.

7.3.2.2 Operational Risk Management (ORM)

ORM is a subset of ERM that focuses on managing the risks in the organization's daily operations. This could include risks associated with personnel, processes, systems, or external events. The goal is to reduce operational losses and enhance operational efficiency.

ORM is a systematic approach to identifying, assessing, managing, and mitigating risks associated with an organization's day-to-day operations [27, 28]. Operational risk can be defined as the risk of loss resulting from inadequate or failed internal processes, people, and systems, or from external events. This definition includes legal risk but excludes strategic and reputational risk. The goal of ORM is to reduce the likelihood and impact of operational failures and improve the efficiency and effectiveness of operations.

Key Steps in the ORM Process

1) **Risk Identification:** Identifying the operational risks that the organization faces. This includes risks associated with personnel, processes, systems, and external events.

2) **Risk Assessment:** After risks are identified, they are assessed in terms of their potential impact and likelihood of occurrence. This often involves the use of risk matrices or other risk assessment tools.

3) **Risk Control/Mitigation:** This involves developing and implementing strategies to manage the identified risks. These could include preventive measures (to reduce the likelihood of a risk) and contingency measures (to reduce the impact of a risk).

4) **Risk Monitoring and Review:** This involves regularly monitoring the identified risks and the effectiveness of risk controls. It also involves reviewing and updating the risk management plan as necessary.

Benefits of ORM

- Enhances the organization's ability to achieve its objectives by reducing the likelihood and impact of operational failures.
- Supports more efficient and effective operations by identifying areas of weakness or inefficiency.
- Helps to protect the organization's reputation by reducing the likelihood of operational failures that could harm its reputation.
- Can lead to cost savings by reducing operational losses and improving operational efficiency.

Challenges of ORM

- Implementing an effective ORM process can be complex and resource-intensive. It requires a clear understanding of the organization's operations and the potential risks associated with them.
- Like all risk management approaches, ORM cannot eliminate all risks. There is always a residual risk that needs to be accepted and managed.
- The effectiveness of ORM is highly dependent on the risk culture within the organization. If the importance of managing operational risk is not recognized and supported by all levels of the organization, the ORM process may not be effective.

Despite these challenges, ORM is a crucial aspect of risk management, particularly for organizations with complex operations or operating in high-risk industries. It helps organizations to manage the risks that are inherent in their operations and to enhance their operational performance and resilience.

7.3.2.3 Strategic Risk Management (SRM)

SRM is a business discipline that focuses on identifying, assessing, and managing the risks that could affect the achievement of an organization's strategic objectives [29, 30]. This could include risks associated with changes in the competitive environment, strategic initiatives, mergers and acquisitions, or changes in customer preferences. SRM provides a structured framework for dealing with the uncertainties that could hinder an organization's ability to execute its strategy.

Strategic risks can arise from numerous sources such as changes in the business environment, disruptive technology, competitive forces, or from within the organization itself. They can also result from decisions concerning an organization's strategic initiatives such as mergers, acquisitions, partnerships, and entering new markets.

Key Steps in the SRM Process

1) **Strategic Planning and Risk Identification:** This involves understanding the organization's strategic objectives, identifying the potential risks that could affect these objectives, and understanding how these risks might impact the overall strategy.

2) **Risk Assessment:** After risks are identified, they are assessed based on their potential impact on the strategic objectives and the likelihood of their occurrence. This helps to prioritize the risks.

3) **Risk Mitigation and Strategy Adjustment:** This involves developing strategies to manage the identified risks. It could involve adjusting the organization's strategy to mitigate the risk,

implementing measures to reduce the risk, transferring the risk, or accepting the risk. The chosen approach should align with the organization's risk appetite.

4) **Monitoring and Review:** This involves regularly monitoring the strategic landscape and the identified risks, reviewing the effectiveness of risk mitigation strategies, and adjusting as necessary.

Benefits of SRM

- SRM enables organizations to proactively address the uncertainties that could affect the achievement of their strategic objectives. This can lead to more effective strategy execution and enhanced organizational performance.
- By providing a structured approach to managing strategic risks, SRM can support better decision-making and risk-informed strategic planning.
- SRM can enhance an organization's resilience by preparing it for potential strategic disruptions.

Challenges of SRM

- Strategic risks can be complex and difficult to predict. They often involve a high degree of uncertainty and can be influenced by numerous external and internal factors.
- Implementing an effective SRM process requires a clear understanding of the organization's strategic objectives and the strategic landscape, as well as the ability to anticipate potential disruptions.
- Like all risk management approaches, SRM cannot eliminate all risks. There is always a residual risk that needs to be accepted and managed.

Despite these challenges, SRM is a crucial aspect of risk management, particularly for organizations operating in volatile and uncertain business environments. It helps organizations to navigate the uncertainties of the strategic landscape and to achieve their strategic objectives.

7.3.2.4 Integrated Risk Management (IRM)

Integrated Risk Management (IRM) is an evolution of ERM that emphasizes a comprehensive and integrated approach to managing risks across all areas of an organization [31, 32]. It emphasizes the interdependencies between risks and the need for a coordinated approach to risk management. The goal of IRM is to manage risk in a manner that not only reduces the likelihood of negative outcomes but also enhances the potential for positive outcomes.

The key principle of IRM is the interconnectedness of risks. In modern organizations, risks are not isolated but often interdependent. Therefore, managing risks in silos can lead to an incomplete understanding of the organization's risk profile. IRM, by taking an integrated approach, aims to address this issue.

Key Steps in the IRM Process

1) **Risk Identification:** IRM begins with identifying risks across all areas of the organization. The identification process should recognize the interconnectedness of risks and the potential for a risk event in one area to impact other areas.
2) **Risk Assessment:** Similar to other risk management approaches, the identified risks are assessed in terms of their potential impact and likelihood of occurrence. However, the assessment process in IRM also considers the interdependencies between risks.
3) **Risk Mitigation:** This involves developing and implementing strategies to manage the identified risks. The risk mitigation strategies should take into account the interconnectedness of risks and the potential for risk mitigation actions in one area to impact risks in other areas.

4) **Risk Monitoring and Reporting:** This involves regular monitoring and reporting of risks to ensure that the risk management strategies are effective and that the organization's risk profile is accurately represented.

Benefits of IRM

- IRM provides a more comprehensive understanding of the organization's risk profile, enabling more informed decision-making and more effective risk management.
- By taking an integrated approach to risk management, IRM can help to manage the interconnected risks that modern organizations face.
- IRM can enhance the organization's resilience by helping it to manage its risk portfolio more effectively.

Challenges of IRM

- Implementing an effective IRM process can be complex and resource-intensive. It requires a clear understanding of the organization's operations and the interdependencies between risks.
- Like all risk management approaches, IRM cannot eliminate all risks. There is always a residual risk that needs to be accepted and managed.
- The effectiveness of IRM is highly dependent on the risk culture within the organization. If the importance of managing risk is not recognized and supported by all levels of the organization, the IRM process may not be effective.

Despite these challenges, IRM is a vital approach in today's complex and interconnected business environment. It provides a more comprehensive and effective approach to managing risk, helping organizations to enhance their resilience and achieve their strategic objectives.

7.3.2.5 Pros and Cons of Contemporary Risk Management Models

Pros

- They provide a more comprehensive and integrated view of the organization's risk profile.
- They align risk management with the organization's strategic objectives.
- They can help to anticipate and manage complex and interrelated risks.
- They can improve decision-making and enhance the organization's resilience.

Cons

- They can be more complex and challenging to implement than traditional approaches.
- They require a change in organizational culture and mindset toward risk.
- They may require more resources and expertise to implement effectively.
- Despite their holistic approach, they may still fail to capture some emergent or unknown risks.

Despite the challenges, contemporary risk management models offer a more effective approach to managing the complex and interrelated risks that modern organizations face. They can help organizations to not just survive in the face of risks, but to thrive and achieve their strategic objectives.

7.3.3 Integrating Risk Management with Strategic Planning

Integrating Risk Management with Strategic Planning involves incorporating risk considerations into strategic planning and decision-making processes. This approach emerged out of the realization that strategic planning and risk management, which were traditionally handled in separate silos, can significantly benefit from being treated as interconnected processes [33, 34].

In the past, risk management was often viewed as a compliance requirement or a necessary evil, and it was typically handled in isolation from the strategic planning process. However, over time, organizations began to realize that risks could significantly impact their ability to achieve their strategic objectives. In response, many organizations started to integrate risk management into their strategic planning processes, leading to the development of the approach we see today.

Here is the framework of this integrated approach:

1) **Strategic Objectives and Risk Appetite**: The first step involves setting the organization's strategic objectives and determining its risk appetite.

 Strategic objectives are the overarching goals that an organization aims to achieve in the long term, typically over a period of 3–5 years. They provide direction for the organization and help guide decision-making at all levels. These objectives should be specific, measurable, attainable, relevant, and time-bound (SMART) to enable the organization to track its progress toward achieving them. Strategic objectives are typically aligned with the organization's mission and vision, and they often involve aspects such as financial performance, market position, innovation, operational efficiency, and social responsibility.

 Risk appetite is the level and type of risk that an organization is willing to take in order to achieve its strategic objectives. It is a reflection of the organization's risk culture and risk capacity, and it should align with its strategic objectives.

 Risk appetite can be expressed qualitatively or quantitatively. A qualitative expression of risk appetite might involve general statements about the organization's willingness to take risks, while a quantitative expression might involve specific metrics such as the maximum acceptable probability of a certain type of risk event occurring, or the maximum acceptable impact if such an event does occur.

 Integrating strategic objectives and risk appetite involves aligning the organization's risk management efforts with its strategic objectives and risk appetite. This means identifying and managing the risks that could impact on the organization's ability to achieve its strategic objectives, within the bounds of its risk appetite.

 For example, if an organization has a strategic objective to enter a new market and has a high-risk appetite, it might be willing to accept a higher level of risk associated with this objective. This could involve accepting the potential for financial loss, reputational damage, or operational disruptions. On the other hand, if the organization has a low-risk appetite, it might choose to mitigate these risks, for example, by entering the new market gradually or in partnership with a local company.

 In conclusion, integrating strategic objectives and risk appetite is a crucial aspect of SRM. It helps to ensure that the organization's risk management efforts are aligned with its strategic objectives and risk appetite, supporting more effective decision-making and enhancing the organization's ability to achieve its strategic objectives.

2) **Risk Identification**: This involves identifying the risks that could impact on the organization's ability to achieve its strategic objectives.

 In Integrating Risk Management with a Strategic Planning approach, risk identification is closely tied to the organization's strategic objectives. This differs from traditional risk identification methods in a few significant ways:

 Focused on Strategic Risks: Risk identification in this approach is specifically focused on identifying risks that could impact the achievement of strategic objectives. This means it needs to consider both internal and external risks that could affect the organization's strategic direction. This includes not just traditional operational and financial risks but also strategic risks like changes in the business environment, market disruptions, and changes in customer behavior.

Forward-Looking and Proactive: Because it is tied to strategic objectives, this approach to risk identification is typically more forward-looking and proactive than traditional approaches. It needs to anticipate future risks, not just identify current ones. This often involves scenario planning, forecasting, and other future-focused risk identification techniques.

Integration with Strategic Planning: In this approach, risk identification is integrated into the strategic planning process. This means that as strategic objectives are being set and strategic decisions are being made, risks are being identified in real-time. This integration allows for a more dynamic and responsive approach to risk identification.

Considering Both Threats and Opportunities: Finally, when integrating risk management with strategic planning, risks are viewed not just as potential threats but also as potential opportunities. This approach recognizes that uncertainty can result in positive outcomes as well as negative ones. Therefore, the risk identification process considers both sides of the coin—the potential downsides and the potential upsides of uncertainty.

By focusing on strategic risks, being forward-looking, integrating with strategic planning, and considering both threats and opportunities, the risk identification process in this approach supports more effective strategic planning and decision-making.

3) **Risk Assessment**: The identified risks are then assessed in terms of their potential impact and the likelihood of their occurrence. This assessment is integrated into the strategic planning process to inform strategic decision-making.

In the approach of Integrating Risk Management with Strategic Planning, the risk assessment process has several unique aspects compared to more traditional risk assessment methods:

Alignment with Strategic Objectives: The risk assessment process evaluates each identified risk in the context of how it could impact the organization's strategic objectives. This is a broader view than traditional risk assessment, which often focuses primarily on operational or financial impacts. By aligning the risk assessment with strategic objectives, organizations can better understand the potential impacts on their broader goals and mission.

Evaluation of Risk Interdependencies: This approach recognizes that risks often do not occur in isolation. They can be interconnected and can influence each other in complex ways. Therefore, risk assessment in this context not only evaluates individual risks but also assesses the interdependencies between risks.

Assessment of Opportunities: Unlike traditional risk assessment methods that focus on negative outcomes, this approach also recognizes that risks can present opportunities. Therefore, the risk assessment process includes an evaluation of the potential upsides of uncertainty, not just the downsides.

Integration with Strategic Decision-Making: In this approach, risk assessment is not a standalone process, but it is integrated into strategic decision-making. The results of the risk assessment inform strategic choices and planning. This helps ensure that strategic decisions are risk-informed and helps the organization navigate uncertainties more effectively.

Dynamic and Ongoing Process: As the strategic landscape is continually evolving, the risk assessment process in this approach is designed to be dynamic and ongoing. It is not a one-time event but a continuous process that is repeated as the strategic context changes and new information becomes available.

By aligning with strategic objectives, evaluating risk interdependencies, assessing opportunities, integrating with strategic decision-making, and being a dynamic and ongoing process, the risk assessment in this approach enables more effective management of strategic risks.

4) **Risk Mitigation**: Strategies are developed to manage the identified risks, which can involve avoiding, reducing, transferring, or accepting the risk. These strategies are integrated into the strategic plan.

In the approach of Integrating Risk Management with Strategic Planning, the risk mitigation task takes on some unique characteristics compared to more traditional risk mitigation methods:

Alignment with Strategic Objectives: Risk mitigation strategies are developed with a clear understanding of the organization's strategic objectives. The aim is not only to reduce potential negative impacts but also to seize opportunities that could facilitate the achievement of strategic objectives.

Strategic Decision Integration: Risk mitigation is deeply integrated with strategic decisions. Instead of being a separate, parallel process, risk mitigation strategies are incorporated into the strategic planning and execution processes. This ensures that risk considerations are not an afterthought, but an integral part of strategy formulation and implementation.

Balancing Risk and Reward: Since both threats and opportunities are considered, risk mitigation does not merely focus on minimizing risks, but on optimizing the risk–reward balance. This involves deciding whether to accept, avoid, transfer, or mitigate risks based on their potential impact on strategic objectives and the organization's risk appetite.

Consideration of Risk Interdependencies: This approach recognizes that risks are interconnected. Hence, when developing risk mitigation strategies, it takes into account how mitigating one risk could impact others. This helps in creating comprehensive and effective strategies that account for the complexity and interdependence of risks.

Continuous Monitoring and Adjustment: Risk mitigation is an ongoing process. Mitigation strategies are regularly reviewed and adjusted based on changes in the strategic landscape, risk profile, or effectiveness of current strategies. This ensures the organization can adapt and respond effectively to a dynamic and uncertain environment.

In conclusion, risk mitigation in this approach is more comprehensive and strategic. It is aligned with strategic objectives, integrated with strategic decision-making, focused on balancing risk and reward, cognizant of risk interdependencies, and continuously monitored and adjusted. This contributes to a more effective SRM that supports the achievement of organizational objectives.

5) **Monitoring and Review**: The organization's risk profile and the effectiveness of risk mitigation strategies are monitored and reviewed regularly, and adjustments are made as necessary.

In Integrating Risk Management with Strategic Planning approach, risk monitoring and review differ from traditional methods in several key ways:

Alignment with Strategic Objectives: Risk monitoring and review are focused on the strategic objectives of the organization. The effectiveness of risk mitigation strategies, changes in the risk landscape, and the impact of risks on strategic objectives are continually monitored and reviewed. This enables a more strategic and focused approach to risk monitoring and review.

Integrated with Decision-Making: The results of risk monitoring and review are integrated into strategic decision-making. This allows the organization to adjust its strategic plans and decisions based on current and updated risk information, enhancing its ability to navigate uncertainties and achieve its strategic objectives.

Proactive and Forward-Looking: Unlike traditional risk monitoring, which is often reactive and focuses on tracking known risks, this approach is proactive and forward-looking. It involves scanning the horizon for emerging risks and changes in the risk landscape that could affect the organization's strategic objectives. This helps the organization stay ahead of risks and take proactive steps to manage them.

Continuous Process: Risk monitoring and review are not one-time or periodic activities but continuous processes. This reflects the understanding that both the risk landscape and the strategic objectives of the organization can change over time. Continuous monitoring and review ensure that the organization can respond quickly and effectively to these changes.

Consideration of Risk Interdependencies: In this approach, risk monitoring and review take into account the interdependencies between risks. This helps the organization to understand the broader risk landscape and to manage risks more holistically.

Evaluation of Opportunities: Risk monitoring and review also consider the potential opportunities associated with risks. This helps the organization to identify and seize opportunities that arise from uncertainty.

By aligning with strategic objectives, integrating with decision-making, being proactive and forward-looking, being a continuous process, considering risk interdependencies, and evaluating opportunities, risk monitoring and review in this approach contribute to more effective SRM.

Benefits of Integration
- **Greater Strategic Alignment**: This approach ensures that risk management efforts are aligned with the organization's strategic objectives.
- **Improved Decision-Making**: By integrating risk considerations into strategic planning, this approach supports more informed and effective decision-making.
- **Enhanced Resilience**: By anticipating and preparing for potential risks, organizations can enhance their ability to respond to disruptions and achieve their strategic objectives.

Challenges of Integration
- **Requires a Cultural Shift**: Integrating risk management with strategic planning requires a shift in organizational culture toward a more risk-aware mindset. This can be challenging to achieve.
- **Requires Expertise and Resources**: Effective integration requires an understanding of both strategic planning and risk management, as well as the resources to implement the integrated approach.
- **Not All Risks Can Be Anticipated**: Despite the best efforts to identify and assess risks, there will always be some risks that are not anticipated.

Despite these challenges, the integration of risk management with strategic planning is a crucial approach in today's volatile and uncertain business environment. It can help organizations to better navigate uncertainties and achieve their strategic objectives.

7.4 Risk Management Techniques

Risk management is a vital cornerstone of any organization's strategic management. This process involves identifying, assessing, mitigating, and controlling risks, which could impede an organization's ability to realize its objectives. Effectual risk management equips organizations to navigate uncertainties, capitalize on opportunities, and sustain a robust and resilient operational framework [3].

In this discourse, we shall delve into an array of intricate methods and techniques that constitute the foundation of efficacious risk management. We will explore the application of these tools to specific tasks, such as risk identification, risk assessment, risk mitigation, and risk control.

This exploration will proffer a practical comprehension of their contribution to the comprehensive risk management process [18].

Understanding and appropriately deploying these techniques can transform risk management from a reactive function into a strategic capability that adds value across all business domains. By doing so, organizations can convert risks into strategic opportunities, enhancing their adaptability and resilience in the face of evolving landscapes [24].

7.4.1 Techniques for Risk Identification

Risk identification techniques are the methods used to uncover, recognize, and describe risks that could affect the achievement of an organization's objectives. A comprehensive set of techniques includes both quantitative and qualitative methods, and they are usually tailored to the nature of the organization, its industry, and the specific risks it faces. Here are some of the most commonly used risk identification techniques:

1) **Brainstorming:** This is a group creativity technique where a team meets to generate a broad list of potential risks. Brainstorming encourages free thinking and the sharing of ideas and can often uncover risks that might be overlooked in a more formal analysis process.
2) **Interviews and Surveys:** Interviews involve one-on-one discussions with experienced individuals who have insight into potential risks, while surveys can gather input from a larger group of individuals. Both techniques can be structured or unstructured depending on the degree of flexibility required.
3) **Delphi Technique:** This is a structured process used to gather expert opinion. The Delphi technique involves multiple rounds of anonymous input and feedback, allowing experts to revise their opinions in light of the views of others.
4) **Checklists:** These are comprehensive lists of known risks based on past experiences or industry standards. Checklists can provide a structured approach to risk identification but must be updated regularly to remain effective.
5) **Risk Registers:** A risk register is a document that lists an organization's key risks, along with information such as their causes, impacts, and potential mitigation strategies. Risk registers are updated continuously as new risks are identified.
6) **Scenario Analysis:** This technique involves identifying potential future scenarios, usually based on certain assumptions or triggering events, and assessing the risks associated with each one. This can be particularly useful for strategic and long-term risk identification.
7) **Root Cause Analysis:** This technique seeks to identify the underlying causes of potential risk events, rather than focusing on the events themselves. Techniques like the "5 Whys" or Fishbone Diagrams can be used to explore root causes.
8) **SWOT Analysis:** By SWOT analysis, an organization can identify risks in its internal (strengths and weaknesses) and external (opportunities and threats) environments.
9) **Historical Data Analysis:** This involves looking at past events and data to identify trends, patterns, and anomalies that could indicate risks.
10) **Process Flow Analysis:** This technique involves creating a detailed flowchart of a process or activity to identify where risks might occur.

These techniques are not mutually exclusive and are often used in combination to ensure a thorough and comprehensive identification of risks. It is also important to note that risk identification should be an ongoing process, as new risks can emerge as circumstances change.

7.4.2 Techniques for Risk Assessment

Risk assessment is the process of understanding the nature of and determining the level of risk. This involves the identification of hazards (or risk factors), the context to which they are relevant, the likelihood that they will occur, and the potential consequences. Here are some of the commonly used risk assessment techniques:

1) **Risk Matrix:** This is a simple tool used to determine the severity of risk by considering the category of probability or likelihood against the category of consequence severity. This is a simple, fast, and easy-to-understand method.

   A risk matrix is a simple grid layout that plots the severity of a risk against its likelihood. Each axis is usually divided into an agreed number of increments, for instance, Low, Medium, and High, or, more granularly, using a numerical scale such as from 1 to 5. Here is an example of a risk matrix populated with numbers, assuming a 5 × 5 matrix (Table 7.3):

Table 7.3 A Sample Risk Matrix

	1. Very Low	2. Low	3. Medium	4. High	5. Very High
1) Very low	1	2	3	4	5
2) Low	2	4	6	8	10
3) Medium	3	6	9	12	15
4) High	4	8	12	16	20
5) Very high	5	10	15	20	25

   In this risk matrix, the Y-axis represents the likelihood of the risk occurring (1 for very low likelihood, 5 for very high likelihood), and the X-axis represents the impact of the risk if it does occur (1 for very low impact, 5 for very high impact). The number in each cell is calculated by multiplying the likelihood and impact scores, providing a risk rating that can be used to prioritize risks.

   For instance, a risk with a likelihood of 4 (High) and impact of 3 (Medium) would have a risk rating of 12 (4×3). Another risk with a likelihood of 2 (Low) and impact of 5 (Very High) would have a risk rating of 10 (2×5). In this case, the first risk would be prioritized for mitigation actions over the second risk, due to its higher risk rating.

2) **Failure Modes and Effects Analysis (FMEA):** This is a step-by-step approach for identifying all possible failures in a design, a manufacturing or assembly process, or a product or service. It is used to identify potential failure modes, determine their effect on the operation of the product, and identify actions to mitigate the failures.

3) **Fault Tree Analysis (FTA):** This is a top-down, deductive analysis method aimed at analyzing the effects of initiating faults and events on a complex system to identify the causes of a specified undesired event.

4) **Event Tree Analysis (ETA):** This is an inductive analytical diagrammatical method where an event is analyzed following the chronological sequence of actual developments or the logic of decisions or a mixture of both.

5) **Monte Carlo Simulation:** This is a quantitative risk analysis technique used to understand the impact of risk and uncertainty in financial, project management, cost, and other forecasting models. A Monte Carlo simulator helps to simulate a variety of possible outcomes in a process that cannot easily be predicted.

6) **Sensitivity Analysis:** This is a risk assessment tool used to determine how different values of an independent variable will impact a particular dependent variable under a given set of assumptions. It is used within specific boundaries that will depend on one or more input variables, such as the effect that changes in interest rates will have on a bond's price.

7) **Bow-Tie Analysis:** This is a simple diagrammatic way of describing and analyzing the pathways of a risk from causes to consequences [35]. It provides a visual image of many plausible accident scenarios that could occur.

8) **SWIFT (Structured What-if Technique):** A team-based qualitative technique for identification of hazards and risks based on structured brainstorming [36].

9) **HAZOP (Hazard and Operability Study):** A systematic and structured team-based method for identifying hazards in a design or process, often used in industries like chemical or process industries [37].

10) **Risk Register:** This is a document that keeps track of all the potential risks and hazards along with information like the nature of the risk, level of impact, and mitigation plan.

These techniques help an organization in identifying potential risks, evaluating their impact, and taking appropriate steps to mitigate them. They can be used individually or in combination depending upon the requirements and circumstances of the organization. The choice of risk assessment technique depends on the type of risk, data availability, and the level of accuracy required.

7.4.3 Quantitative and Qualitative Risk Analysis

Risk analysis can be divided into two main types: quantitative and qualitative. Both types have their own strengths and are often used in conjunction with each other to get a comprehensive view of risks.

Qualitative risk analysis involves a subjective assessment of the potential risks based on the experience, expertise, and intuition of the people involved. This type of analysis usually employs a risk matrix to categorize risks based on their probability (likelihood of occurrence) and impact (the extent of damage if the risk does occur).

Qualitative risk analysis often involves ranking the identified risks in terms of their likelihood of occurrence and potential impact. Here are some techniques commonly used:

1) **Risk Matrix or Risk Register:** This is a common tool for recording and prioritizing risks. It typically includes columns for risk description, likelihood, impact, and mitigation strategies. The risks are often color-coded based on severity.

2) **SWOT Analysis:** This technique helps identify internal and external factors that could impact project success, categorized into Strengths, Weaknesses, Opportunities, and Threats.

3) **Expert Judgment:** Leveraging the experience and intuition of industry experts or experienced project team members can provide valuable insights in assessing risk.

4) **Delphi Technique:** A panel of experts assesses risk in multiple rounds of anonymous surveys. After each round, a facilitator provides an anonymous summary of the experts' forecasts and their reasons for their judgments from the previous round. This process continues until a consensus is reached.

5) **Checklists:** Checklists derived from historical data or best practices can be used to identify potential risks.

Quantitative risk analysis, on the other hand, uses numerical values and probabilities. It attempts to assign numeric values to risks, their impacts, and the probability of their occurrence.

Methods used in quantitative risk analysis include decision tree analysis, Monte Carlo simulation, sensitivity analysis, and EMV analysis.

For example, consider the risk of a machine breakdown in a production line. A qualitative risk analysis might rate the risk as "high" because the machine is old, and its breakdown would severely impact production. A quantitative risk analysis might further estimate that the machine has a 30% chance of breaking down in the next year and that the breakdown would lead to US$50,000 in lost productivity.Quantitative risk analysis attempts to assign numeric values to risks for further analysis. Here are some popular techniques:

1) **Sensitivity Analysis:** This technique involves changing variables within a model to see how the changes affect the overall risk. It is often depicted using a tornado diagram.
2) **Expected Monetary Value (EMV) Analysis:** This technique involves calculating the average outcome when the future includes scenarios that may or may not happen (i.e., analyzing under conditions of uncertainty). EMV is calculated by multiplying the value of each possible outcome by their probability and summing them.
3) **Decision Tree Analysis:** A decision tree is a graphic model of various alternative courses of action and their possible outcomes. It allows one to map out the risks and rewards of different options and thus choose the most beneficial course.
4) **Monte Carlo Simulation:** This is a mathematical technique that uses statistical sampling to estimate possible outcomes of uncertain events. This technique is particularly useful in estimating the potential impact of high-risk activities.
5) **Fault Tree Analysis (FTA):** A top-down, deductive failure analysis in which an undesired state of a system is analyzed using Boolean logic to combine a series of lower-level events.

These techniques are often used in combination to provide a robust and comprehensive risk analysis. It is also worth noting that the nature and complexity of a project or business activity often dictate which risk analysis techniques are most suitable.

In conclusion, qualitative risk analysis is a good starting point and helps identify which risks need further quantitative analysis. Quantitative risk analysis can then be used for those major risks to get a more precise analysis. Together, these two methods can provide a comprehensive understanding of risks, which allows for effective risk mitigation strategies.

7.5 Technology and Risk Management

In our rapidly evolving digital age, technology plays an increasingly significant role in the field of risk management. It provides tools and capabilities that not only enhance efficiency and accuracy but also deliver unprecedented possibilities in understanding and addressing risk scenarios. This section will delve into three crucial aspects of this intertwining relationship. Firstly, we will explore the overarching role of technology in risk management, highlighting the transformative impact it has on the risk landscape. Secondly, we will focus on the growing importance of automation and AI in risk assessment, exploring how these innovative technologies are redefining our ability to identify and evaluate potential risks. Finally, we will discuss the indispensable role of data analytics in risk prediction and management, underlining the profound influence data-driven decision-making has on creating proactive and resilient risk management strategies. By understanding these facets, we can better appreciate the exciting opportunities and challenges that technology presents in our continuous pursuit of effective risk management.

7.5.1 Role of Technology in Risk Management

In the past, risk management relied heavily on manual processes and intuition. Risk identification and assessment were largely performed by individuals or teams drawing upon their experience and expertise, often involving tedious, time-consuming tasks. This process was not only labor-intensive but also prone to human error and oversight.

With the advent of digital technology, the landscape of risk management began to change. Initially, this was largely manifested in the form of database management systems and simple analytics tools that automated data collection and basic analysis. This marked the first significant shift in risk management, enabling a more systematic approach to risk identification and assessment, reducing human error, and freeing up time for more strategic tasks.

The advent of the internet, followed by the rise of cloud computing, further transformed risk management. These developments ushered in a new era of connectivity and data availability, allowing for real-time monitoring and management of risks facilitating improved collaboration and data sharing among risk management teams.

7.5.1.1 Current Role of Technology in Risk Management

In the present day, the role of technology in risk management has expanded and deepened significantly. It is not only about automating processes or improving efficiency; technology now plays a strategic role in how risks are identified, assessed, and managed. Here is how:

1) **Risk Identification:** Advanced data analytics, machine learning, and AI are used to trawl through large volumes of data to identify potential risks. These technologies can uncover patterns and trends that might be missed by the human eye, allowing for proactive risk identification.

Example 7.2 Financial Fraud Detection

Banks and other financial institutions deal with large volumes of transactions daily. Manually monitoring these transactions for fraudulent activities is not only labor-intensive but practically impossible, given the sheer volume of data. Here is where advanced data analytics, machine learning, and AI come into play.

Advanced data analytics tools can process and analyze huge datasets in real time, allowing financial institutions to quickly detect anomalous transactions that could indicate fraud [38]. For instance, if a customer who typically makes small, local purchases suddenly makes a large purchase in a foreign country, the system can flag this as a potential risk.

Machine learning algorithms can be trained to recognize patterns associated with fraudulent transactions. Over time, these algorithms learn from the data and continually improve their detection capabilities. They can identify subtle patterns that might be missed by humans, making them highly effective in detecting fraudulent activities.

AI takes this a step further by not only detecting potential fraud but also learning and adapting to new, evolving forms of fraudulent activity. AI can analyze transaction patterns, track user behavior, and make informed decisions about whether a particular transaction might be fraudulent. It can also use natural language processing to analyze customer complaints or inquiries to identify potential risks that may not be captured in transaction data.

For instance, MasterCard uses AI and machine learning for its Decision Intelligence technology, which provides real-time, predictive analytics to assess and score every transaction for potential fraud. This system helps reduce false declines, improves the efficiency of fraud detection, and enhances customer experience.

This example illustrates the power of advanced data analytics, machine learning, and AI in identifying potential risks. By leveraging these technologies, businesses can proactively manage risks, improve their decision-making, and enhance their operational efficiency.

2) **Risk Assessment:** Technologies such as predictive analytics and simulation models enable a more precise and forward-looking risk assessment. They allow organizations to model different risk scenarios and their potential impacts, enabling more informed decision-making.

Example 7.3   Predictive Analytics in Healthcare Risk Management

In the healthcare sector, predictive analytics and simulation models are used to foresee and mitigate potential risks, leading to improved patient outcomes and efficient resource allocation.

Consider the application of predictive analytics in patient readmission risk. Hospitals around the world are continuously working to minimize patient readmissions, which often indicate suboptimal care and result in higher healthcare costs. By using predictive analytics, hospitals can assess which patients are more likely to be readmitted based on a multitude of factors like age, disease type, treatment method, and social determinants of health.

For instance, the University of Louisville Hospital in Kentucky implemented a predictive analytics platform that used machine learning algorithms to predict patient readmission risks [39]. The platform considered numerous variables such as patients' medical history, demographics, and even the medications they were prescribed upon discharge. The resulting risk scores helped medical professionals identify high-risk patients and develop personalized care plans to reduce the likelihood of readmission. This predictive model not only improved patient outcomes but also helped the hospital manage its resources more effectively.

Similarly, simulation models are used in healthcare to assess the potential impacts of various risk scenarios. For instance, during the COVID-19 pandemic, simulation models were extensively used to predict the spread of the virus, its impact on healthcare resources, and the potential effects of various mitigation strategies. These models played a crucial role in informing public health strategies and allocating resources effectively.

These examples demonstrate how predictive analytics and simulation models enable a more precise and forward-looking risk assessment, allowing organizations to proactively manage risks and make more informed decisions.

---

3) **Risk Monitoring:** Technologies such as dashboards and real-time analytics provide organizations with the tools to monitor risks continuously. This enables timely detection of any changes in risk exposure and allows for quick action.
4) **Risk Reporting and Communication:** Technology facilitates efficient and effective risk reporting and communication. Dashboards, visualizations, and collaborative tools provide a clear, concise, and up-to-date picture of the organization's risk profile, making it easier to communicate with stakeholders.

5) **Risk Mitigation:** Technology can also aid in the mitigation of risks. For instance, cybersecurity technologies can help mitigate cyber risks, while advanced analytics can support decision-making to manage strategic or operational risks.

In conclusion, the role of technology in risk management has evolved from being a tool for efficiency to a strategic partner that enables proactive and informed risk management. With the rapid advancement of technologies such as AI, machine learning, and big data analytics, this role is set to deepen further, opening up new opportunities and challenges for risk management in the digital age.

7.5.2 Automation and Artificial Intelligence in Risk Assessment

Automation and AI have rapidly changed the landscape of risk assessment over the past few decades. Initially, automation was utilized in risk management for repetitive and labor-intensive tasks, such as data collection, data cleaning, and the generation of standard risk reports. The main purpose was to improve efficiency, reduce human error, and allow risk management professionals to focus more on tasks that required expert judgment.

In parallel, early applications of AI in risk management were focused on pattern recognition tasks, such as identifying fraudulent credit card transactions or detecting anomalous patterns that could signify operational risks. These were rule-based systems that were programmed to flag transactions or operations that deviated from predefined norms.

However, as computing power increased and machine learning algorithms improved, AI systems started evolving beyond rule-based systems. They began learning from data, improving their accuracy over time, and making predictions about future risks.

7.5.2.1 State of the Art as of Now

As of now, automation and AI have significantly enhanced and transformed risk assessment in multiple ways:

1) **Automated Data Collection and Analysis:** Automation tools can gather and process massive amounts of data much faster than humans, offering real-time insights into potential risks. AI can then be applied to this data to extract meaningful insights, identify patterns, and make accurate predictions about future risks.
2) **Predictive Analytics:** With machine learning, AI can now predict potential risks even before they occur. This is particularly helpful in industries like finance and healthcare, where predicting risks such as loan defaults or patient readmissions can result in significant cost savings and improved outcomes.
3) **AI-powered Risk Modeling:** Advanced AI algorithms can now model complex risk scenarios that consider multiple interrelated factors. For instance, in the insurance industry, AI models are being used to assess the risk of insuring a particular individual or property, considering a vast number of variables simultaneously.

Example 7.4   Auto Insurance Pricing

One of the major auto insurance providers, Progressive, has leveraged AI and machine learning to streamline their underwriting process and accurately assess the risk associated with insuring a particular individual or vehicle. This system is known as Snapshot, and it is a usage-based insurance program that observes the driver's behavior [40].

When enrolled in the Snapshot program, customers are provided with a device that plugs into their car. This device collects data related to driving behavior such as miles driven, time of day, and hard braking incidents. More recently, the program has been extended to a mobile app that can also collect data on phone usage while driving.

This vast amount of data, when processed and analyzed with machine learning algorithms, allows Progressive to accurately assess the risk of insuring a particular individual. The model considers a multitude of variables, such as the time the vehicle is typically on the road, the number of hard brakes per trip, and the total miles driven. Using AI, they can identify patterns and correlations that would be nearly impossible for human analysts to detect.

The result is a personalized insurance rate that accurately reflects the driver's risk, rewarding safer drivers with lower premiums and penalizing risky driving behaviors with higher rates. This AI-driven approach to risk assessment not only leads to more accurate pricing but also encourages safer driving behavior.

This use case demonstrates the power of AI and machine learning in risk assessment, providing a more precise, data-driven evaluation of risk that can lead to better decision-making and more efficient operations.

4) **Natural Language Processing (NLP):** AI's ability to understand and generate human language has opened up new avenues in risk assessment. For example, NLP can be used to analyze customer complaints, social media chatter, or news articles to identify emerging risks that may not be captured in structured data.

Example 7.5   Case Example: Sentiment Analysis in Finance

In the financial industry, companies like Bloomberg and Reuters use NLP to perform sentiment analysis on news articles, social media posts, and financial documents to predict market movements and identify potential risks. This analysis can inform trading decisions and risk management strategies [41].

For example, if negative sentiments about a specific company are increasingly expressed in news articles or social media posts, it could be an indicator of potential financial risk for that company. This might prompt risk managers to re-evaluate the risk profile of investments in that company.

NLP can also be used to analyze customer complaints for financial institutions. For instance, if there is a sudden surge in complaints about a specific issue, it might signify an operational risk that needs immediate attention. Similarly, analyzing the text of customer reviews can help identify potential reputational risks.

A notable case of NLP application is JPMorgan's LOXM program. LOXM is a self-learning trading program that uses NLP and machine learning to decipher high volumes of data [42]. The program was designed to automate trading tasks traditionally handled by humans. By understanding trader communications and observing the procedures they took to react to specific market scenarios, LOXM was able to learn the best trading strategies under different conditions.

This application of NLP and AI allows financial institutions to identify and respond to risks more quickly than traditional methods. It also exemplifies how AI can learn and adapt to complex tasks by learning from human actions, making it a powerful tool for modern risk management.

These examples illustrate how NLP can be utilized to monitor and analyze a vast range of data sources, providing a comprehensive view of potential risks that might otherwise be missed in structured data alone.

5) **Real-Time Risk Monitoring:** AI algorithms can constantly monitor various data streams for potential risk indicators. This enables organizations to respond to potential risks more quickly and proactively.

6) **AI in Cybersecurity:** AI is playing a crucial role in risk assessment in the field of cybersecurity. It can identify suspicious activities, detect anomalies, and predict future attacks based on patterns in the data.

However, while AI and automation bring significant benefits to risk assessment, it is important to remember they also introduce new risks, including algorithmic bias, over-reliance on automated systems, and new cybersecurity threats. These risks must be managed as part of a comprehensive risk management strategy.

In conclusion, the state of the art in AI and automation has fundamentally transformed risk assessment, allowing for real-time, predictive, and comprehensive risk analysis. The future will likely see further integration of these technologies into risk management, offering even more sophisticated and nuanced risk assessment capabilities.

7.5.3 Data Analytics for Risk Prediction and Management

The use of data in risk management is not a new concept. Traditionally, organizations have used historical data and statistical methods to identify and assess risks. This often involved simple data analysis techniques like trend analysis, ratio analysis, and scenario analysis.

However, with the advent of digital technology and the proliferation of data, the ability to capture and analyze data has significantly expanded. This has led to the development of more sophisticated data analytics techniques that can process larger volumes of data, handle different types of data (including unstructured data), and deliver insights in real time.

At the same time, the focus of risk management has shifted from simply understanding and managing known risks to predicting future risks and identifying unknown risks. This has resulted in a greater emphasis on predictive analytics and machine learning techniques that can identify patterns and make predictions about future events.

Today, data analytics plays a pivotal role in risk prediction and management, enabling organizations to identify, assess, mitigate, and monitor risks more effectively. Here are some ways data analytics is currently being applied:

1) **Predictive Analytics:** Organizations use machine learning algorithms to analyze historical data and identify patterns that can help predict future outcomes. This is particularly useful for predicting operational risks, financial risks, and cyber risks.

2) **Real-Time Risk Monitoring:** Advanced data analytics tools can process and analyze data in real-time, providing instant insights into potential risks. This enables organizations to respond to risks more quickly and proactively.

Example 7.6 Real-Time Risk Assessment in Healthcare

Cerner, a global healthcare technology company, has developed a system called St. John Sepsis Agent that utilizes real-time data analytics to identify patients at risk of developing sepsis—a potentially life-threatening condition [43].

Sepsis is difficult to diagnose in its early stages, and delays in treatment can lead to serious complications. To tackle this problem, Cerner developed a machine learning model that analyzes a variety of patient data in real-time, such as vital signs, lab results, and clinical notes.

The model has been trained to identify patterns in the data that may indicate a patient is at risk of developing sepsis. When such patterns are detected, the system sends an alert to the medical staff, enabling them to take immediate action.

This use of real-time data analytics has made a significant difference. A study involving 6000 patients found that the St. John Sepsis Agent reduced sepsis-related mortality rates by 21%.

This case shows the potential of real-time data analytics in identifying and managing risks. It also highlights the importance of real-time data processing capabilities for modern risk management. The ability to instantly analyze and act upon data can not only save lives in healthcare but also prevent operational disruptions, financial losses, and other adverse outcomes in various industries.

3) **Risk Modeling:** Organizations use data analytics to build more sophisticated and accurate risk models. For instance, in the financial sector, companies use data analytics to build credit risk models that can predict the likelihood of default based on a multitude of factors.

4) **Stress Testing and Scenario Analysis:** Data analytics enables organizations to simulate different risk scenarios and assess their potential impact. This helps organizations prepare for different risk events and develop effective mitigation strategies.

Example 7.7 Stress Testing in Banks

After the financial crisis of 2008, banks and other financial institutions are required to perform regular stress tests under regulations such as the Dodd-Frank Act in the U.S. These stress tests are designed to evaluate how banks would fare under adverse economic conditions [44].

To conduct these stress tests, banks use data analytics to simulate various risk scenarios, such as a sharp increase in unemployment, a steep decline in property prices, or a major change in interest rates. They then assess the potential impact of these scenarios on their financial health, considering factors like credit risk, market risk, liquidity risk, and operational risk.

For instance, JPMorgan Chase, one of the largest banking institutions in the world, uses complex data analytics models to simulate thousands of economic scenarios and their potential impacts on the bank's portfolio. They use historical data, economic indicators, and machine learning algorithms to generate these scenarios and predict their outcomes.

These stress tests help banks identify vulnerabilities in their operations and make informed decisions about risk mitigation strategies. They also provide regulators with assurance that banks can withstand adverse economic conditions.

This case illustrates the power of data analytics in simulating risk scenarios and assessing their potential impact. By leveraging data analytics, organizations can prepare for a wide range of risk events and make more informed decisions about risk management.

---

5) **Risk Identification:** With the help of data analytics, organizations can trawl through large volumes of data to identify potential risks. This includes unstructured data, such as text data from social media or news articles, which can be analyzed using NLP techniques.

6) **Risk Mitigation:** Data analytics can also help organizations identify the most effective risk mitigation strategies. For instance, machine learning algorithms can be used to identify the risk mitigation strategies that have had the greatest impact in the past.

Example 7.8   Cybersecurity Risk Mitigation

Cybersecurity threats are an omnipresent concern for organizations, and data analytics plays an instrumental role in identifying and mitigating such risks.

One instance is IBM's QRadar, a Security Information and Event Management (SIEM) system that uses advanced data analytics to detect, prioritize, and manage potential threats [45]. The system collects data from across an organization's IT infrastructure and uses machine learning algorithms to identify unusual activity that could signify a security threat.

QRadar goes beyond simple threat detection; it also assists in risk mitigation. The system analyses past incidents and the organization's response to those incidents. Using this data, QRadar can recommend the most effective mitigation strategies for specific types of threats. It can also suggest improvements to the organization's overall security posture, based on patterns and trends identified in the data.

For instance, if QRadar identifies that certain types of phishing attacks have been particularly successful in the past, it might recommend increased staff training on phishing awareness, or it might suggest implementing stronger email filtering solutions.

This real-world example underscores how data analytics can not only identify risks but also help organizations pinpoint the most effective risk mitigation strategies. By analyzing past data and learning from it, organizations can continuously improve their risk management practices.

---

7) **Risk Reporting:** Advanced data visualization tools can help organizations communicate risks more effectively. Interactive dashboards and real-time reports can provide stakeholders with a clear and up-to-date view of the organization's risk landscape.

In conclusion, the state of the art in data analytics has revolutionized risk prediction and management, allowing organizations to be more proactive and strategic in managing risks. However, like all technologies, data analytics brings its own risks, including data privacy and security risks, which need to be managed carefully.

7.6   Resilience and Business Continuity

Resilience and business continuity serve as two paramount facets of an organization's risk management strategy, fortifying entities to absorb disturbances, adapt, and persist with operations under challenging conditions [46]. This discourse examines the importance of nurturing resilience within organizations, the criticality of business continuity planning, and the compelling requirement for robust disaster recovery and emergency response mechanisms.

These elements are not standalone but intimately interlace with each other, constituting a comprehensive strategy that protects the organization's interests and ensures sustainability and success amidst unanticipated crises and challenges [47]. In the ensuing subsections, we will delve into the nuances of fostering organizational resilience, unravel the complexities of business continuity planning, and comprehend the importance of disaster recovery and emergency response [48].

The synthesis of these elements into a well-rounded risk management strategy is crucial for navigating an increasingly volatile and uncertain business environment. This strategy not only enhances an organization's capacity to weather adverse conditions but also equips it to emerge stronger, more adaptable, and better prepared for future crises [49].

7.6.1 Cultivating Resilience in Organizations

The concept of building resilience within organizations has grown increasingly pertinent over recent decades [50], particularly in the face of escalating global challenges such as cybersecurity threats, climate change, and, most recently, the COVID-19 pandemic. Organizational resilience is defined as the capacity of an enterprise to endure, adapt to, and recuperate from disruptions while continuing to operate effectively [46].

7.6.1.1 Historical Context and Evolution

In bygone eras, organizations concentrated primarily on risk management, intending to prevent and mitigate adverse events. However, as the intricacy and interconnectedness of business operations have expanded, it has become evident that it is unfeasible to predict and avert all potential risks [49]. Consequently, the focus has shifted toward cultivating resilience—an approach acknowledging that adverse events will transpire and emphasizing the necessity for organizations to be primed to respond and recover effectively [51].

This shift has been propelled by numerous factors, including the transformation of the global economy, technological advancements, and growing recognition of systemic risks such as climate change and pandemics [52]. The process of building resilience has gradually evolved from an oft-overlooked aspect of business strategy into a central tenet of contemporary organizational design and culture [53].

7.6.1.2 Current Approaches to Building Resilience

Resilience in an organization is not simply about having robust systems and processes in place. It involves a combination of various factors, including but not limited to:

1) **Adaptive Capacity**: Organizations must have the ability to adapt quickly to changing circumstances. This requires flexible systems and processes, as well as a culture that encourages innovation and learning.

    Adaptive Capacity refers to an organization's ability to adjust and evolve its strategies, processes, and operations in response to changes in its external and internal environments. It is a crucial aspect of organizational resilience as it ensures the organization is not just robust in the face of shocks, but can also evolve over time in response to long-term changes and challenges. Here are some key facets of adaptive capacity:

    - **Flexibility:** Resilient organizations are flexible in their structures, strategies, and operations. They are not rigidly attached to existing ways of doing things and are willing to change their methods when circumstances require it. This might involve adjusting business strategies in response to market changes, modifying operational processes due to regulatory shifts, or implementing new technologies to enhance efficiency.
    - **Innovation:** Adaptive capacity goes hand in hand with innovation. Organizations need to continuously innovate to stay ahead of changing market conditions, technological advancements, regulatory changes, and evolving customer expectations. This involves fostering a culture of innovation where employees are encouraged to think creatively, take calculated risks, and learn from their failures.
    - **Learning Orientation:** Adaptive organizations are learning organizations. They view challenges and failures as learning opportunities and use these experiences to improve and evolve. They promote a culture of continuous learning and improvement, investing in training and development, encouraging knowledge sharing, and embracing a mindset of curiosity and exploration.

- **Responsiveness:** Adaptive capacity also involves the ability to respond quickly and effectively to changes. This requires robust monitoring and alert systems to identify changes in real-time, effective decision-making processes to respond to these changes, and efficient execution capabilities to implement the necessary actions.
- **Diversity and Redundancy:** Diversifying operations, suppliers, and markets can provide more options when responding to disruptions, reducing dependence on a single entity. Redundancy, in terms of multiple methods for achieving the same outcome, can provide fallback options when a preferred method is disrupted.

  Enhancing adaptive capacity often involves a balance between efficiency and resilience. While certain practices such as maintaining redundant systems or diversifying suppliers may seem less efficient in the short term, they enhance an organization's adaptive capacity and contribute to its long-term resilience.

2) **Robust Design**: Resilient organizations are designed to withstand shocks. This might involve diversifying supply chains to reduce dependence on a single supplier, implementing redundant systems to ensure business continuity in case of a failure, or designing physical infrastructure to withstand natural disasters [54, 55].

   Robust Design in the context of organizational resilience refers to the creation and implementation of systems, processes, and structures within an organization that are resistant to disruptions and are capable of maintaining function despite adverse conditions. It underlines the principle of "designing for failure," where it is assumed that disruptions will inevitably occur, and thus the focus is on designing systems that can continue to function even in the face of such disruptions.

   Key elements of robust design include:

- **Redundancy**: This is about having backup systems or processes that can be used if the primary system fails. For instance, a robust IT system might include redundant servers that can continue to provide services to customers if the primary server fails.
- **Diversity**: Diversity in supply chains, personnel, technology, and more can reduce the risk of a single point of failure. For example, diversifying suppliers can help an organization avoid disruptions if a single supplier faces issues.
- **Modularity**: This involves designing systems such that they are composed of independent, interchangeable parts or modules. If a module fails or needs to be upgraded, it can be replaced without disrupting the entire system.
- **Fail-Safe Mechanisms:** These are designed to ensure that if a system fails, it fails in a way that minimizes harm. An example could be a circuit breaker in an electrical system that cuts off power if an overload is detected, preventing a possible fire.
- **Flexibility:** Designing systems and processes that can be easily adapted to changing circumstances, allowing the organization to respond more effectively to disruptions.
- **Safety Margins**: These are buffers built into systems to account for uncertainty and variability. For instance, keeping extra inventory to account for fluctuations in demand or supply would be an example of a safety margin.

  While implementing a robust design might require additional resources in the short term, the long-term benefits include reduced downtime, greater resilience to disruptions, and less likelihood of catastrophic failure. Thus, investing in robust design is a strategic decision that can significantly enhance an organization's resilience.

3) **Effective Leadership and Culture**: Resilience must be embedded in an organization's culture, driven by effective leadership. Leaders play a crucial role in promoting a resilience mindset, demonstrating adaptability, fostering open communication, and making informed decisions during crises.

4) **Preparedness and Planning**: Organizations must have plans in place to manage potential crises. This includes business continuity plans, disaster recovery plans, and emergency response procedures.

5) **Learning from Experience**: Resilient organizations learn from past disruptions, using these experiences to improve their strategies and become better prepared for future crises.

Learning from experience is a critical component of building resilience within an organization. It is the process of reflecting on past events, both positive and negative, to gain insights that can be used to enhance the organization's ability to handle future challenges and risks.

Key aspects of learning from experience include:

- **Incident Analysis:** This involves the detailed examination of events that have led to disruptions or failures in the past. The goal is to understand what happened, why it happened, and what actions could have been taken to prevent it. It is about dissecting incidents to extract lessons and actionable insights.

- **Feedback Loops**: Organizations should create mechanisms for capturing and sharing lessons learned across all levels. This might include debriefing sessions after major projects or events, a system for recording and sharing lessons learned, and processes to incorporate feedback into strategic planning and operational decision-making.

- **Continuous Improvement**: Learning from experience should drive continuous improvement. Insights gained from past experiences should be used to enhance systems, processes, strategies, and decision-making methods.

- **Fostering a Learning Culture:** It is not enough to simply analyze incidents and identify lessons. For an organization to truly learn from its experiences, it must foster a learning culture. This means creating an environment where employees are encouraged to share their experiences and insights, where failures are viewed as opportunities for learning rather than causes for blame, and where continuous learning and improvement are valued and rewarded.

- **Updating Risk Management Strategies**: Learning from past experiences should directly inform the organization's risk management strategies. This could mean updating risk assessment procedures to account for newly identified risks, re-evaluating risk mitigation strategies based on what has worked in the past, or adjusting risk tolerance levels based on lessons learned.

In essence, learning from experience is about turning past failures and successes into a valuable resource for enhancing organizational resilience. This requires a commitment to learning and improvement at all levels of the organization, from frontline employees to top leadership.

In today's VUCA world, building organizational resilience is no longer a luxury but a necessity. Organizations that prioritize resilience are better equipped to navigate through disruptions, recover more quickly, and emerge stronger than before. The ability to adapt and recover from shocks will continue to be a critical factor in an organization's long-term success.

7.6.1.3 Building Resilience through Complexity Theory

Complexity theory indeed provides an invaluable lens through which to understand organizational vulnerabilities and guide redesign toward enhanced resilience. This theory is a scientific approach focusing on how parts of a system interact to produce overall system behavior, particularly relevant to organizations since they are complex adaptive systems. These organizations consist of interconnected and interdependent elements that adjust in response to changes in their environment [56].

Key Insights from Complexity Theory

Complexity theory offers several key insights that can inform organizational design:

- **Emergent Behavior**: Complex systems give rise to emergent behavior at the system level, unpredictable from simply comprehending individual elements. This underscores the importance of understanding relationships and interactions within an organization, rather than focusing solely on individual components [57].
- **Nonlinearity**: Due to the nonlinear interactions between system elements, minor changes can occasionally lead to substantial effects (and vice versa). Organizations, therefore, should be alert to small signals that may denote emerging risks or opportunities [58].
- **Adaptation and Co-evolution**: Complex systems adapt—they evolve in response to their environment. Additionally, they co-evolve with their environment, meaning both the system and its environment influence each other's evolution. Organizations should aim for flexibility and adaptability, allowing for evolution in response to alterations in their external and internal environments [59].
- **Self-organization and Decentralization**: Order and structure emerge from local interactions between system elements, rather than being imposed from above in complex systems—a phenomenon known as self-organization. A degree of decentralization could enhance an organization's adaptability and resilience. Decision-making authority, information, and resources should be dispersed across the organization to enable swifter and more localized responses to changes [60].
- **Edge of Chaos**: Complex systems are most adaptable and innovative at the "edge of chaos"—a state poised between utter order and complete disorder. Striking a balance between stability (required for efficient operation) and flexibility (necessary for adaptation and innovation) is crucial for organizations [61].

To identify vulnerabilities, complexity theory emphasizes scrutinizing the system's structure and interdependencies since vulnerabilities frequently emerge from the nonlinear, interconnected nature of complex systems. For example, a failure in one segment of the organization can spread throughout the system due to these interdependencies [62].

By recognizing these factors related to complexity, organizations can not only detect their vulnerabilities but also adapt their structures and processes to boost resilience in the face of disruptions and unexpected events [63].

7.6.2 Business Continuity Planning

Business Continuity Planning (BCP) [64, 65]is an essential aspect of organizational resilience that ensures the continuity of operations during a disruption or emergency. The concept has evolved significantly over time, as organizations have faced increasing complexities and a variety of new challenges.

Historically, BCP was often viewed as a relatively straightforward disaster recovery effort, mostly focusing on IT and data recovery. The objective was to ensure the organization could continue to operate in case of a significant system or technology failure. In this era, the primary threats considered were natural disasters, power outages, and other localized incidents.

However, the increasing interconnectivity and globalization of business, along with the emergence of complex, systemic risks such as pandemics and cyberattacks, have led to a broader understanding of BCP. The evolution of threats, coupled with regulatory changes and higher stakeholder expectations, has led organizations to develop more comprehensive and holistic BCPs.

The state of the art of BCP, as of now, incorporates a far broader perspective that takes into account all critical business functions, key stakeholders, and encompasses both internal and external threats.

The modern approach to BCP generally involves several key steps:

1) **Business Impact Analysis (BIA):** This involves identifying the most critical business functions and understanding the impact if these functions were to be interrupted. It is crucial to assess not just the financial implications, but also impacts on reputation, regulatory compliance, customer satisfaction, and other factors.
2) **Risk Assessment:** This step involves identifying the various risks that could disrupt business operations, assessing their likelihood and potential impact, and prioritizing them based on these factors.
3) **Business Continuity Strategies:** Based on the BIA and risk assessment, the organization develops strategies to maintain or quickly resume critical business functions in the face of a disruption. This might involve diversifying supply chains, developing redundant systems, creating crisis management teams, or implementing robust data backup solutions.
4) **Plan Development:** This involves creating a detailed plan that lays out the specific actions that will be taken in the event of a disruption, including roles and responsibilities, communication protocols, and recovery procedures.
5) **Testing and Maintenance:** The BCP should be tested regularly to ensure it is effective and updated as necessary to reflect changes in the business environment or internal operations.

In terms of best practices, it is now recognized that BCP should be a continuous, cyclical process, rather than a one-time effort. Furthermore, BCP should be aligned with the overall risk management framework and integrated with other related processes, such as incident management, IT disaster recovery, and emergency response.

BCP also increasingly incorporates considerations related to supply chain resilience, reputational risk, and stakeholder communication. In addition, there is a growing focus on building overall organizational resilience, such as by enhancing the adaptability and flexibility of business processes, rather than simply planning for specific identified threats.

In summary, BCP has evolved to become a more strategic, holistic, and integrated process, focused not just on recovering from disruptions, but on enabling the organization to adapt and thrive in the face of uncertainty and change.

7.6.3 Disaster Recovery and Emergency Response

Disaster Recovery and Emergency Response are integral parts of a broader organizational risk management strategy, designed to prepare for and respond to major incidents that can disrupt operations or cause significant harm to stakeholders.

Historically, disaster recovery primarily focused on recovery and restoration of IT systems and data after a major disruption, such as a natural disaster, a fire, or a power outage. Emergency response, on the other hand, was more oriented toward immediate actions required to protect life, property, and the environment during and immediately after an emergency event. Both areas, however, have seen significant evolution in their scope and sophistication over time.

Over the years, the emergence of new technologies and the increasing interconnectivity of systems have changed the disaster recovery landscape. Organizations today are much more reliant on digital infrastructure and data, and disruptions can have far-reaching impacts. Consequently, disaster recovery has evolved to address a wider range of scenarios, including cyberattacks, large-scale data breaches, and systems failure due to software bugs or human error.

The state of the art for disaster recovery now involves comprehensive planning and implementation of measures to ensure not just the recovery of data and IT systems, but the continuity of business operations. This involves:

1) **Data Backup and Recovery:** Regular and comprehensive backups of critical data, and strategies for recovering this data in the event of loss or corruption.
2) **IT Systems and Infrastructure Recovery:** Strategies and plans for restoring IT systems and infrastructure to their normal function after a disaster.
3) **Alternate Processing Sites:** In the event of a physical disaster that renders a primary site unusable, having a secondary site ready to take over processing can be critical.
4) **Testing and Maintenance:** Regular testing of disaster recovery plans to ensure they are effective, and updating the plans as necessary to reflect changes in technology, business operations, or the risk environment.

Emergency response, similarly, has seen a significant shift in focus. While the immediate safety of individuals is always a priority, organizations now recognize the need to manage a wider range of potential emergencies and to engage in more proactive planning. Today's emergency response strategies often include:

1) **Emergency Preparedness:** Developing comprehensive plans for responding to a range of potential emergency scenarios, including natural disasters, fires, medical emergencies, terrorist attacks, or cyber incidents.
2) **Incident Response Teams:** Establishing dedicated teams with clearly defined roles and responsibilities to manage and respond to emergencies.
3) **Communication and Coordination:** Strategies for communicating effectively with employees, first responders, authorities, and other stakeholders during an emergency.
4) **Training and Drills:** Regular training for employees to understand emergency response procedures, and conducting drills to practice these procedures.
5) **Learning and Improvement:** Post-incident reviews to learn from each emergency and continually improve emergency response capabilities.

The evolution of these fields has seen a more proactive approach to risk management, with the aim of preventing disasters wherever possible, and mitigating their impact when they do occur. The increasing use of technology, such as automated monitoring and alerting systems, predictive analytics, and advanced simulation models, also enables organizations to detect and respond to incidents more quickly and effectively. However, the human element remains crucial in both disaster recovery and emergency response, highlighting the importance of leadership, training, and a strong organizational culture of safety and resilience.

References

1 Haimes, Y.Y. (2005). *Risk Modeling, Assessment, and Management.* Wiley.
2 Dionne, G. (2013). Risk management: history, definition, and critique. *Risk Management and Insurance Review* 16 (2): 147–166.
3 Aven, T. (2016). Risk assessment and risk management: review of recent advances on their foundation. *European Journal of Operational Research* 253 (1): 1–13.

4 Lam, J. (2014). *Enterprise Risk Management: From Incentives to Controls*. Wiley.

5 Head, B.W. (2016). Toward more "evidence-informed" policy making? *Public Administration Review* 76 (3): 472–484.

6 Borge, D. (2002). *The Book of Risk*. Wiley.

7 Patrucco, A.S. and Kähkönen, A.K. (2021). Agility, adaptability, and alignment: new capabilities for PSM in a post-pandemic world. *Journal of Purchasing and Supply Management* 27 (4): 100719.

8 Mikes, A. and Kaplan, R.S. (2015). When one size doesn't fit all: evolving directions in the research and practice of enterprise risk management. *Journal of Applied Corporate Finance* 27 (1): 37–40.

9 Schulman, P. and Roe, E. (2016). *Reliability and Risk: The Challenge of Managing Interconnected Infrastructures*. Stanford University Press.

10 Hokstad, P., Utne, I.B., and Vatn, J. (2012). *Risk and Interdependencies in Critical Infrastructures*. London: Springer.

11 Rasmussen, J. and Suedung, I. (2000). *Proactive Risk Management in a Dynamic Society*. Swedish Rescue Services Agency.

12 Ganguly, A., Nilchiani, R., and Farr, J.V. (2011). Identification, classification, and prioritization of risks associated with a disruptive technology process. *International Journal of Innovation and Technology Management* 8 (02): 273–293.

13 Berkhout, F. (2012). Adaptation to climate change by organizations. *Wiley Interdisciplinary Reviews: Climate Change* 3 (1): 91–106.

14 Linnenluecke, M. and Griffiths, A. (2010). Beyond adaptation: resilience for business in light of climate change and weather extremes. *Business & Society* 49 (3): 477–511.

15 Aven, T. (2016). Risk assessment and risk management: review of recent advances on their foundation. *European Journal of Operational Research* 253 (1): 1–13.

16 Haimes, Y.Y. (2005). *Risk Modeling, Assessment, and Management*. Wiley.

17 Taleb, N.N. (2007). *The Black Swan: The Impact of the Highly Improbable*. Random House.

18 Kaplan, R.S. and Mikes, A. (2012). Managing risks: a new framework. *Harvard Business Review* 90 (6): 48–60.

19 Borodzicz, E.P. (2005). *Risk, Crisis and Security Management*. Chichester: Wiley.

20 McNeil, A.J., Frey, R., and Embrechts, P. (2015). *Quantitative Risk Management: Concepts, Techniques and Tools-Revised Edition*. Princeton University Press.

21 Purdy, G. (2010). ISO 31000: 2009—setting a new standard for risk management. *Risk Analysis: An International Journal* 30 (6): 881–886.

22 Hopkin, P. (2018). *Fundamentals of Risk Management: Understanding, Evaluating and Implementing Effective Risk Management*. Kogan Page Publishers.

23 Aven, T. (2016). Risk assessment and risk management: Review of recent advances on their foundation. *European Journal of Operational Research* 253 (1): 1–13.

24 Nocco, B.W. and Stulz, R.M. (2022). Enterprise risk management: theory and practice. *Journal of Applied Corporate Finance* 34 (1): 81–94.

25 Gates, S., Nicolas, J.L., and Walker, P.L. (2012). Enterprise risk management: a process for enhanced management and improved performance. *Management Accounting Quarterly* 13 (3): 28–38.

26 Fraser, J.R., Quail, R., and Simkins, B. (ed.) (2021). *Enterprise Risk Management: Today's Leading Research and Best Practices for Tomorrow's Executives*. Wiley.

27 Araz, O.M., Choi, T.M., Olson, D.L., and Salman, F.S. (2020). Data analytics for operational risk management. *Decision Science* 51 (6): 1316–1319.

28 Azvine, B., Cui, Z., Majeed, B., and Spott, M. (2007). Operational risk management with real-time business intelligence. *BT Technology Journal* 25 (1): 154–167.

29 Frigo, M.L. and Anderson, R.J. (2011). What is strategic risk management? *Strategic Finance* 92 (10): 21.

30 Calandro, J. (2015). A leader's guide to strategic risk management. *Strategy & Leadership* 43 (1): 26–35.

31 Hillson, D. (2006). Integrated risk management as a framework for organizational success. In: *Proceedings of the PMI Global Congress*, Seattle, Washington (14–21 October 2006). https://www.researchgate.net/profile/David-Hillson/publication/265288286_Integrated_Risk_Management_As_A_Framework_For_Organisational_Success/links/54aa51c10cf2eecc56e6d5e4/Integrated-Risk-Management-As-A-Framework-For-Organisational-Success.pdf (accessed 29 August 2023).

32 Wu, J. and Wu, Z. (2014). Integrated risk management and product innovation in China: the moderating role of board of directors. *Technovation* 34 (8): 466–476.

33 Sax, J. and Andersen, T.J. (2019). Making risk management strategic: integrating enterprise risk management with strategic planning. *European Management Review* 16 (3): 719–740.

34 Pierce, E.M. and Goldstein, J. (2018). ERM and strategic planning: a change in paradigm. *International Journal of Disclosure and Governance* 15: 51–59.

35 Book, G. (2012). Lessons learned from real world application of the bow-tie method. In: *SPE Middle East Health, Safety, Security, and Environment Conference and Exhibition*, Abu Dhabi, UAE (2–4 April 2012), SPE-154549. Society of Petroleum Engineers.

36 Card, A.J., Ward, J.R., and Clarkson, P.J. (2012). Beyond FMEA: the structured what-if technique (SWIFT). *Journal of Healthcare Risk Management* 31 (4): 23–29.

37 Dunjó, J., Fthenakis, V., Vílchez, J.A., and Arnaldos, J. (2010). Hazard and operability (HAZOP) analysis. A literature review. *Journal of Hazardous Materials* 173 (1–3): 19–32.

38 Ngai, E.W., Hu, Y., Wong, Y.H. et al. (2011). The application of data mining techniques in financial fraud detection: a classification framework and an academic review of literature. *Decision Support Systems* 50 (3): 559–569.

39 Woeste, M.R., Strothman, P., Jacob, K. et al. (2022). Hepatopancreatobiliary readmission score out performs administrative LACE+ index as a predictive tool of readmission. *The American Journal of Surgery* 223 (5): 933–938.

40 A.O.S. OFFER (2013). The influence of telematics on customer experience: case study of progressive's snapshot program. *Progressive* 59 (31): 89.

41 Man, X., Luo, T., and Lin, J. (2019). Financial sentiment analysis (FSA): a survey. In: *2019 IEEE International Conference on Industrial Cyber Physical Systems (ICPS)*, 617–622. IEEE.

42 Terekhova, M. (2017). JPMorgan takes AI use to the next level. Business Insider (2 August). https://www.businessinsider.com/jpmorgan-takes-ai-use-to-the-next-level-2017-8

43 Honeyford, K., Nwosu, A.P., Lazzarino, R. et al. (2023). Prevalence of electronic screening for sepsis in National Health Service acute hospitals in England. *BMJ Health & Care Informatics* 30 (1).

44 Geithner, T.F. (2015). *Stress Test: Reflections on Financial Crises*. Crow.

45 Chakrabarty, B., Patil, S.R., Shingornikar, S. et al. (2021). *Securing Data on Threat Detection by Using IBM Spectrum Scale and IBM QRadar: An Enhanced Cyber Resiliency Solution*. IBM Redbooks.

46 Linnenluecke, M.K. (2017). Resilience in business and management research: a review of influential publications and a research agenda. *International Journal of Management Reviews* 19 (1): 4–30.

47 Hiles, A. (2010). *The Definitive Handbook of Business Continuity Management*. Wiley.

48 Sawalha, I.H.S. (2015). Managing adversity: understanding some dimensions of organizational resilience. *Management Research Review* 38 (4): 346–366.

49 Williams, T.A., Gruber, D.A., Sutcliffe, K.M. et al. (2017). Organizational response to adversity: fusing crisis management and resilience research streams. *Academy of Management Annals* 11 (2): 733–769.

50 Sutcliffe, K.M. and Vogus, T.J. (2003). Organizing for resilience. In: *Positive Organizational Scholarship: Foundations of a New Discipline* (ed. K.S. Cameron, J.E. Dutton, and R.E. Quinn), 94–110. Berrett-Koehler Publishers.

51 Välikangas, L. (2010). *The Resilient Organization: How Adaptive Cultures Thrive Even when Strategy Fails*. New York: McGraw Hill.

52 Van Der Vegt, G.S., Essens, P., Wahlström, M., and George, G. (2015). Managing risk and resilience. *Academy of Management Journal* 58 (4): 971–980.

53 Bhamra, R., Dani, S., and Burnard, K. (2011). Resilience: the concept, a literature review and future directions. *International Journal of Production Research* 49 (18): 5375–5393.

54 Jen, E. (ed.) (2005). *Robust Design: A Repertoire of Biological, Ecological, and Engineering Case Studies*. Oxford University Press.

55 Carlson, J.M. and Doyle, J. (2002). Complexity and robustness. *Proceedings of the National Academy of Sciences of the United States of America* 99 (suppl_1): 2538–2545.

56 Uhl-Bien, M., Marion, R., and McKelvey, B. (2007). Complexity leadership theory: shifting leadership from the industrial age to the knowledge era. *The Leadership Quarterly* 18 (4): 298–318.

57 Anderson, P. (1999). Perspective: complexity theory and organization science. *Organization Science* 10 (3): 216–232.

58 Holland, J.H. (1992). Complex adaptive systems. *Daedalus* 121 (1): 17–30.

59 Kauffman, S.A. (1993). *The Origins of Order: Self-Organization and Selection in Evolution*. New York, USA: Oxford University Press.

60 Heylighen, F. (2013). *Self-Organization in Communicating Groups: The Emergence of Coordination, Shared References and Collective Intelligence*, 117–149. Berlin Heidelberg: Springer.

61 Waldrop, M.M. (1993). *Complexity: The Emerging Science at the Edge of Order and Chaos*. Simon and Schuster.

62 Buldyrev, S.V., Parshani, R., Paul, G. et al. (2010). Catastrophic cascade of failures in interdependent networks. *Nature* 464 (7291): 1025–1028.

63 Dooley, K.J. (1997). A complex adaptive systems model of organization change. *Nonlinear Dynamics, Psychology, and Life Sciences* 1: 69–97.

64 Cerullo, V. and Cerullo, M.J. (2004). Business continuity planning: a comprehensive approach. *Information Systems Management* 21 (3): 70–78.

65 Lindström, J., Samuelsson, S., and Hägerfors, A. (2010). Business continuity planning methodology. *Disaster Prevention and Management: An International Journal* 19 (2): 243–255.

8

Emerging Organizational Changes in the 21st Century

The narrative of organizational structures is one of perpetual evolution, spurred by technological progress, societal dynamics, and economic fluctuations [1]. These evolutions span from rigid hierarchies to agile, networked frameworks, with goals of increased productivity, fostering innovation, and promoting employee satisfaction. As we delve deeper into the 21st century, the pace of these transformations appears to be accelerating, posing both trials and opportunities for contemporary businesses and professionals.

From the era of the Second Industrial Revolution, firms were designed into inflexible, hierarchical configurations, often depicted as traditional pyramid structures. This structure was efficient for coordinating large-scale undertakings, preserving a transparent chain of command, and retaining accountability. However, with the emergence of a knowledge-based economy, the downsides of these hierarchical organizations became apparent. The insular thinking and lack of agility inherent in these structures were incompatible with an increasingly intricate and swiftly changing business landscape [2].

This prompted a shift toward flatter organizational structures, notably among technology companies and startups. These organizations stressed open dialogue, prompt decision-making, and a stronger focus on innovation. Yet, while bringing significant enhancements, these models also introduced their own challenges, such as ambiguity in roles and responsibilities and potential conflict arising from an unclear chain of command.

The advent of the digital age heralded the era of networked and agile organizations, characterized by cross-functional collaboration, employee empowerment, and an emphasis on swiftness and customer-centricity [3]. Presently, as technologies like Artificial Intelligence (AI), big data, and Industry 4.0 increasingly impact businesses, organizations are encountering another wave of transformation. This is seen in the rise of the gig economy, remote work, and the need for heightened flexibility, steering organizational structures toward more decentralized and distributed models. These new organizational forms are exemplified by concepts such as holacracy and models like Haier's Rendanheyi [4].

Simultaneously, societal changes are injecting additional complexity. Work-life boundaries are becoming increasingly blurred, and alterations in lifestyles, work habits, thought processes, and social interactions are occurring rapidly. There is a growing demand for work-life integration, personalized experiences, and purpose-driven work, all contributing to shaping the organizations of the future [5].

Looking ahead, it is evident that these trends will persistently reshape our workplaces over the next two decades. For quality professionals, these shifts offer both challenges and opportunities.

Quality in the Era of Industry 4.0: Integrating Tradition and Innovation in the Age of Data and AI,
First Edition. Kai Yang.

The rise of AI and big data compels a deeper comprehension of these technologies and their repercussions for quality management. The shift toward more decentralized and distributed structures implies that quality professionals must adeptly navigate ambiguity, handle change, and collaborate effectively across networks. Furthermore, the emphasis on human skills, such as creativity, complex problem-solving, and emotional intelligence, underscores the importance of these skills in the quality profession.

This chapter will further investigate these emerging alterations in organizational structures, examine their origins and implications, and contemplate their impact on the role of quality professionals. Through this, it aims to offer insights and guidance for navigating the ever-evolving organizational landscape of the 21st century.

8.1 The Continuously Shifting Landscape of Organizational Structures

The unfolding story of organizational structures is intrinsically intertwined with technological progress and sociocultural shifts. As manufacturing and production processes grew more complex during the Second Industrial Revolution, the imperative for structured coordination gave rise to hierarchical, or pyramid, structures. These traditional configurations were underscored by clear chains of command, uniform procedures, and rigorously outlined roles and responsibilities [6].

The post-World War II era signaled a new wave of change in the organizational landscape with the advent of Information Technology (IT). This revolution demanded more swift decision-making, enhanced adaptability, and a superior level of innovation [7]. Concurrently, Japan's economic ascent, coupled with the innovation of quality control methodologies such as Lean and Just-in-Time, spurred Western corporations to reassess their operational and organizational practices [8].

The digital revolution at the turn of the century introduced additional impetus for change. Technological breakthroughs such as the Internet, personal computing, and mobile devices necessitated a critical reevaluation of traditional models, promoting the emergence of more flexible and flat structures. Organizations recognized the merit of empowering employees, fostering open communication, and dismantling silos to facilitate cross-functional collaboration [9].

Now, in the era of Industry 4.0, characterized by AI, big data, Internet of Things (IoT), and automation, the nature of work and the subsequent organizational structures are experiencing a further transformative wave. This new industrial revolution is driving organizations toward even more flexible, agile, and networked configurations.

In the subsequent sections, we will delve into these historical shifts more deeply, examining the evolution from conventional pyramid structures to modern organizational frameworks. We will focus on understanding the genesis and impact of flexible and flat structures. These insights will illuminate the path of organizational evolution, providing a glimpse into what the future may hold as we journey further into the 21st century.

8.1.1 Evolution from Traditional Pyramid to Contemporary Organizational Structures

The journey from traditional pyramid structures to contemporary organizational models mirrors the broader transformation of our economies and societies. This evolution has been shaped by technological advancements, societal shifts, and changes in economic thinking.

8.1.1.1 Traditional Pyramid Structures

During the Second Industrial Revolution and into the early 20th century, the dominant organizational structure was the traditional pyramid or hierarchical model. This structure was characterized by a clear chain of command from top (executives) to bottom (frontline employees), well-defined roles and responsibilities, and centralized decision-making.

This structure was particularly suited to the manufacturing-oriented economy of the era. The assembly-line model of production, epitomized by Henry Ford's Model T production, required strict control, efficiency, and standardization, all of which the hierarchical model provided. The clear chain of command minimized confusion and the separation of strategic decision-making (top-level) from execution (lower levels) streamlined operations. Organizations like General Motors embodied this structure. They achieved great success due to their ability to maintain order, coordinate large-scale operations, and ensure clear communication channels.

8.1.1.2 The Move to Matrix Structures

Following World War II, technological innovations began to reshape the business landscape, prompting organizations to seek greater flexibility to swiftly respond to evolving market dynamics. The 1960s and 1970s witnessed the emergence of matrix structures. These models blended functional and divisional lines of authority, enabling employees to report to two managers: a functional manager and a product or project manager. This structure facilitated resource sharing across projects and promoted cross-functional collaboration. Early adopters of this model included companies like Philips and Texas Instruments, though the approach was not without its challenges, such as potential managerial conflicts and employee confusion [10].

In parallel, the late 20th century heralded the Total Quality Management (TQM) movement, popularized by Japanese manufacturers. TQM principles emphasized employee empowerment, continuous improvement, and customer focus, thereby challenging the strict hierarchy and rigidity of the traditional pyramid model [11].

8.1.1.3 The Flat and Horizontal Organizations

Throughout the 1980s and 1990s, rapid advancements in IT catalyzed an accelerated pace of change, prompting organizations to flatten their structures. The resultant flat organizations, delineated by fewer managerial layers, facilitated direct communication between executives and frontline employees, encouraged an innovative culture, and demonstrated greater adaptability. Tech companies and startups, notably Google and Amazon, became exemplars of this organizational structure [12].

8.1.1.4 Contemporary Organizational Structures

As the 21st century dawned, the pervasiveness of the Internet revolution and digital technology triggered organizations to pivot toward more networked, flatter structures. Tech industry pioneers like Google and Valve adopted these flat structures, hallmarked by low power distance, open communication, and empowered employees. The decentralization of decision-making equipped these organizations to foster rapid innovation and remain competitive in the swiftly changing tech landscape [13].

This transition from traditional pyramid models to contemporary designs reflects the cumulative influence of technological advancements, societal evolution, and economic fluctuations. Markedly, this journey has underscored an escalating emphasis on flexibility, collaboration, and employee empowerment. As we gaze forward, these trends are anticipated to persist, fueled by continued technological progress, shifting societal expectations, and the mounting complexity of the business environment [14].

8.1.2 The Emergence of Flexible and Flat Structures

Flexible and flat structures have emerged as a response to the changing demands of the modern business landscape, characterized by rapid technological changes, evolving customer expectations, and the increasing need for innovation. Let us delve into this trend in detail.

The flat organizational structure emerged in response to the perceived rigidity and inefficiency of traditional hierarchical models. Flat organizations are characterized by few or no levels of middle management between staff and executives. This move toward flatter structures started gaining traction in the 1980s and 1990s, as businesses sought to be more responsive and adaptive to market changes.

Several factors have contributed to this shift toward flat structures:

- **Technological Advancements:** Developments in IT have facilitated easy access to information and improved communication, reducing the need for multiple layers of management.
- **Speed and Innovation:** The increasing pace of business and the need for rapid innovation have driven companies to reduce bureaucratic hurdles and empower employees to make decisions.
- **Employee Empowerment:** There is a growing recognition that employees closest to problems often have the best solutions. Flat structures empower these employees to take decisions and contribute ideas.
- **Cost Efficiency:** By eliminating levels of middle management, organizations can reduce costs and increase operational efficiency.

Flat structures offer several benefits:

- **Speed and Agility:** With fewer layers of management, decisions can be made and implemented more quickly, allowing the organization to be more agile and responsive.
- **Empowerment and Engagement:** Employees in flat structures are typically given more responsibility, which can lead to increased job satisfaction and engagement.
- **Communication and Collaboration:** With less hierarchical barriers, communication improves, fostering collaboration and innovation.

There are several notable examples of organizations that have adopted flat structures:

Example 8.1   W.L. Gore & Associates

W.L. Gore & Associates, the makers of the waterproof and breathable fabric Gore-Tex, provides an insightful example of a successful flat organizational structure [15]. The company, founded in 1958 by Bill Gore, has consistently featured on lists of the best companies to work for and has built a reputation for innovation and high-quality products.

Organizational Structure

Bill Gore designed the organization with a unique, lattice-like structure to encourage hands-on innovation and problem-solving. The company famously operates without traditional managerial roles. Instead of being assigned to a boss, associates (as employees are called) commit to projects that align with their skills and interests.

Leadership in the company is determined not by appointment but by "followership." In other words, leaders emerge according to their ability to attract followers, gain trust, and inspire their peers. This kind of leadership is not cemented in a title but can ebb and flow with different projects and tasks.

The "Waterline" Principle

One of the guiding principles at Gore is the "waterline" concept. The idea is that it is fine for associates to make decisions on their own as long as the potential mistake would not cause the company to sink. However, if the decision could potentially hit above the waterline, causing serious harm to the company, associates are encouraged to consult with their peers before proceeding.

Teams and Communication

The organization is also known for its small, cross-functional teams, usually capped around 8–12 members. This is based on Bill Gore's belief in the concept of "dunbar's number"–the idea that there is a cognitive limit to the number of people with whom one can maintain stable social relationships. Keeping teams small encourages direct communication, cooperation, and collective problem-solving.

Culture of Innovation

The lack of rigid hierarchies and the emphasis on associate autonomy fosters a culture of innovation at Gore. Associates have the freedom to spend 10% of their work time on projects of their own choosing, similar to Google's famous "20% time." This has led to numerous product innovations, from medical devices to guitar strings, demonstrating the power of intrinsic motivation and autonomy.

Challenges and Criticisms

Despite its successes, Gore's model is not without its critics or challenges. Some have pointed out that the lack of traditional hierarchy can lead to a lack of clarity about decision-making and responsibility. Also, not everyone thrives in such an environment; it requires a high degree of self-motivation and comfort with ambiguity.

In summary, W.L. Gore & Associates exemplifies a flat structure that encourages innovation, autonomy, and natural leadership. Its success illustrates the potential benefits of such a structure: increased innovation, high levels of employee satisfaction, and the ability to adapt to change. However, it also highlights the need for organizations adopting such a model to be aware of the potential challenges and foster a culture supporting autonomy, collaboration, and mutual trust.

Example 8.2 Valve Corporation

Valve Corporation, a video game development company known for blockbuster titles like "Half-Life," "Portal," and "Counter-Strike," and the gaming platform "Steam," is renowned for its nonhierarchical, flat organizational structure [16]. Founded in 1996 by Gabe Newell and Mike Harrington, former Microsoft employees, Valve has been an intriguing case study in innovative organizational design.

Organizational Structure

In Valve, there are no job titles, no formal bosses, and employees can work on projects they choose. The organizational structure is based on the idea that in creative work, autonomy, and the freedom to innovate without managerial constraints are crucial. This model is designed to create an environment where the employees can be fully immersed in the creative process with minimal administrative distractions.

In essence, Valve operates as a "boss-less" organization. It is been described as anarchy in the sense that there is no formal authority, but it is an anarchy that is organized around the work to be done.

Mobility and Projects

One of the unique aspects of Valve's structure is the mobility of employees. Their offices are designed to be adaptable, with desks fitted with wheels to allow employees to move freely to different teams or projects. This physical manifestation of their flat structure emphasizes their commitment to flexibility, employee choice, and cross-functional collaboration.

Projects at Valve typically start with one or two people coming up with an idea and then recruiting others within the organization to join them. The value of the project is determined not by a managerial decree but by the ability to attract coworkers to the idea. If a project fails to draw interest, it is likely it will be dropped, acting as a form of peer review.

The Employee Handbook

Valve's unique approach to organizational structure is best encapsulated in its employee handbook. It states: "Of all the people at this company who aren't your boss, Gabe is the MOST not your boss, if you get what we're saying." This playful and bold assertion underscores the commitment to a flat structure from the top down.

Challenges

While Valve's structure has contributed to its innovation and success in the gaming industry, it is not without its challenges. Some former employees have commented on the difficulty of navigating within the organization, as it may lead to a lack of clarity around decision-making processes and career progression.

In conclusion, Valve Corporation's flat structure represents a bold experiment in workplace organization that prioritizes employee autonomy and innovation. While this model may not be appropriate for every industry or company, it serves as a noteworthy example of an alternative to traditional hierarchies in the creative industry. It underscores the potential advantages of a flat structure but also highlights potential drawbacks that should be carefully considered.

Example 8.3   Spotify

Spotify, a Swedish audio streaming and media services provider, stands out in its application of an innovative organizational structure [17]. Though not entirely flat, Spotify's "squad" model embodies a modern approach to organizational structure that emphasizes autonomy, cross-functionality, and agility.

Organizational Structure

Spotify operates with what they call the "Squad" model. The basic unit in this model is a Squad, a small cross-functional team with a specific mission. Each squad is somewhat like a mini-startup within the Spotify ecosystem, equipped with all the necessary competencies to design, develop, test, and release to production.

Squads are grouped into Tribes, larger organizational units that encapsulate squads working in related areas. Each tribe is led by a Tribe Lead, who works to ensure that all squads are in alignment and have the resources they need.

The Model: Squads, Tribes, Chapters, and Guilds

The Squad model is complemented by Chapters and Guilds. A Chapter is a horizontal organization of specialists working in the same area, across different squads. For example, a Chapter Lead for Backend Development would have regular meetings with all backend developers, irrespective of their Squad or Tribe.

Guilds are more informal and expansive, cutting across the entire organization. They are essentially communities of interest, where employees share knowledge, tools, and practices.

Impact and Results

Spotify's model has facilitated the handling of vast, complex tasks by breaking them down into manageable, autonomous units (Squads). The Chapter and Guild system encourages knowledge sharing and mentoring, fostering a culture of continuous learning.

This organizational structure has been pivotal in promoting innovation, efficiency, and adaptability at Spotify. It has enabled the company to stay agile and responsive to market changes, maintaining its leading position in the dynamic tech industry.

Challenges and Criticism

The Squad model, while innovative, has its challenges. Maintaining alignment and communication between squads can be difficult, especially as the organization scales. Moreover, squads could potentially become too siloed, causing inconsistencies in the product or user experience.

To address these challenges, Spotify has to constantly evolve its structure, demonstrating the company's commitment to adaptability and continuous improvement.

In conclusion, Spotify's Squad model offers a unique approach to modern organizational structure, effectively balancing autonomy and alignment, innovation, and efficiency. Its successes and challenges provide valuable insights for companies looking to adopt more agile, flexible organizational structures.

However, it is important to note that flat structures are not without their challenges. They require a high level of mutual trust and responsibility, clear communication, and often involve navigating ambiguity in roles and responsibilities.

Overall, the move toward flexible and flat structures is a reflection of the evolving business landscape. As the pace of change continues to accelerate, these models offer a way for organizations to stay nimble, innovative, and adaptive.

8.2 Impact of Technological Advances on Organizational Structures

In the rapidly evolving business landscape of the 21st century, the most profound changes in organizational structures can be traced back to technological advances. The advent of technologies such as AI, Big Data, and the onset of Industry 4.0 have brought about seismic shifts in the ways organizations are structured and operate. Technology has not only disrupted traditional business models, but it has also reshaped workplace dynamics, team compositions, communication methods, and decision-making processes. This has led to the emergence of flexible, adaptive, and increasingly flat organizational structures, better equipped to navigate the complexities of the digital age. The subsequent sections delve into the specific impacts of AI, Big Data, and Industry 4.0 on organizational structures, providing a comprehensive understanding of the transformational power of these technologies.

8.2.1 Impact of Artificial Intelligence

AI represents one of the most transformative technologies of our time. Its ability to learn from data, predict outcomes, automate tasks, and facilitate decision-making has had significant implications for organizational structures.

1) **Streamlining Hierarchies**
   AI has the potential to simplify and streamline organizational hierarchies. By automating routine tasks, AI reduces the need for supervisory roles and layers of management traditionally required to oversee these tasks. The result is a flatter, more efficient organizational structure.
2) **Facilitating Decision-Making**
   With AI's capability to analyze vast amounts of data and provide insights, decision-making can be made faster and more accurate. This can empower employees at all levels, reducing the need for top-down decision-making and fostering a more democratic and inclusive organizational structure.
3) **Enhancing Collaboration**
   AI-driven tools and platforms can enhance collaboration across different teams and departments. By breaking down communication barriers and facilitating knowledge sharing, AI can foster a more interconnected and networked organizational structure.

8.2.1.1 AI Technologies and Their Impact
Several AI technologies are driving these changes:

- **Machine Learning:** By enabling computers to learn from data, Machine Learning can automate decision-making processes and reduce the need for human intervention, leading to flatter structures.

- **Natural Language Processing (NLP):** NLP can automate and enhance customer service, reducing the need for large customer service teams and management.
- **Robotic Process Automation (RPA):** By automating routine tasks, RPA can reduce the need for supervisory roles and create more streamlined, efficient organizational structures [18].
  RPA is a type of AI technology designed to automate high-volume, repetitive, routine tasks that humans previously performed. This can include tasks such as data entry, processing transactions, responding to simple customer service queries, and more.

  RPA works by creating software robots, or "bots," that can be programmed to perform specific tasks. These bots can interact with digital systems and software just like a human user would, following predefined rules and workflows. They can log into applications, enter data, calculate and complete tasks, and then log out.

  Here is how RPA impacts organizational structures:

  - **Efficiency and Productivity:** RPA can significantly improve efficiency and productivity as software bots can work 24/7 without breaks, and they can complete tasks more quickly and accurately than humans. This can result in a leaner, more efficient organizational structure, as fewer employees are needed to perform routine tasks.
  - **Reshaping Roles and Responsibilities:** The automation of routine tasks frees up employees to focus on higher-value tasks that require human skills such as problem-solving, creativity, and interpersonal communication. This can lead to a shift in roles and responsibilities within the organization, emphasizing more strategic and creative roles.
  - **Reducing Hierarchical Layers:** With fewer routine tasks to manage, there is less need for supervisory roles and the layers of management that were previously required to oversee these tasks. This can lead to flatter and more agile organizational structures.

  Deutsche Bank provides a real-world example of RPA's impact on organizational structure. The bank has employed RPA to automate thousands of jobs that were previously performed by humans, particularly in operations and back-office roles. This has resulted in a leaner, more efficient organizational structure, with human employees freed up to focus on more complex, higher-value tasks.

  In summary, RPA is a powerful technology that can bring significant changes to organizational structures, leading to increased efficiency, reshaped roles and responsibilities, and flatter organizational hierarchies.

- **AI-Driven Analytics:** AI-driven analytics utilizes technologies like machine learning and natural language processing to scrutinize large and complex datasets, generating actionable insights, enabling organizations to make informed decisions, and guiding their structural evolution. It has several noteworthy impacts on organizational structures:

  - **Data-Driven Decision-Making:** AI-driven analytics supports data-driven decision-making across organizational hierarchies, encouraging a more democratic and inclusive structure [19].
  - **Enhanced Customer Understanding:** AI's proficiency in analyzing copious amounts of customer data can offer profound insights into customer behavior, preferences, and trends, which could prompt structural adaptations to accentuate customer-oriented roles and teams [20].
  - **Predictive Analytics:** Leveraging historical data, AI-driven predictive analytics enable organizations to anticipate future trends and outcomes, fostering proactive strategies and potentially inspiring the creation of roles centered on predictive analysis and strategic foresight [21].

Streaming giant Netflix exemplifies the effective use of AI-driven analytics in business strategy. By analyzing extensive data on user behavior, Netflix gains insights into viewer preferences and viewing patterns, guiding decisions on content production and acquisition, as well as personalizing viewer recommendations. This data-driven approach has shaped Netflix's organizational structure, with substantial resources dedicated to data analytics teams and decision-making authority distributed throughout the organization [22].

In summary, AI-driven analytics can revolutionize organizational structures by facilitating data-driven decision-making, enhancing customer understanding, and supporting predictive analytics. As AI-driven analytics adoption continues, we can anticipate further evolution toward more data-centric, predictive, and customer-focused structures. AI-driven analytics can provide real-time insights, aiding decision-making processes and fostering a more data-driven and inclusive organizational structure.

Example 8.4 Google's DeepMind

A prominent example of an organization utilizing AI to impact its structure is Google's DeepMind. DeepMind has developed AlphaGo, an AI program that demonstrated the potential of AI to solve complex problems by defeating the world champion in the complex board game Go. This underscores DeepMind's approach to problem-solving: instead of relying on hierarchical decision-making, the organization leverages AI's predictive and analytic capabilities to inform strategic decisions. This has allowed DeepMind to maintain a flatter, more flexible organizational structure where innovation and collaboration thrive.

In conclusion, the impact of AI on organizational structures cannot be overstated. As these technologies continue to evolve, they will further shape the organizations of the future, pushing them toward more flexible, inclusive, and efficient structures.

8.2.2 The Role of Big Data

In the age of digital transformation, Big Data plays a crucial role in shaping organizational structures. It refers to the vast amounts of structured and unstructured data generated by digital technologies that can be analyzed for insights. Big Data analytics helps in decision-making, identifying trends, and finding ways to streamline operations, all of which have significant implications for organizational structures.

1) **Data-Driven Decision-Making**
   Big Data allows for more data-driven decision-making. Access to large amounts of data and the ability to analyze it means decisions no longer have to be made based on intuition or limited information. This can flatten hierarchies, as decisions are not restricted to the top tiers of the organization, and employees at all levels can make informed decisions.
2) **Specialized Roles and Teams**
   The rise of Big Data has led to the creation of new roles and teams within organizations, such as data scientists, data analysts, and dedicated data teams. These roles are crucial for managing, analyzing, and deriving insights from the organization's data.
3) **Improved Customer Understanding**
   Through the analysis of Big Data, organizations can gain a better understanding of their customers' needs, behaviors, and preferences. This can inform the development of products and services, and influence the organization's structure to become more customer-centric.

8.2.2.1 Applications of Big Data and Their Impact

There are several key applications of Big Data that have shaped organizational structures:

- **Predictive Analytics:** Using Big Data to predict future outcomes and trends can help organizations to strategize and plan effectively. This can lead to the creation of roles focused on predictive analytics and strategic planning.
- **Personalization:** Big Data can be used to personalize customer experiences, from marketing messages to product recommendations. This necessitates roles and teams focused on customer engagement and personalization.
- **Risk Management:** Big Data can help identify potential risks and threats, enabling organizations to be proactive in risk management. This could lead to a more distributed risk management structure, with employees at all levels involved in identifying and managing risks.

Example 8.5   Amazon

Amazon is a prime example of an organization that has leveraged Big Data to shape its organizational structure [23]. The company uses Big Data to personalize customer experiences, from product recommendations to tailored marketing messages. This has influenced Amazon's structure, with teams dedicated to data analytics and customer engagement. Additionally, the use of Big Data in decision-making processes has flattened hierarchies and fostered a culture of data-driven decision-making.

In conclusion, the role of Big Data in shaping organizational structures is significant. As organizations continue to embrace Big Data, we can expect further evolution toward more data-driven, customer-centric, and adaptive structures.

8.2.3   Effects of Industry 4.0

Industry 4.0, also known as the Fourth Industrial Revolution, refers to the integration of digital technologies such as the IoT, cloud computing, AI, and big data into industrial processes. These technological advancements have significant impacts on organizational structures.

1) **Decentralized Decision-Making**
   Industry 4.0 technologies allow for real-time data collection and analysis, enabling quicker and more informed decision-making processes. This promotes a more decentralized decision-making structure as employees at all levels can make data-informed decisions.
2) **Enhanced Collaboration**
   The interconnectedness of Industry 4.0 technologies facilitates better collaboration and communication among different departments, leading to a more networked and less siloed organizational structure.
3) **Creation of New Roles**
   Industry 4.0 leads to the creation of new roles and teams, such as IoT specialists, data analysts, AI experts, and cybersecurity teams. These roles are necessary to manage and leverage the new technologies effectively.

8.2.3.1 Industry 4.0 Technologies and Their Impact

Several key technologies underpin Industry 4.0 and are shaping organizational structures:

- **Internet of Things (IoT):** The IoT connects devices and machines, enabling real-time data collection and analysis. This leads to a more data-driven and decentralized decision-making structure.

- **Cloud Computing:** Cloud computing enables data and systems to be accessed from anywhere, promoting remote work and a more flexible organizational structure.
- **Artificial Intelligence and Machine Learning:** AI and Machine Learning allow for automation and predictive analysis, leading to a more streamlined and proactive organizational structure.
- **Cyber-Physical Systems:** These systems integrate physical processes with digital technologies, necessitating new roles and teams to manage and leverage these systems effectively.

Siemens, a multinational industrial manufacturing company, provides a good example of how Industry 4.0 technologies are reshaping organizational structures. The company has leveraged IoT, AI, and cloud computing to optimize its manufacturing processes, leading to a more data-driven, networked, and decentralized structure. They have also created new roles focused on managing these technologies and analyzing the data they produce.

In conclusion, the effects of Industry 4.0 on organizational structures are significant. As more organizations embrace these technologies, we can expect further evolution toward more decentralized, collaborative, and data-driven structures.

8.3 Emerging Organizational Models in the 21st Century

The 21st century continues to usher in organizational adaptations to an ever-changing business landscape, influenced by technological advancements, shifting societal values, and novel work modalities. This engenders innovative organizational models that contest conventional hierarchies and bureaucratic systems. This section elucidates several of these emerging models:

- **The Networked Organization**, interconnected and collaborative;
- **The Holacracy Model**, favoring self-management;
- **The Agile Organization**, marked by flexibility and responsiveness;
- **Virtual and Remote Organizations**, leveraging geographical dispersion;
- **The Platform Model**, emphasizing ecosystem-centricity; and
- **The Rendanheyi Model**, a unique blend of multiple approaches.

Each model exemplifies different responses to the opportunities and challenges borne from our rapidly evolving world, presenting diverse ways for organizations to structure their operations and relationships. By understanding these models, we gain insights into the potential future of organizational structures and their potential impacts on business performance, employee satisfaction, and societal well-being [24, 25].

8.3.1 The Networked Organization

A networked organization is an organizational model that is characterized by less hierarchical structure, greater flexibility, and more emphasis on interconnectivity [26]. Instead of a traditional top-down hierarchy, networked organizations emphasize lateral communication and collaboration, with a more interconnected and less centralized structure.

8.3.1.1 Structure of a Networked Organization
The structure of a networked organization is less about defined roles and fixed positions, and more about fluid, project-based teams. These teams are cross-functional, bringing together diverse skillsets needed for specific projects or tasks. The members of these teams may change based on the project requirements.

Communication and collaboration are key in a networked organization. Information flows freely across the organization, not just up and down a hierarchical chain. This fluid communication enables faster decision-making and problem-solving, as well as fostering innovation and creativity.

8.3.1.2 Reasons for Adopting a Networked Structure

Organizations may choose a networked structure for several reasons:

1) **Flexibility:** Networked organizations can adapt quickly to changing market conditions, customer needs, and technological advancements.
2) **Efficiency:** By breaking down silos and encouraging cross-functional collaboration, networked organizations can enhance efficiency and productivity.
3) **Innovation:** The free flow of information and ideas across the organization can foster innovation.
4) **Employee Engagement:** Networked structures can enhance employee engagement by empowering employees to participate in decision-making and problem-solving.

In a networked organization, the assignment of projects and tasks often happens in a more organic and collaborative manner, rather than through a top-down directive.

- **Project and Work Assignment**: Projects or tasks might be determined by strategic objectives or customer needs and are often assigned to cross-functional teams that have the necessary skills and expertise. In some cases, employees in networked organizations may have the flexibility to choose or suggest projects they wish to work on, based on their interests and skills. This can lead to higher motivation and engagement.
- **Performance Appraisal**: Evaluating performance in a networked organization can be a complex process. Traditional metrics such as meeting individual targets might be less relevant in a structure where collaboration and collective output are the focus. Therefore, new metrics are often required. These can include:
  Team Performance: The success of projects or initiatives the team has worked on can be a major factor in performance appraisal.
  Collaboration and Contribution: Individuals might be evaluated based on their contribution to the team or project, and how effectively they collaborate with others.
  Innovation and Problem-Solving: Given the emphasis on innovation in networked organizations, the ability to contribute new ideas and solve problems can be key performance indicators.
  Recognition and Reward: In a networked organization, recognition and rewards often go beyond simple financial incentives. They might involve acknowledging contributions in team meetings, offering opportunities for further learning and development, or providing increased autonomy and decision-making power.

It is worth noting that transparency is crucial in networked organizations. Clear communication about how projects are assigned and how performance is evaluated can help to maintain trust and engagement within the team. Feedback should be ongoing and constructive, with the goal of continuous improvement and learning. This collaborative, open approach to work distribution and appraisal aligns with the overall philosophy of the networked organization, helping to foster a culture of flexibility, innovation, and shared responsibility.

Example 8.6   Google

Google is a well-known example of a successful networked organization. Known for its collaborative and open culture, Google encourages cross-functional teaming and the free flow of ideas. Employees at Google often work in project teams that are formed to solve specific problems or develop new products. These teams are fluid and can change based on the task at hand.

Google's networked structure has contributed to its success by fostering innovation, enhancing efficiency, and enabling the company to adapt quickly to changing market conditions. This structure has also played a role in attracting and retaining top talent, as it provides a dynamic and empowering work environment.

In summary, networked organization is an emerging organizational model that offers flexibility, efficiency, and innovation. As the business landscape continues to evolve, we can expect to see more organizations exploring and adopting this model.

8.3.2   The Holacracy Model

Holacracy is a method of decentralized management and organizational governance in which authority and decision-making are distributed throughout a holarchy of self-organizing teams rather than being vested in a management hierarchy [27]. The term "Holacracy" is derived from the term "holarchy," coined by Arthur Koestler in his 1967 book "The Ghost in the Machine." A "holarchy" is a connection between holons—where a holon is both a part and a whole.

In a holacracy, there is a clear organizational structure and everyone knows who is responsible for what, but there is no managerial hierarchy. Instead, power is distributed among roles (not individuals) which are grouped in circles (not departments) that are interconnected. Each role has a specific purpose and set of accountabilities, and a person can hold multiple roles. The structure is flexible and can be modified regularly to reflect the evolving needs of the organization.

Organizations might adopt a holacracy for various reasons:

1) **Efficiency:** With roles and responsibilities clearly defined, there is less confusion and overlap, leading to increased efficiency.
2) **Adaptability:** Since the structure can be easily modified, it allows for quick adaptation to changing circumstances.
3) **Empowerment:** By distributing authority, individuals are empowered to make decisions and take actions related to their roles, fostering a sense of ownership and engagement.
4) **Transparency:** There is high transparency in decision-making processes, as all decisions are made in a structured process in which all relevant role-fillers are invited to participate.

In a holacracy, roles and responsibilities are defined by the team, or "circle," and are regularly updated through a collective governance process. Individuals can fill multiple roles, and roles can evolve as needed.

Example 8.7   Zappos

Online shoe and clothing retailer Zappos is one of the most famous examples of a company implementing holacracy [28]. Zappos transitioned to a holacracy in 2013, replacing its traditional hierarchy with a system of overlapping, self-organizing teams. Despite facing initial challenges during the transition, the company believes that this model has improved their ability to adapt to change and fostered innovation. They maintain that holacracy has enabled them to stay agile, empowering their employees to act as entrepreneurs within their roles and contributing to their ongoing success.

In conclusion, holacracy represents a radical departure from traditional organizational models. While not suitable for every organization, it offers a compelling framework for companies that seek to enhance efficiency, adaptability, and employee empowerment in a fast-paced and dynamic business environment.

8.3.3 The Agile Organization

Agile organizations are characterized by their capacity to swiftly respond to changes in the market environment and customer needs [29]. Originally derived from agile methodologies in software development, the concept of agility has transcended its roots and is now applied to organizational design and management practices.

The term "Agile" first emerged in the realm of software development with the Manifesto for Agile Software Development in 2001 [30]. The manifesto proposed values and principles that focus on customer satisfaction through early and continuous delivery of valuable software, embracing changing requirements, and promoting sustainable development.

The success of the agile approach in software development sparked interest in its potential application to other areas of business. Companies began to see the benefit of adapting quickly to change, being customer-focused, and fostering a collaborative environment. This led to the birth of the "agile organization."

Agile organizations typically feature flat structures, with minimal hierarchy and bureaucracy. Teams are cross-functional, self-managed, and organized around products, projects, or services rather than traditional departments or functions.

In an agile organization, work is broken down into smaller, manageable parts known as iterations or sprints. These sprints allow for constant feedback and adjustment to ensure that the work is aligned with customer needs and business goals.

Organizations might adopt an agile structure for several reasons:

1) **Speed and Adaptability:** Agile organizations are better equipped to respond quickly to changes in their environment, be it customer preferences, market conditions, or technological innovations.
2) **Customer-Centric:** By focusing on delivering value to the customer continuously, agile organizations are better aligned with customer needs and can foster stronger customer relationships.
3) **Empowerment and Engagement:** With self-managing teams, individuals are empowered to make decisions, fostering higher levels of engagement and job satisfaction.

In an agile organization, roles and responsibilities are often fluid and based on the needs of the project or product. Team members take on different roles based on the tasks at hand, their skills, and interests. Leadership is shared and often rotates among team members.

Music streaming service Spotify is a well-known example of an agile organization. Spotify is organized into "squads," "tribes," "chapters," and "guilds"—their own unique interpretation of an agile organization. Each squad is a self-managing team responsible for a specific aspect of the product. Squads with similar areas of focus are grouped into tribes. Chapters are horizontal structures across different squads for individuals with similar skills, while guilds are informal, cross-tribe organizations for sharing knowledge.

This structure has allowed Spotify to rapidly innovate and stay ahead in a highly competitive market. It fosters a culture of ownership, collaboration, and continuous learning and improvement.

In summary, the agile organization model offers a dynamic and customer-centric approach to business, ideal for environments that are characterized by rapid change and uncertainty.

8.3.4 Virtual and Remote Organizations

Virtual and remote organizations represent a growing trend in modern business. These organizations are characterized by geographically distributed teams that rely on digital communication and collaboration tools to work together.

The advent of the Internet and advances in communication technology laid the groundwork for the evolution of virtual and remote organizations. As more robust tools for collaboration and communication became available, it became possible to coordinate work across vast distances, overcoming the geographical limitations of traditional office-based work.

This trend accelerated rapidly in the wake of the COVID-19 pandemic. Many organizations were compelled to shift to remote work almost overnight. As a result, the number of fully virtual organizations has increased significantly.

The structure of a virtual or remote organization can vary widely, from entirely flat structures to more traditional hierarchies. The common thread is that most or all employees work remotely, and the organization may not have a physical headquarters or office space.

Virtual teams can be organized around projects, functions, or products, and individuals may be part of multiple teams. Decision-making can be centralized or distributed, depending on the organization's culture and nature of work.

Organizations may choose a virtual or remote structure for several reasons:

1) **Access to Global Talent:** A virtual structure allows organizations to recruit from a global talent pool, not limited by geographical boundaries.
2) **Cost Savings:** Without the need for physical office space and related costs, organizations can realize significant savings.
3) **Flexibility:** Remote work offers greater flexibility to employees, which can lead to higher job satisfaction and retention.

Certain circumstances are particularly conducive to virtual and remote work:

1) **Knowledge Work:** Work that relies primarily on knowledge and information, such as software development, design, consulting, and research, can often be performed remotely with little loss of productivity.
2) **Global Markets:** Organizations that operate in global markets may benefit from having a distributed workforce that can provide local insights and connections.
3) **Flexible Work Arrangements:** If an organization values flexibility and work-life balance for its employees, a virtual or remote structure can be a good fit.

Roles and responsibilities in virtual and remote organizations are often defined by the work to be done rather than the location. Performance is usually evaluated based on output or results rather than time spent working.

Automattic, the company behind http://WordPress.com, is a notable example of a successful remote organization [31]. With over 1000 employees across more than 70 countries, Automattic operates entirely remotely, relying on blogs, Slack, and other tools for communication and collaboration. This remote structure has allowed Automattic to tap into a global talent pool and maintain a flexible, adaptable organization.

In conclusion, virtual and remote organizations represent a powerful adaptation to the realities of the modern, digital, and increasingly global economy. With the right tools and practices, they can offer significant advantages in terms of flexibility, access to talent, and cost efficiency.

8.3.5 The Platform Model

In the digital age, the platform model has emerged as a dominant business structure [32]. In this model, organizations operate platforms that facilitate exchanges between two or more interdependent groups, usually consumers and producers. Rather than creating value through traditional linear processes, platform businesses create value by facilitating connections and interactions.

The concept of a "platform" has long existed in various forms. However, the proliferation of digital technology and the Internet has enabled the platform model to reach unprecedented scales. With the advent of Web 2.0 technologies in the early 2000s, online platforms began to emerge that facilitated the exchange of goods, services, and even social interactions.

Over the past decade, the platform model has been adopted by many of the most successful and disruptive companies in the digital economy, revolutionizing industries from retail and transportation to hospitality and entertainment.

Platform organizations have a unique structure that is designed to facilitate interactions. The core of the platform is the technology that allows for these exchanges, and the organization is often lean, focusing primarily on developing, maintaining, and improving this technology.

On either side of the platform are the user groups—typically, one group that provides a product, service, or content, and another group that consumes or uses it. The organization does not own the means of production; rather, it facilitates the exchange.

Organizations might adopt a platform structure for several reasons:

1) **Scalability:** Once a platform is built, it can be scaled up to accommodate a vast number of interactions with relatively little incremental cost.
2) **Network Effects:** The value of the platform increases as more users join, creating a positive feedback loop that can lead to rapid growth.
3) **Access to Data:** Platforms generate a wealth of data on user behavior, which can be used to improve the platform, personalize experiences, and drive strategic decision-making.

The platform model works particularly well in circumstances where:

1) **There are fragmented resources that can be brought together efficiently on a platform.**
2) **Network effects can be harnessed, and there is potential for rapid growth.**
3) **Digital technology can facilitate interactions that were previously difficult or impossible.**

8.3.5.1 Assigning Roles and Responsibilities

In a platform organization, roles and responsibilities often revolve around developing, managing, and improving the platform, as well as attracting, supporting, and retaining users. Some employees may focus on technical development, while others focus on user experience, marketing, community management, or data analysis.

Airbnb is a prime example of a successful platform model. The platform connects hosts, who have available space to rent, with guests seeking accommodations. Airbnb does not own any properties; instead, it facilitates exchanges between hosts and guests, taking a commission on each booking.

The platform model has allowed Airbnb to scale rapidly, offering millions of listings in countries worldwide. Its success has been driven by network effects—as more hosts and guests join the platform, the value of the platform to all users increases.

In conclusion, the platform model represents a significant shift in the way businesses operate, driven by advances in digital technology. With their potential for scalability, network effects, and data generation, platforms are reshaping industries and challenging traditional business models.

8.3.6 Rendanheyi Model

The Rendanheyi model is a unique organizational model that originated in the Haier Group, a multinational consumer electronics and home appliances company based in Qingdao, China [33]. It represents a significant departure from traditional hierarchical organizations, designed instead around self-managed microenterprises that operate almost like startups within the larger organization.

The term "Rendanheyi" is a compound word that combines two Chinese phrases: "Ren" (人) stands for "employees," "Dan" (单) for "user orders," and "Heyi" (合一) means "integration." Together, "Rendanheyi" signifies the integration of employees' value creation with user needs.

This model was first introduced by Haier's CEO Zhang Ruimin around 2005, as part of a radical transformation to deal with the challenges of the Internet era and drive innovation within the company.

Under the Rendanheyi model, the organization is divided into thousands of self-managing units called microenterprises (MEs). Each ME, typically consisting of 10–15 employees, is essentially a mini-startup that operates independently. They have the autonomy to make decisions regarding their business, including product development, marketing, and sales.

At the same time, each ME is accountable for its profits and losses. The performance of each ME and its employees is tied directly to the market response, creating a strong alignment of incentives.

The Rendanheyi model was adopted to foster entrepreneurship, innovation, and customer orientation within the organization. By granting autonomy to MEs, the model encourages experimentation and fast iteration, enabling the organization to adapt quickly to market changes.

By linking performance directly to market response, it also encourages MEs to stay focused on customer needs and create value for customers. This reduces bureaucracy and improves efficiency, making the organization more competitive in the rapidly evolving consumer electronics industry.

The Rendanheyi model is well-suited to industries characterized by rapid change, high uncertainty, and the need for continuous innovation. It works well in environments where customer preferences are evolving rapidly, and organizations need to adapt quickly to stay relevant.

In the Rendanheyi model, roles and responsibilities are not assigned in a top-down manner but are largely determined by the MEs themselves based on their business needs. Each ME defines its own roles and tasks, and employees have the freedom to take on multiple roles and move between different MEs.

Since implementing the Rendanheyi model, Haier has reported increased innovation, agility, and customer satisfaction. It has grown into one of the world's leading home appliance manufacturers, with products sold in more than 100 countries.

Example 8.8	**From a Micro Enterprise to a Market Leader in the Gaming Laptop Industry**

In May 2013, Lu Kailin and his three colleagues at Haier initiated a revolutionary idea: to create a laptop specifically designed for online gaming. They believed traditional, business-oriented laptops fell short of meeting the rigorous demands of hardcore gaming. Their first order of business was to meticulously analyze over 30,000 online reviews of gaming PCs. From this research, Lu's team identified 13 recurring customer pain points [33].

Taking their findings, Lu's team sought assistance from the head of a Haier "Platform," a division that provides support to various businesses, including laptop manufacturing. Initially met with skepticism, the team eventually secured a modest seed funding of US$270,000. It was understood that any additional funding would be contingent on the successful market testing of their product.

With the aid of external resources, by December 2013 — just seven months after their initial idea — this newly formed microenterprise introduced its first batch of 500 vibrantly colored and boldly designed gaming laptops online. In an impressive display, they sold out in just five days. A few weeks later, a subsequent batch of 3000 units sold out in a staggering 20 minutes online.

Buoyed by this success, the team prepared a comprehensive business plan. In April 2014, they received additional funding from Haier, and they also reinvested a significant portion of their profits back into the burgeoning enterprise.

Fast forward just over three years, and they have successfully established their own brand, *Thunderobot*. Now valued at US$180 million with a team of 80 staff members, Thunderobot leads the gaming laptop market in China and is rapidly expanding its footprint in other Asian markets.

The success of the Rendanheyi model has also led other organizations to explore similar approaches, signifying a potential shift in organizational thinking in the era of digital disruption.

In conclusion, the Rendanheyi model represents an innovative approach to organizing for the digital age. By decentralizing decision-making and aligning incentives with market response, it fosters entrepreneurship, customer orientation, and agility, enabling the organization to thrive in a rapidly changing environment.

8.4 Future of Organizational Structures

Organizational structures, which dictate the allocation of roles, responsibilities, and power within companies, have been traditionally regarded as fundamental to any enterprise's effective operation. However, as we traverse deeper into the 21st century, these frameworks stand at the precipice of dramatic transformation. Factors such as digital technology proliferation, shifting demographic and societal expectations, global interconnectedness, and the continuous quest for sustainable business models are redefining the customary norms around organizing and managing work. The potential magnitude and profundity of these transformations have far-reaching implications, affecting not only the efficiency and competitiveness of individual businesses but also influencing the nature of work, employee wellbeing, societal inequality, and global economic resilience [34, 35].

Subsequent sections will further dissect these imminent transformations:

Section 8.4.1 explores the primary factors driving changes in organizational structures and projects the likely paths of these transformations. Section 8.4.2 discusses an array of obstacles organizations may face in adapting to these changes and suggests possible strategies to overcome these challenges. Finally, Section 8.4.3 considers how advances in technology—ranging from AI and blockchain to virtual reality—could disrupt organizational structures further and offer new opportunities for reimagining work.

The journey ahead promises to be a challenging one, yet filled with unprecedented possibilities.

8.4.1 Predicted Trends and Patterns

As we delve into this section, we shall explore four key trends that are predicted to shape the future of organizational structures: the rise of remote and flexible work, increased emphasis on cross-functional teams, flattening hierarchies, and the evolution toward purpose-driven organizations.

1) **Rise of Remote and Flexible Work:** This trend has been significantly accelerated by the COVID-19 pandemic and the subsequent realization of the viability of remote work in many sectors. With advancements in technology and digital communication tools, organizations have realized that work can be conducted virtually anywhere. The shift toward more remote and flexible work arrangements can lead to lower operating costs for businesses and offer employees a better work-life balance. However, it also brings challenges related to team cohesion, organizational culture, and employee engagement that need to be addressed.

2) **Increased Emphasis on Cross-Functional Teams:** As the speed of business and the pace of change continue to accelerate, traditional, siloed organizational structures are becoming less effective. Many organizations are moving toward more cross-functional teams, where individuals from different disciplines come together to work on specific projects or objectives. This trend reflects the growing complexity of business problems, which often require diverse expertise to solve. It promotes collaboration, fosters innovation, and enables more rapid decision-making but demands new management skills and approaches.

3) **Flattening Hierarchies:** This trend is driven by changing societal attitudes toward authority, the increasing value placed on diversity and inclusion, and the recognition of the limitations of hierarchical decision-making. Flatter organizational structures can facilitate more open communication, enhance employee engagement, and foster a culture of shared responsibility. However, they also require changes to leadership styles, which may bring new decision-making and accountability challenges.

4) **Evolution Toward Purpose-Driven Organizations:** Today's employees, especially younger generations, are increasingly seeking work that aligns with their personal values and contributes to a broader societal good. In response, many organizations are seeking to become more purpose-driven, integrating societal and environmental considerations into their strategic objectives, organizational culture, and daily operations [36]. This trend can improve employee motivation, enhance corporate reputation, and drive long-term sustainable success, but it also requires authentic and transparent leadership. As we delve into Section 8.4.1, we shall explore four key trends that are predicted to shape the future of organizational structures: the rise of remote and flexible work, increased emphasis on cross-functional teams, flattening hierarchies, and the evolution toward purpose-driven organizations.

In summary, the future of organizational structures will likely be marked by more flexibility, collaboration, flatness, and purposefulness. While each of these trends presents its own unique opportunities and challenges, they all point toward a broader shift in how we conceptualize and organize work, reflecting the changing needs and expectations of both businesses and their employees in the 21st century.

5) **Incorporation of Human-in-the-Loop Artificial Intelligence**

The incorporation of "human-in-the-loop" (HITL) AI into organizational structures is indeed a transformative trend that holds significant implications for the future of work. This model of AI, which integrates human decision-making into automated systems, is likely to have several key impacts [37, 38]:

- **Creation of Small, Powerful Enterprises:** HITL AI could enable small organizations or even single individuals to achieve what, in the past, only large organizations could do. For instance, AI could handle routine tasks and data processing, leaving humans to focus on creative, strategic, and interpersonal activities. This could enable small enterprises to operate with unprecedented efficiency and competitiveness. For instance, a small design firm could use AI to handle administrative tasks, customer inquiries, and market analysis, freeing up human workers to focus exclusively on creative design work.

- **Revival of Craftsmanship:** By automating routine and mundane tasks, HITL AI could free up time and resources for humans to engage in more artisanal and high-skill work. In this way, HITL AI could contribute to a revival of craftsmanship in various fields. For example, a clothing designer could use AI to automate fabric cutting and sewing, allowing them to focus on designing unique and high-quality garments.

- **New Job Roles and Skills:** As AI takes on routine tasks, new roles that combine human expertise with AI capabilities could emerge. This includes AI trainers, who teach AI systems how to perform tasks; AI explainers, who interpret AI outputs and decisions; and AI sustainers, who ensure AI systems operate ethically and transparently.

- **Reskilling and Upskilling:** As HITL AI becomes more pervasive, workers will likely need to acquire new skills to work effectively with AI systems. This could involve technical skills, such as understanding AI capabilities and limitations, as well as soft skills, such as adaptability, problem-solving, and interpersonal communication.

- **Ethical and Societal Considerations:** HITL AI also brings ethical and societal challenges. For instance, there may be concerns around job displacement due to automation, or issues related to data privacy and bias in AI systems. These challenges will need to be proactively managed to ensure the benefits of HITL AI are realized in a fair and responsible manner.

In sum, HITL AI is poised to reshape organizational structures and the nature of work, potentially enabling smaller, more efficient enterprises and a revival of craftsmanship. However, it also demands significant changes in terms of skills development, job roles, and ethical considerations. These issues will need to be carefully managed to ensure a positive and inclusive transition to the future of work.

8.4.2 Potential Challenges and Solutions

The transformation of organizational structures in response to the "Predicted Trends and Patterns" will undeniably bring about several challenges. While these challenges are significant, there are also potential solutions and strategies organizations can employ to navigate these changes effectively.

1) **Challenge: Maintaining Culture and Cohesion in Remote Work:** As remote and flexible work becomes more common, organizations may struggle to maintain a strong company culture and ensure team cohesion. Without physical interaction, it can be difficult to build and sustain the interpersonal relationships and shared norms that underpin effective teamwork.

Solution: Digital Collaboration Tools and Virtual Team Building: Organizations can use digital tools to facilitate communication and collaboration, and organize virtual team-building activities to foster a sense of community. Additionally, they can emphasize and reinforce their organizational values and norms in their communications with remote employees to sustain their corporate culture.

2) **Challenge: Managing Cross-Functional Teams:** Cross-functional teams bring together individuals with diverse expertise, which can lead to difficulties in communication, decision-making, and conflict resolution.

    Solution: Training and Support for Team Leaders: Organizations can provide training and support to team leaders to equip them with the skills needed to manage diverse teams effectively. This might include training in areas such as conflict resolution, facilitation, and intercultural communication.

3) **Challenge: Implementing Flatter Hierarchies:** Flatter hierarchies can lead to confusion about roles and responsibilities, difficulties in decision-making, and potential power struggles.

    Solution: Clear Communication and Role Definition: Organizations can overcome these challenges by clearly communicating about the changes and defining roles and responsibilities. This might include using tools such as role charts or matrices and organizing regular check-ins to discuss issues and make decisions collaboratively.

4) **Challenge: Becoming a Purpose-Driven Organization:** Transitioning to a more purpose-driven organization can be complex, requiring changes to strategic objectives, operational practices, and organizational culture. There is also the risk of perceived "purpose washing" if the change is not perceived as authentic.

    Solution: Authentic Leadership and Stakeholder Engagement: Leaders need to demonstrate a genuine commitment to the organization's purpose, and all stakeholders should be engaged in the process of defining and implementing the organization's purpose. This might involve open discussions, consultations, or participatory decision-making processes.

5) **Challenge: Integrating HITL AI:** The integration of AI into organizational structures can lead to concerns around job displacement, data privacy, and ethical considerations.

    Solution: Responsible AI Practices and Workforce Reskilling: Organizations can implement responsible AI practices, such as transparency in decision-making and respect for data privacy. They can also invest in reskilling their workforce to prepare them for new roles and collaborations with AI systems.

In conclusion, while the future of organizational structures holds significant challenges, these can be addressed through thoughtful strategies, practices, and tools. The successful navigation of these challenges will be critical for organizations to leverage future opportunities and achieve sustainable success.

8.4.3 Impact of Future Technologies

As we examine the "Impact of Future Technologies" on organizational structures, we'll discuss several advancements that have the potential to reshape the way businesses operate. These technologies not only include HITL AI but also Blockchain, Augmented Reality (AR), Virtual Reality (VR), and IoT.

1) **Human-In-The-Loop AI:** As we have previously discussed, HITL AI combines human decision-making with automated systems, creating a powerful synergy that allows for greater efficiency and precision. It has the potential to automate repetitive tasks, freeing up human resources for more strategic, creative, and complex tasks. Moreover, HITL AI could enable real-time learning and adaptation within organizations, as the AI system learns from the human

user and vice versa. However, successful integration of HITL AI will require businesses to address issues of workforce reskilling, data privacy, and ethical AI usage.

2) **Blockchain:** Blockchain technology could fundamentally alter organizational structures by providing a secure, transparent, and decentralized method of recording transactions. This could reduce the need for intermediaries and streamline processes in industries such as finance, supply chain, and Human Resource (HR). By establishing trust and transparency, it could also promote more collaborative and peer-to-peer organizational structures. However, the adoption of blockchain technology comes with challenges such as regulatory uncertainties, integration difficulties, and energy consumption concerns.

3) **Augmented Reality (AR) and Virtual Reality (VR):** AR and VR technologies can enhance collaboration, especially in a remote work setting. They can create immersive virtual environments for team meetings, training sessions, and even product development, breaking down geographical barriers and fostering stronger cooperation. However, widespread use of AR and VR will require addressing challenges like the cost of equipment, technical issues, and potential health effects such as cyber sickness.

4) **Internet of Things (IoT):** IoT refers to the network of physical devices connected via the Internet. In an organizational context, IoT can streamline operations, improve decision-making through real-time data, and enhance customer experiences. For example, in a manufacturing firm, IoT sensors can monitor equipment performance and predict maintenance needs, reducing downtime. However, the adoption of IoT also raises concerns about data security, privacy, and the need for new skills to manage and interpret large volumes of data.

In summary, future technologies have the potential to significantly reshape organizational structures, shifting them toward more flexibility, efficiency, and collaboration. However, the successful adoption and integration of these technologies will require organizations to address a range of technical, regulatory, ethical, and skills-related challenges. The ability of businesses to navigate these changes will be a key determinant of their success in the future landscape of work.

8.5 The Impact on Quality Professionals

As the spheres of technology, organizations, and society continue their relentless evolution, the impact on Quality Professionals—the stewards of organizational excellence—is inevitable and far-reaching. The traditional roles of these professionals, centered around quality control and assurance, are being challenged and transformed by a trifecta of forces: technological advancements like AI and automation, shifting organizational structures, and societal changes reflecting evolving attitudes toward work, leadership, and environmental responsibility. These transformations hold profound implications for Quality Professionals, altering their roles, necessitating new approaches to quality management, and presenting challenges and opportunities around remote work.

In the subsequent sections, we will delve into these impacts in more depth. "Role Shifts and Adaptation" will scrutinize how the roles of Quality Professionals are shifting in response to these forces and outline the adaptations required to thrive in this new landscape. Following that, "New Quality Management Approaches" will illuminate the emerging strategies and methodologies that are redefining quality management in the age of digital transformation and societal shifts. Lastly, "Impact of Remote Working on Quality Management" will examine the implications of the rise in remote work—a trend further cemented by the COVID-19 pandemic—on the processes, practices, and challenges associated with ensuring quality. As we navigate this period of intense change, the journey of Quality Professionals epitomizes the dynamic and adaptive spirit required for success in our evolving world.

8.5.1 Role Shifts and Adaptation

In this section, we will explore two major shifts in the roles of Quality Professionals: the transition from a focus on quality control to quality assurance and the need for increased technological and data literacy.

1) **Evolution from Quality Control to Comprehensive Quality Assurance**: Traditionally, Quality Professionals' role was confined to quality control—testing products or services post-production to ensure adherence to set standards. Technological advancements, however, are automating many of these tasks. This shift is redefining the role of Quality Professionals to encompass comprehensive quality assurance—proactively managing processes and systems to prevent defects, drive value creation, and foster seamless collaboration across teams. Thus, the focus is not just on maintaining quality standards but also contributing to the organization's overall value proposition.

   Adaptation: To accommodate this shift, Quality Professionals will need to cultivate a thorough understanding of their organization's processes and systems and acquire skills in areas like risk management, process improvement, and value creation. The transformation further necessitates the enhancement of communication and collaboration skills, as modern quality assurance involves working closely with various teams to collectively identify and address potential quality issues, thereby driving value for the organization.

2) **Increased Technological and Data Literacy:** The digital transformation of businesses is generating vast amounts of data that can be used to identify trends, predict quality issues, and drive continuous improvement. At the same time, technologies like AI and IoT are becoming increasingly important in quality management. This is changing the role of Quality Professionals, who now need to be able to use these technologies and analyze data to inform their work.

   Adaptation: Quality Professionals will need to develop skills in areas such as data analysis, data visualization, and predictive analytics. They may also need to learn about specific technologies used in their industry, such as AI or IoT. Continuous learning and adaptation will be key, as the technological landscape continues to evolve rapidly.

   In conclusion, the roles of Quality Professionals are shifting significantly in response to technological advancements and changes in organizational structures and practices. These shifts require Quality Professionals to adapt by developing new skills and knowledge and adopting a more proactive and data-driven approach to their work. This will be critical in enabling them to continue to ensure quality in a rapidly changing business environment.

   Along with the shifts toward proactive quality assurance and increased technological and data literacy, the roles of Quality Professionals are further transforming toward value creation and fostering cross-functional collaboration.

3) **Transition to Value Creation:** As organizations strive to stay competitive in the ever-evolving business landscape, Quality Professionals are not just seen as gatekeepers for maintaining standards but also as strategic contributors to creating value for the organization. This includes identifying opportunities for product or service improvements, streamlining operations, and influencing strategic decisions to enhance customer satisfaction and business performance.

   Adaptation: This transition requires Quality Professionals to broaden their perspective and develop a deep understanding of their organization's strategic goals, customer needs, and market trends. It also necessitates enhancing skills in areas such as strategic thinking, innovation, and customer-focused design. Furthermore, Quality Professionals will need to develop their ability to communicate the value and impact of quality initiatives to stakeholders within and outside the organization.

4) **From Siloed Professionals to Collaborative Partners:** Traditionally, Quality Professionals often worked in specific quality departments, separate from other functional teams. However, with the emergence of flatter and more cross-functional organizational structures, Quality Professionals are increasingly working as integrated partners within diverse teams. They collaborate with various functions—from operations to marketing to HR—to ensure quality is embedded in all processes and decisions.

 Adaptation: This shift toward collaboration requires Quality Professionals to develop strong interpersonal and teamwork skills. They will also need to become proficient in collaboration tools and platforms, especially in a remote or hybrid work context. Moreover, they will need to understand the roles, perspectives, and priorities of different functions within the organization and be able to negotiate and build consensus around quality issues.

In sum, the roles of Quality Professionals are undergoing profound transformations as they shift from quality control to value creation and from siloed professionals to collaborative partners. These changes require Quality Professionals to continuously adapt and develop new competencies, thus positioning themselves as pivotal players in driving organizational excellence and competitiveness in the future.

8.5.2 New Quality Management Approaches

As Quality Professionals undergo these significant role shifts, the approaches to quality management are also being reinvented. These new methods capitalize on advanced technology, data analysis, and collaborative processes to proactively ensure quality, create value, and drive continuous improvement.

1) **Data-Driven Quality Management:** The rise of Big Data and advanced analytics is transforming quality management. Organizations are moving from sporadic, manual quality checks to continuous, data-driven quality monitoring. Advanced analytics allow Quality Professionals to identify trends, predict issues, and make data-informed decisions.

 Basis: The proliferation of data and advancements in data processing capabilities provide the raw materials and tools needed for this shift. Quality Professionals need to leverage these tools to turn data into actionable insights that can drive quality improvements.

2) **Quality 4.0:** Quality 4.0 refers to the application of Industry 4.0 technologies to quality management. These technologies include IoT, AI, machine learning, and blockchain. For instance, IoT sensors can monitor production processes in real-time, AI can predict quality issues, and blockchain can ensure traceability and transparency.

 Basis: The ongoing digital transformation of industries provides a platform for Quality 4.0. As these technologies become more prevalent and accessible, they present new opportunities for enhancing quality management.

3) **Integrated Quality Management Systems (IQMS):** Organizations are moving away from siloed quality systems toward IQMS. These systems link quality data and processes across different functions, providing a holistic view of quality and enabling more coordinated and effective quality management.

Basis: The trend toward more cross-functional and collaborative organizational structures supports this shift. Moreover, the availability of more advanced and interoperable technology systems facilitates the integration of quality management systems.

4) **Sustainability and Quality:** Quality management is also increasingly linked with sustainability. Organizations are recognizing that ensuring quality is not just about meeting technical specifications but also about minimizing environmental impact, ensuring ethical practices, and creating social value.

   Basis: This trend is driven by growing societal awareness and concern about environmental and social issues, as well as regulatory changes. Quality Professionals will need to integrate sustainability considerations into their quality management approaches.

In conclusion, new quality management approaches are emerging in response to advances in technology, data availability, and societal changes. These approaches are more data-driven, technology-enabled, integrated, and sustainability-oriented. They represent a significant shift in how quality is managed and provide new opportunities for Quality Professionals to create value and drive organizational excellence.

8.5.3 Impact of Remote Working on Quality Management

The shift toward remote work, precipitated by the COVID-19 pandemic and facilitated by advancing technology, has considerable implications for quality management. The "Impact of Remote Working on Quality Management" encompasses changes to collaboration, communication, and process control methods.

1) **Collaboration in Remote Environments:** Ensuring quality in remote work setups requires heightened collaboration among team members and between different teams. Cross-functional collaboration becomes critical for maintaining process integrity and quality output.

   Basis: Remote work inherently lacks the physical proximity that often aids spontaneous communication and problem-solving. Hence, Quality Professionals must use digital tools effectively and maintain open lines of communication to collaborate on maintaining quality standards.

2) **Quality of Digital Communication:** As communication transitions to digital platforms, Quality Professionals must ensure that these platforms are reliable, secure, and facilitate clear communication.

   Basis: Miscommunication or misunderstandings due to poor digital communication infrastructure can lead to quality issues. Therefore, ensuring the quality of these systems is as critical as ensuring the quality of products or services.

3) **Remote Process Control and Monitoring:** With remote work, monitoring and controlling processes to ensure quality can be more complex, given that work is happening across disparate locations.

   Basis: Traditional methods of process control often rely on on-site monitoring. However, remote work necessitates new methods for overseeing processes. This could include remote monitoring technologies, real-time data analysis, or developing robust self-monitoring procedures for remote workers.

4) **Human Factors in Remote Work:** Factors like work-life balance, mental health, and ergonomics can affect quality in remote work setups.

   Basis: In remote work, these human factors can significantly influence workers' productivity and the quality of their output. Quality Professionals must consider these factors and work collaboratively with HR and management to address them.

5) **Data Security and Privacy:** As more data is shared and accessed remotely, ensuring data security and privacy becomes critical to maintaining quality.

   Basis: Breaches in data security or privacy can have severe consequences, from regulatory penalties to loss of customer trust. Hence, quality management in a remote work context must incorporate data security and privacy protocols.

In conclusion, the rise of remote work presents both challenges and opportunities for quality management. Adapting to this new way of working requires innovative approaches, a strong understanding of digital tools and systems, and an appreciation for the human factors that can impact quality. By addressing these considerations, Quality Professionals can help their organizations navigate the remote work landscape while maintaining, and even enhancing, quality outcomes.

8.6 Required Skills and Knowledge for Quality Professionals in the Future

As we traverse deeper into the 21st century, the landscape for Quality Professionals continues to evolve in response to technological advancements, shifting organizational structures, and societal changes. In turn, the skill set and knowledge required to succeed in the quality field are being redefined. Increasingly, it is not just about understanding traditional quality management techniques but also about being adept in using advanced technology, agile methodologies, and appreciating the human dimensions of quality.

In the subsequent sections, we will explore the key competencies that will be critical for Quality Professionals in the future. First, we will delve into the importance of "Emphasizing Data Literacy" as data becomes an integral tool in quality management. Then, we will discuss the growing necessity for "Proficiency in AI and Machine Learning," followed by the "Understanding of Agile and Lean Methodologies" that are increasingly becoming standard in many industries. The fourth section will illuminate the importance of the "Understanding Human Side of Quality," signifying the role of people-centric approaches in quality management. Finally, we will examine the significance of an "Understanding Holistic View of Quality" as sustainability and societal values become embedded in the concept of quality. As the contours of the quality field shift, so too must the skills and knowledge of those who navigate it.

8.6.1 Emphasizing Data Literacy

In an age characterized by the exponential growth of data, the ability to interpret and utilize data effectively, often termed as "Data Literacy," has become indispensable for Quality Professionals. Data literacy encompasses several interconnected competencies that help professionals harness the power of data for improved quality management.

1) **Understanding of Data Fundamentals:** At its core, data literacy requires a solid grasp of basic data concepts. This includes understanding different data types, data structures, and the principles of data management. This fundamental knowledge serves as a foundation for more advanced data work.

2) **Statistical Proficiency:** Quality professionals need a robust understanding of statistical methods to analyze data effectively. This includes understanding statistical concepts such as variability, correlation, regression, and hypothesis testing and being able to use these techniques to make sense of data.

3) **Data Visualization Skills:** Quality professionals need to be adept at presenting data in a visually understandable manner. This includes understanding how to use charts, graphs, and other visual aids to clearly communicate complex data to various stakeholders.

4) **Data-Driven Decision-Making:** Data literacy also involves the ability to use data to inform decisions. This requires not just understanding the data itself but also understanding the context in which it is used, the limitations of the data, and the potential implications of different decisions.

5) **Data Ethics Understanding:** With the growing reliance on data, it is crucial for Quality Professionals to understand the ethical considerations related to data use. This includes respecting privacy, ensuring data security, and using data responsibly.

The rise of Big Data, AI, and advanced analytics has significantly increased the volume and complexity of data that organizations deal with. As such, data literacy has become a vital skill for Quality Professionals. They must be able to understand, analyze, visualize, and use this data effectively to ensure quality, identify opportunities for improvement, and make informed decisions. Furthermore, they must do so in an ethical manner that respects privacy and security considerations. The growing emphasis on data literacy not only reflects these technological and societal shifts, but also a broader trend toward more data-driven decision-making in organizations.

Two further areas of data literacy critical for Quality Professionals are the understanding and management of data quality and data-related risk:

6) **Data Quality Management:** Data quality is paramount when it comes to making reliable, data-driven decisions. Quality Professionals should be well-versed in data quality dimensions such as accuracy, completeness, consistency, timeliness, and relevance. They should be capable of implementing data validation processes, cleaning data, and managing metadata to ensure that the data used in quality management is trustworthy and fit for purpose.

7) **Data Risk Management:** As data becomes a more critical asset in organizations, managing the associated risks becomes a key aspect of a Quality Professional's role. They need to understand potential risks such as data breaches, loss of data, incorrect data interpretation, or misuse of data. Furthermore, they should be familiar with relevant data protection regulations and standards. Quality Professionals should be able to identify and assess data-related risks and to develop and implement strategies to mitigate these risks.

In essence, Quality Professionals' data literacy should not only enable them to understand and use data effectively but also to ensure the quality of the data and manage the risks associated with its use. They play a critical role in fostering a culture of data quality and risk awareness in the organization, which is fundamental for reliable and ethical data usage. Their skills in these areas directly impact the organization's ability to make data-informed decisions and use data to drive quality improvements.

8.6.2 Proficiency in AI and Machine Learning

As AI and Machine Learning (ML) become increasingly integral to a multitude of industries, Quality Professionals will need to gain proficiency in these areas. The depth and breadth of knowledge required will depend on the role and industry, but there are some key competencies that will be broadly applicable.

1) **Fundamental Understanding of AI and ML Concepts:** Quality Professionals should understand the basic principles of AI and ML, including key concepts such as supervised learning, unsupervised learning, reinforcement learning, neural networks, and natural language

processing. Understanding these principles will help them to appreciate the capabilities and limitations of AI and ML and to engage in meaningful conversations with technical experts.

2) **Application of AI and ML to Quality Management:** Quality Professionals need to understand how AI and ML can be applied in their specific field to enhance quality management. This could involve using ML algorithms for predictive quality control, leveraging AI to automate quality checks, or using these technologies to analyze customer feedback and improve products or services.

3) **Ethical Considerations:** As with data, AI and ML bring a host of ethical considerations that Quality Professionals need to be aware of. This includes issues around bias in AI algorithms, transparency of AI decisions, and the impact of AI on jobs and society.

4) **AI and ML in Data Analysis:** AI and ML are critical tools in modern data analysis. Quality Professionals need to understand how these tools can be used to analyze complex datasets, identify patterns, and make predictions. This could involve understanding the basics of how to train an ML model, or how to use AI-powered data analysis tools.

5) **Risk Management:** Just as with data, AI and ML come with certain risks, from algorithmic bias to security vulnerabilities. Quality Professionals need to understand these risks and be able to implement strategies to mitigate them.

While it is not essential for Quality Professionals to become expert AI and ML practitioners, a certain level of proficiency is becoming increasingly important. They need to understand the potential and pitfalls of these technologies and how to apply them effectively and ethically in their work. This will enable them to harness the power of AI and ML for quality management and to ensure that their organizations use these technologies responsibly.

8.6.3 Understanding of Agile and Lean Methodologies

Agile and Lean methodologies, initially developed in the context of software development and manufacturing, respectively, have now found their place in a wide range of industries, including quality management. The adoption of these methodologies by Quality Professionals signals a shift toward more flexible, efficient, and customer-focused approaches to ensuring quality.

1) **Understanding of Agile Principles:** Quality Professionals need a solid understanding of Agile principles and how they can be applied to quality management. This includes concepts such as iterative development, self-organizing teams, regular feedback loops, and a focus on delivering value to the customer. The application of Agile principles can help Quality Professionals to respond more quickly to changes and to continuously improve quality processes.

2) **Knowledge of Agile Techniques:** Quality Professionals also need to be familiar with specific Agile techniques that can be used in quality management. For instance, techniques such as Scrum or Kanban can be used to manage quality projects and processes. These techniques can help to improve transparency, foster collaboration, and ensure that work is prioritized effectively.

3) **Understanding of Lean Principles:** Lean principles focus on minimizing waste and maximizing value. Quality Professionals need to understand these principles and how they can be applied to improve quality processes. This includes techniques such as value stream mapping, Just-in-Time (JIT) production, and continuous improvement (Kaizen).

4) **Lean Tools and Techniques:** Quality Professionals should also be familiar with specific Lean tools and techniques, such as 5S (Sort, Set in order, Shine, Standardize, and Sustain), Six Sigma, or TQM. These tools can help to identify and eliminate waste, reduce variation, and improve process efficiency.

5) **Integration of Agile and Lean:** Some organizations are adopting hybrid approaches that combine Agile and Lean, such as Lean-Agile or SAFe (Scaled Agile Framework). Quality Professionals need to understand how these integrated methodologies work and how they can be used to enhance quality management.

The adoption of Agile and Lean methodologies reflects a broader trend toward more flexible, efficient, and customer-centric approaches in organizations. By understanding and applying these methodologies, Quality Professionals can help their organizations to adapt more quickly to changes, improve efficiency, and ensure that quality processes deliver maximum value to the customer. They can also foster a culture of continuous improvement, where quality is not just checked but is embedded in the way work is done.

8.6.4 Understanding the Human Side of Quality

Understanding the "Human Side of Quality" is a critical facet of a Quality Professional's role, particularly in our increasingly technology-driven age. This human-centric approach recognizes that the ultimate measure of quality is the value and satisfaction it brings to the people who use a product or service. It embodies Robert Pirsig's philosophy of "Quality" as an experience, as described in his book "Zen and the Art of Motorcycle Maintenance," and Christopher Alexander's focus on "life-enhancing" quality in "The Nature of Order."

1) **Quality as an Experience:** Building on Pirsig's work, Quality Professionals need to appreciate that quality is not just about meeting technical specifications but about the experience a product or service offers. This includes how well it meets customer needs, how intuitive it is to use, and how it makes the user feel. Quality Professionals need to be adept at gathering and interpreting customer feedback and should work closely with design and development teams to ensure that the customer's experience is at the heart of quality considerations.
2) **Life-Enhancing Quality:** Inspired by Christopher Alexander's ideas, Quality Professionals should strive for "life-enhancing" quality. This means creating products and services that not only fulfill their function but also contribute positively to people's lives and the wider society. This could involve considering factors such as sustainability, ethics, and social impact in quality management.
3) **Empathy and Communication Skills:** The human side of quality requires Quality Professionals to have strong empathy and communication skills. They need to be able to understand and relate to the needs, values, and experiences of a diverse range of stakeholders, including customers, employees, and society at large. They also need to be able to communicate effectively about quality issues and to facilitate collaboration between different stakeholders.
4) **Quality of Work Life:** The human side of quality also involves considering the quality of work life for employees. Quality Professionals can play a key role in creating a work environment where quality is valued, where employees have the resources and support they need to maintain high-quality work, and where they are involved in quality improvement efforts.
5) **Ethics and Social Responsibility:** Quality Professionals need to understand the ethical implications of their work and the social responsibility of their organization. This includes considering the impact of products and services on society and the environment and ensuring that quality management practices are ethical and sustainable.

In essence, understanding the human side of quality requires a shift in perspective from seeing quality as a technical challenge to seeing it as a human-centered endeavor. By focusing on the experience of quality, the life-enhancing potential of products and services, the quality of work life, and the ethical and social implications of their work, Quality Professionals can contribute to a more holistic and meaningful approach to quality management.

Christopher Alexander's concept of "Quality Without a Name" (QWAN) is a significant part of understanding the human side of quality. Alexander describes QWAN as the quality found in spaces and objects that feel most alive, a sense of profound comfort and belonging. It is a quality that is elusive and hard to pin down but deeply meaningful. Incorporating this philosophy into quality management might look like the following:

6) **Quality Without a Name (QWAN):** The philosophical idea of QWAN necessitates that Quality Professionals strive to infuse this intangible, yet deeply impactful quality into their products or services. This often involves prioritizing aspects that may not be easily quantified but contribute significantly to user satisfaction. It could mean enhancing the aesthetic appeal of a product, focusing on the usability of a service, or designing processes in a way that feels intuitively right.

Quality Professionals would need to develop a sense of this type of quality, which might involve deepening their understanding of their customers and users beyond surface-level needs and wants. They might need to use qualitative research methods, like interviews or ethnographic studies, to gain insights into what QWAN means in their specific context.

Additionally, instilling the QWAN in products and services is often a collaborative and iterative process. Quality Professionals will need to work closely with designers, engineers, and other stakeholders, iterating and refining until they have achieved a result that has this nameless quality.

In essence, the QWAN represents a kind of ultimate quality goal, one that goes beyond technical specifications and satisfies at a deeper, more holistic level. This concept resonates with the shift toward more human-centric, experience-driven notions of quality, and Quality Professionals will need to develop the skills and sensitivities to pursue this kind of quality in their work.

8.6.5 Understanding Holistic View of Quality

The term "holistic view of quality" refers to comprehending and applying quality principles that transcend specific operational elements or departments and encompassing the entire organization and even the whole value chain. It includes the interdependencies among various components of an organization and how they all contribute to overall quality. This perspective recognizes that quality is not just the absence of defects or the meeting of specifications but a complex system that involves people, processes, products, and relationships, both within and outside the organization.

1) **Systems Thinking:** Quality Professionals should adopt a systems thinking approach, seeing the organization and its processes as a complex, interrelated system. Every change to one part of the system will affect other parts. This perspective enables them to anticipate the wider impacts of quality issues or improvements and to develop solutions that enhance the overall system, not just its individual parts.

2) **Cross-Functional Collaboration:** To holistically manage quality, Quality Professionals must collaborate effectively across different functions and departments. They need strong communication and influencing skills to promote quality principles throughout the organization and ensure that all functions align with its quality objectives.

3) **Customer and Stakeholder Focus:** A holistic view of quality also means considering the needs and expectations of all stakeholders, including customers, employees, suppliers, investors, and society at large. Quality Professionals need to understand these stakeholder perspectives and consider them in their quality management practices.

4) **Value Chain Understanding:** Quality Professionals need to understand the entire value chain, from suppliers through to the end customers. They need to consider how each element of the value chain impacts quality and work with suppliers and partners to ensure that they meet the organization's quality standards.

5) **Balanced Metrics:** While it is important to measure quality performance, a holistic view of quality requires balanced metrics that consider different aspects of quality, from operational performance to customer satisfaction to employee engagement. Quality Professionals need to be skilled in developing and using such metrics to guide quality management decisions.

6) **Sustainability and Ethics:** A holistic view of quality also means considering the longer-term sustainability and ethical implications of products, services, and operations. Quality Professionals need to incorporate sustainability and ethics into their quality management practices, considering factors such as environmental impact, social responsibility, and ethical conduct.

In essence, a holistic view of quality requires Quality Professionals to see beyond the technical aspects of quality and to consider how all elements of the organization and its context contribute to quality. It requires a shift from a reactive, problem-solving approach to a proactive, system-wide approach that aims to prevent quality issues from arising in the first place and to continuously improve the entire system. This perspective is key to achieving excellence in quality management, and to ensuring that quality contributes to the overall success and sustainability of the organization.

References

1 Burns, T. and Stalker, G.M. (1961). *The Management of Innovation*. London: Tavistock Publications.

2 Drucker, P. (1993). *Post-Capitalist Society (First)*. Butterworth Heinemann.

3 Kotter, J.P. and Sorensen, C. (2017). *Accelerate*. Recorded Books.

4 Frynas, J.G., Mol, M.J., and Mellahi, K. (2018). Management innovation made in China: Haier's Rendanheyi. *California Management Review* 61 (1): 71–93.

5 Pink, D.H. (2011). *Drive: The Surprising Truth About What Motivates Us*. Penguin.

6 Chandler, A.D. Jr. (1969). *Strategy and Structure: Chapters in the History of the American Industrial Enterprise*, vol. 120. MIT Press.

7 Beniger, J. (2009). *The Control Revolution: Technological and Economic Origins of the Information Society*. Harvard University Press.

8 Womack, J.P., Jones, D.T., and Roos, D. (2007). *The Machine that Changed the World: The Story of Lean Production—Toyota's Secret Weapon in the Global Car Wars That Is Now Revolutionizing World Industry*. Simon and Schuster.

9 Castells, M. (2000). *The Information Age. Vol. 1: The Rise of the Network Society Blackwell*. Wiley-Blackwell.

10 Galbraith, J.R. (1971). Matrix organization designs how to combine functional and project forms. *Business Horizons* 14 (1): 29–40.

11 Deming, W.E. (1986). *Out of the Crisis*. MIT Press. Reprint, ISBN-13.

12 Birkinshaw, J. and Ridderstråle, J. (1999). Fighting the corporate immune system: a process study of subsidiary initiatives in multinational corporations. *International Business Review* 8 (2): 149–180.

13 Laloux, F. and Wilber, K. (2014). *Reinventing Organizations: A Guide to Creating Organizations Inspired by the Next Stage of Human Consciousness*, vol. 360. Brussels: Nelson Parker.

14 Morgan, J. (2014). *The Future of Work: Attract New Talent, Build Better Leaders, and Create a Competitive Organization*. Wiley.

15 Grønning, T. (2016). *Working Without a Boss: Lattice Organization With Direct Person-to-Person Communication at WL Gore & Associates, Inc. In SAGE Business Cases.* SAGE Publications: SAGE Business Cases Originals.

16 Foss, N.J. and Dobrajska, M. (2015). Valve's way: wayward, visionary, or voguish? *Journal of Organization Design* 4 (2): 12–15.

17 Mankins, M. and Garton, E. (2017). How Spotify balances employee autonomy and accountability. *Harvard Business Review* 95 (1): 134–139.

18 Van der Aalst, W.M., Bichler, M., and Heinzl, A. (2018). Robotic process automation. *Business & Information Systems Engineering* 60: 269–272.

19 Davenport, T.H. and Ronanki, R. (2018). Artificial intelligence for the real world. *Harvard Business Review* 96 (1): 108–116.

20 Bughin, J., Hazan, E., Ramaswamy, S. et al. (2017). *How Artificial Intelligence Can Deliver Real Value to Companies.* McKinsey Global Institute.

21 Siegel, E. (2013). *Predictive Analytics: The Power to Predict Who Will Click, Buy, Lie, or Die.* Wiley.

22 Amatriain, X. (2013). Mining large streams of user data for personalized recommendations. *ACM SIGKDD Explorations Newsletter* 14 (2): 37–48.

23 Mazzei, M.J. and Noble, D. (2017). Big data dreams: a framework for corporate strategy. *Business Horizons* 60 (3): 405–414.

24 Uhl-Bien, M. and Arena, M. (2017). *Complexity Leadership: Enabling People and Organizations for Adaptability.* Organizational Dynamics.

25 Lee, M.Y. and Edmondson, A.C. (2017). Self-managing organizations: exploring the limits of less-hierarchical organizing. *Research in Organizational Behavior* 37: 35–58.

26 Sproull, L., Kiesler, S., and Kiesler, S.B. (1991). *Connections: New Ways of Working in the Networked Organization.* MIT Press.

27 Robertson, B.J. (2015). *Holacracy: The New Management System for a Rapidly Changing World.* Henry Holt and Company.

28 Kumar, S.V. and Mukherjee, S. (2018). Holacracy–the future of organizing? The case of Zappos. *Human Resource Management International Digest* 26 (7): 12–15.

29 Brosseau, D., Ebrahim, S., Handscomb, C., and Thaker, S. (2019). *The Journey to an Agile Organization,* 14–27. McKinsey & Company, May, 10.

30 Manifesto, A., 2001. Manifesto for Agile Software Development. http://www.agilealliance.org.

31 Vecchi, A., Della Piana, B., Feola, R., and Crudele, C. (2021). Talent management processes and outcomes in a virtual organization. *Business Process Management Journal* 27 (7): 1937–1965.

32 Rahman, K.S. and Thelen, K. (2019). The rise of the platform business model and the transformation of twenty-first-century capitalism. *Politics & Society* 47 (2): 177–204.

33 Hamel, G. and Zanini, M. (2018). The end of bureaucracy. *Harvard Business Review* 96 (6): 50–59.

34 Daft, R.L., Murphy, J., and Willmott, H. (2010). *Organization Theory and Design.* Cengage Learning EMEA.

35 Teece, D.J. (2018). Business models and dynamic capabilities. *Long Range planning* 51 (1): 40–49.

36 Rey, C., Bastons, M., and Sotok, P. (2019). *Purpose-Driven Organizations: Management Ideas for a Better World,* 138. Springer Nature.

37 Mosqueira-Rey, E., Hernández-Pereira, E., Alonso-Ríos, D. et al. (2023). Human-in-the-loop machine learning: a state of the art. *Artificial Intelligence Review* 56 (4): 3005–3054.

38 Tsiakas, K. and Murray-Rust, D. (2022). Using human-in-the-loop and explainable AI to envisage new future work practices. In: *Proceedings of the 15th International Conference on PErvasive Technologies Related to Assistive Environments,* 588–594. ACM (Association for Computing Machinery).

Index

Quality in the Era of Industry 4.0: Integrating Tradition and Innovation in the Age of Data and AI,
First Edition. Kai Yang.
© 2024 John Wiley & Sons, Inc. Published 2024 by John Wiley & Sons, Inc.